Technological Learning and Competitive Performance

NEW HORIZONS IN THE ECONOMICS OF INNOVATION

General Editor: Christopher Freeman, *Emeritus Professor of Science Policy, SPRU – Science and Technology Policy Research, University of Sussex, UK*

Technical innovation is vital to the competitive performance of firms and of nations and for the sustained growth of the world economy. The economics of innovation is an area that has expanded dramatically in recent years and this major series, edited by one of the most distinguished scholars in the field, contributes to the debate and advances in research in this most important area.

The main emphasis is on the development and application of new ideas. The series provides a forum for original research in technology, innovation systems and management, industrial organization, technological collaboration, knowledge and innovation, research and development, evolutionary theory and industrial strategy. International in its approach, the series includes some of the best theoretical and empirical work from both well-established researchers and the new generation of scholars.

Titles in the series include:

Industrial Diversification and Innovation
An International Study of the Aerospace Industry
François Texier

Trade Specialisation, Technology and Economic Growth
Theory and Evidence from Advanced Countries
Keld Laursen

Firm Size, Innovation and Market Structure
The Evolution of Industry Concentration and Instability
Mariana Mazzucato

Knowledge Flows in National Systems of Innovation
A Comparative Analysis of Sociotechnical Constituencies
in Europe and Latin America
Edited by Roberto E. López-Martínez and Andrea Piccaluga

The Anatomy and Evolution of Industries
Technological Change and Industrial Dynamics
Orietta Marsili

Managing New Product Development and Innovation
A Microeconomic Toolbox
Hariolf Grupp and Shlomo Maital

Technological Learning and Competitive Performance
Paulo N. Figueiredo

Research and Innovation Policies in the New Global Economy
An International Comparative Analysis
Philippe Laredo and Philippe Mustar

The Strategic Management of Innovation
A Sociological and Economic Theory
Jon Sundbo

Innovation and Small Enterprises in the Third World
Edited by Miene Peter van Dijk and Henry Sandee

Technological Learning and Competitive Performance

Paulo N. Figueiredo

Adjunct Professor of Technology and Innovation Management in the Brazilian School of Public Administration at the Getulio Vargas Foundation (EBAP–FGV), Brazil

NEW HORIZONS IN THE ECONOMICS OF INNOVATION

Edward Elgar

Cheltenham, UK • Northampton, MA, USA

Published by
Edward Elgar Publishing Limited
Glensanda House
Montpellier Parade
Cheltenham
Glos GL50 1UA
UK

Edward Elgar Publishing, Inc.
136 West Street
Suite 202
Northampton
Massachusetts 01060
USA

A catalogue record for this book
is available from the British Library

Library of Congress Cataloguing in Publication Data
Figueiredo, Paulo N., 1963–
    Technological learning and competitive performance / Paulo N. Figueiredo.
        p. cm. — (New horizons in the economics of innovation)
    Based on the author's doctoral thesis at the University of Sussex.
    Includes bibliographical references and index.
    1. Organizational learning. 2. Knowledge management. 3. Technological innovations—Management. 4. New business enterprises—Effect of technological innovations on. 5. Competition. 6. Steel industry and trade—Technological innovations—Brazil—Case studies. I. Title. II. Series.

HD58.82 .F54 2001
658.4'038—dc21   2001023564

ISBN 1 84064 656 X

Printed and bound in Great Britain by MPG Books Ltd, Bodmin, Cornwall

*To my parents who over the years have
inspired me to renew my own learning process*

Contents

Figures

Tables

Boxes

Abbreviations

ABM	Brazilian Association of Materials and Metallurgy
ACESITA	Companhia de Aços Especiais Itabira SA
AÇOMINAS	Aços Minas Gerais SA
ADEP	Appropriability of project engineering design system
AHMSA	Alto Hornos de Mexico SA
API	American Petroleum Institute
ASTM	American Society for Testing and Materials
BAHINT	Booz-Allen and Hamilton International
BF	Blast furnace
BNDE	National Bank for Economic Development
BNDES	National Bank for Economic and Social Development
BOF	Basic oxygen furnace
CAD	Computer-aided design
CAM	Computer-aided manufacturing
CC	Continuous casting
CI	Continuous improvement
CIM	Computer integrated manufacturing
COBRAPI	Companhia Brasileira de Projetos Industriais
COG	Coke oven gas
CONSIDER	National Non-Ferrous and Ferrous Council
COSIPA	Compahia Siderúrgica Paulista SA
CDP	Personnel Development Centre
CSBM	Companhia Siderúrgica Belgo-Mineira SA
CSN	Companhia Siderúrgica Nacional SA
CST	Companhia Siderúrgica de Tubarão SA
CUT	Workers' Central Unit
CVRD	Companhia Vale do Rio Doce SA
DRAM	Dynamic Access Random Memory
ETPC	Pandiá Calógeras Technical School
EXIMBANK	Export and Import Bank
FEM	Fábrica de Estruturas Metálicas
FGV	Getulio Vargas Foundation
FINEP	Financier of Studies and Projects
FUBRAE	Brazilian Foundation for Education Support
FUGEMMS	General Edmundo Macedo Soares Foundation
GCIS	Consulting Group for the Steel Industry
GHH	Gute Hoffnungs Huette Sterkrade AG

HIC	Hydrogen-induced cracking
HSM	Hot strip mill
IBRD	International Bank for Reconstruction and Development
IBS	Brazilian Steel Institute
IF	Interstitial free
IISI	International Iron and Steel Institute
ILAFA	Latin American Iron and Steel Institute
INMETRO	National Institute of Metrology
IPT	Technological Research Institute
IS	Import substitution
ISO	International Standardisation Organisation
JIT	Just in time
LCL	Latecomer company literature
LD	Linz and Donawitz
MBO	Management by objectives
MERCOSUR	Common Market of the South Cone
MRP	Material requirements planning
Nm^3	Newton per cubic meters
NSC	Nippon Steel Corp.
OBM	Own brand manufacturing
ODM	Own design manufacturing
OEM	Original equipment manufacturing
OIM	Operating income margin
OJT	On-the-job training
PAQ	Continuous quality improvement process
PBQP	The Brazilian Programme for Quality and Productivity
PCI	Pulverised coal injection
PDCA	Plan, do, check, action
PETROBRAS	Petroleo Brasileiro SA
PC	Personal computer
PEG	Process enhancement groups
PICE	Industrial and Foreign Trade Policy
POP	Plant optimisation plan
POSCO	Pohang Iron and Steel Company
PPC	Production planning and control
PSAM	Problem solving analysis and method
QCC	Quality control circles
QS	Quality system
R&D	Research and development
RFFSA	Brazilian Federal Railways Co.
SAE	Society for Automotive Engineers
SENAI	National Service of Industrial Apprenticeship
SIDERBRAS	Siderúrgicas Brasileiras SA
SIDOR	Siderúrgica Orinoco SA
SIPGP	Integrated System for Production Planning and Control
SM	Siemens-Martin

SPC	Statistical process control
TA	Technical assistance
TAG	Technology Absorption Group
TFCL	Technological Frontier Company Literature
TIC	Technical information centre
Tpy	Tonnes per year
TQM	Total quality management
TR	Test and research
TWI	Training within industry
UEC	United States Steel Engineers and Consultants Co.
UFF	Fluminense Federal University
UFMG	Minas Gerais Federal University
UFRJ/IE	Federal University of Rio de Janeiro
UFSCAR	São Carlos Federal University
ULSAB	Ultra light steel auto body
UMSA	Usiminas Mecânica SA
UNICAMP	University of Campinas
USIMEC	Usiminas Mecânica SA
USIMINAS	Usinas Siderúrgicas de Minas Gerais SA
USS	United States Steel Corp.
ZD	Zero defect

Preface

This book results from the research for my PhD thesis completed at SPRU – Science and Technology Policy Research at the University of Sussex, UK. I am profoundly grateful to my supervisors Martin Bell and John Bessant for their superb guidance and tremendous encouragement. The learning I have acquired from this fruitful relationship will influence my whole research career in the years to come.

I am also deeply grateful to Chris Freeman and Keith Pavitt at SPRU and Sanjaya Lall at the University of Oxford for their very positive and constructive comments on my manuscript. I regard their comments as an enormous encouragement for my future research work.

The motivation to undertake the research underlying this book is rooted in my career path. During my first degree training in business management I picked up the issue of technology and innovation management within companies. After experiencing the management task within companies for some years, I decided to pursue a master programme in the Brazilian School of Public Administration at the Getulio Vargas Foundation (EBAP-FGV) in Brazil. The contact with the works of the remarkable Argentinian researcher Jorge Sabato greatly influenced my interest in technology and innovation management within companies that operate in late-industrialising countries. From that time, I became even more committed to generating new understanding of how the management of technological learning can improve the competitive performance of late-industrialising companies. In the light of this commitment, I engaged in my doctorate work at SPRU that led to this book.

Technological Learning and Competitive Performance is addressed to managers – who are engaged in daily efforts to build up and sustain the technological capabilities of their companies in order to compete in international markets – and to academics – involved with teaching and research on the practical implications of learning processes and technological capability-accumulation for the improvement of competitive performance within companies.

When I had the research – and the book – completed I realised how indebted I am to a number of other people who were involved with this project, in different ways. I am deeply grateful to all the operators, foremen, technicians, engineers, and managers who participated in the fieldwork for this research in the steel companies. They patiently welcomed me into their units and answered all my questions, providing invaluable information for the successful completion of this project. In particular, I would like to express my special gratitude to Marcus R. Lemos, Renato Dietrich, José A. Penna, Ricardo Leal and Wilson Martins. These managers demonstrated an enormous interest in this research by providing additional comments and interpretations on the data. They also provided substantial support during my stay in their companies. During the stages of the fieldwork for this research I also had the privilege of meeting key leaders in the steel industry in Brazil: Amaro Lanari Jr., Sylvio Coutinho, Antonio Polanczyk, Marco Pepino and André R. Vicente. I very much appreciated their interest in this research and their enthusiastic and inspiring comments.

I am also grateful for the co-operation of the Brazilian Metallurgy and Materials Association (ABM) and the Brazilian Steel Institute (IBS). In particular, I am grateful to Cátia MacCord, Rita Clemente, Vicente Mazzarella and Vicente Chiaverini. Drawing on their deep expertise, they provided helpful comments on the research topic and permitted my access to the archival records of their institutions. The sharing of experience with all these professionals has made me even more interested in engaging in further research within the steel industry in late-industrialising countries.

Several other people have contributed to this research. Over the past few years I have benefited from talks with numerous doctoral students at SPRU, at the Centre for Research in Innovation Management (CENTRIM), at the Institute of Development Studies (IDS) and other units at Sussex University. I would like to express my gratitude to Pari Patel and Reinaldo Diniz for their helpful comments on the procedures I used to deal with some of the operational performance indicators. I wish to thank SPRU staff for their valuable co-operation during the research work. I am grateful to Cynthia Little for quickly proof reading the initial manuscript. Sarah and Jason at the Sussex University Computing Centre gave me a great help with the use of software to solve practical research problems.

I am indebted to the faculty members of EBAP-FGV – most of them my former professors and currently my colleagues – particularly Armando Cunha, Bianor Cavalcanti, Deborah Zouain, Fernando Tenório, Hermano Cherques, Paulo Motta and Sylvia Vergara, for the encouragement they gave me, in different ways. I am particularly grateful to two other faculty members: Enrique Saravia, who very especially supported my decision to engage in the doctorate work at SPRU and Paulo Reis Vieira – who has accumulated decades of studies on organisational behaviour – for his interest in reading my final typescript and making such encouraging comments. I wish to express my gratitude to José C. Barbieri in the São Paulo School of Business Administration at the Getulio Vargas Foundation (EAESP-FGV) not only for influencing my interest in technology management during my early years at university, but also for his suggestions during the fieldwork for this research. Also, I am personally most grateful to Tânia Fischer at the Federal University of Bahia (UFBA) in Brazil for her support of this project.

In addition, this project would not have been achieved without the funding provided by CAPES – the government agency for postgraduate support – at the Brazilian Ministry of Education.

During the process of preparing the camera-ready version of the manuscript I relied on the excellent team at Edward Elgar Publishing Ltd, especially Dymphna Evans, Matthew Pitman, Julie Leppard, Alison Edwards and Edward himself who gave me their very best support to get this book published.

Finally, I wish to thank my family. Although far away, they shared with me key moments of this undertaking. Their love and unfailing encouragement were crucial for me to overcome many of the difficulties that arose during this project. Also, I wish to thank my dearest friend Diana Garland-Wells (Dee-Dee) for kindly welcoming me into her house during the key years of this research and for her daily care about my personal and professional activities.

Although several people have been involved with this project, any inaccuracies in this book are my own responsibility. Seeking to continuously improve my future writing, I would be the most grateful if readers could provide me with their feedback about their reading of this book.

1. Introduction

This book is concerned with how and why latecomer companies differ in the way and rate at which they accumulate technological capability over time. The focus is on how key features of learning processes influence the paths of technological capability-accumulation and, in turn, operational performance improvement. This set of relationships is examined during the lifetime of two large steel companies in Brazil.

Being centred on latecomer companies, the focus of analysis in this book differs from most of the recent studies on knowledge and capabilities in technological frontier companies. In the latter companies innovative technological capabilities already exist. Latecomer companies, however, move into a business on the basis of technology they have acquired from other companies in other countries. Therefore during their start-up time they lack even the basic technological capabilities. To become competitive and catch up with technological frontier companies they first have to acquire knowledge to build up and accumulate their own technological capabilities. In other words, they need to engage in a process of technological 'learning'. The term technological 'learning' is usually understood in two alternative senses.

The first sense refers to the trajectory or path along which the accumulation of technological capability proceeds. The way paths proceed may change over time: technological capability may be accumulated in different directions and at differing rates. The second sense refers to the various processes by which knowledge is acquired by individuals and converted into the organisational level. In other words, it deals with the processes by which individual learning is converted into organisational learning.

Learning in this book is addressed in the second of the two senses outlined above. Hereafter learning will be understood as a *process* that permits the company to accumulate technological capability over time. Technological capability is defined here as the resources needed to generate and manage improvements in processes and production organisation, products, equipment, and engineering projects. They are accumulated and embodied in individuals (skills, knowledge, and experience) and organisational systems (Bell and Pavitt, 1995). Thus this book focuses on the relationship between learning processes, technological capability-accumulation paths and operational performance improvement, in other words, how technological learning influences firms' competitive performance.

The book describes the paths of technological capability-accumulation in the two companies. This is followed by a description of the learning processes underlying these paths. Learning is decomposed here into two distinct processes: (1) knowledge acquisition; and (2) knowledge conversion. These are then decomposed into external and internal-knowledge acquisition and knowledge-socialisation and knowledge-codification processes. The focus of the analysis is centred on how these learning processes have worked over the lifetime of the two companies.

There are two key reasons why the research for this book has been undertaken. First, the book moves further in relation to a pioneering and inspiring set of studies on technological capability-accumulation paths and learning processes undertaken in the late 1970s and early 1980s (e.g. Katz, 1976, 1987; Katz et al., 1978; Dahlman and Fonseca, 1978; Maxwell, 1981; Lall, 1987; Bell et al. 1982 among others). These studies have made an enormous contribution to the latecomer literature. They have suggested the need for deeper studies, particularly on the basis of inter-firm comparisons. This book seeks to cover some of the issues not explored in those studies.

Second, most of the studies in the 1990s, of technological capability in the latecomer context, have focused on the description of paths, particularly at the industry and country levels. Studies investigating the relationship between paths and the learning processes underlying them have been scarce. Some more recent revealing studies have tackled these issues within individual companies (e.g. Kim, 1995, 1997a; Dutrénit, 2000), but comparative studies are still scarce. Additionally, during the 1990s there has been a dissemination of studies on 'knowledge' and 'capabilities', particularly within technological-frontier companies. Many of the studies specify 'what' companies have learnt. Others provide a conceptual treatment of learning processes. Only a few studies are concerned with 'how' these learning processes work (e.g. Leonard-Barton, 1990, 1992a, 1992b, 1995; Iansiti, 1998). Much less clear in those studies are the practical implications of learning processes for the rate of technological capability-accumulation and operational performance improvement.

The research underlying this book is concentrated on steel companies. Steel is the basis and symbol of industrialisation. Technological capability development in steel producers is a key issue in the industry. Technological capability has played a substantial role in the development of the technology and the industry in different countries over time. World-wide, the steel industry is passing through a series of transformations. These are associated with the emergence of new process and product technologies and the demand for thinner, lighter, and more resistant steel for a wide range of applications: from car-making to the manufacture of surgical instruments and implants. In addition, there is a call for new practices for continuous improvements in process control.

During the 1990s, steel has been deregulated and privatised across several late-industrialising countries. These factors have led to additional competitive pressures and the need for technological capability development and continuous performance improvement. The steel industry in Brazil is the eighth largest in the world and ranks first in Latin America. In the early-1990s, the industry was deregulated and privatised. Since then it has been seeking to compete in world markets. Therefore steel companies in Brazil offer a rich reservoir of empirical evidence for the researcher to explore. This book extends and builds on previous empirical studies on technological capability focusing on steel, though from different perspectives (e.g. Dahlman and Fonseca, 1978; Maxwell, 1981, 1982; Bell et al., 1982; Viana, 1984; Lall, 1987; Pérez and Pinache, 1987; Piccinini, 1993; Bell et al., 1995; Shin, 1996).

This study combines qualitative and quantitative elements. The strategy is based on in-depth comparative case-studies. They draw on empirical evidence gathered through detailed fieldwork. The strategy has permitted examination of inter-firm differences in the three issues of the book: technological capability-accumulation paths, learning processes, and operational performance improvement. The qualitative elements are reflected in the way most of the empirical evidence is presented: in the form of histories. These histories are structured on a chronological, analytical and comparative basis. The quantitative elements are reflected in the analyses of cross-company differences, particularly operational performance improvement. This book has been structured to answer three questions:

1. How different were the paths of technological capability-accumulation followed by two large steel companies in Brazil, over time?
2. To what extent can those differences be explained by the key features of the various processes by which knowledge is acquired by individuals and converted into the organisational level – the underlying learning processes?
3. What are the implications of the technological capability-accumulation paths for operational performance improvement in these companies?

The book is explicitly concerned with the accumulation of capabilities which are 'technological'. Other types of capabilities (e.g. marketing or financial) are not explored. The accumulation of these technological capabilities is associated with particular functions: investments, process and production organisation, products, and equipment. These functions are associated with the main production flow in an integrated steel company and some production support units (e.g. quality control, research, engineering, planning). Other operational areas (e.g. energy generation, exploitation of iron mines) and other corporate areas (e.g. finance, sales, accounting) are not explored. As far as performance is concerned, the book focuses on indicators that are operational in the first place. Nevertheless, reference is made to some of the implications of operational performance improvement for financial performance.

Technological Learning and Competitive Performance is organised in three parts. Part I provides the background and framework for firm-level empirical research on technological capability-accumulation paths and learning processes in latecomer companies. Chapter 2 reviews some empirical studies on technological capability-accumulation and learning processes. The review covers both the latecomer company literature (LCL) and the technological frontier-company literature (TFCL). The primary objective of this chapter is to demonstrate some of the reasons for undertaking this research.

Chapter 3 focuses on the frameworks within which the empirical evidence on technological capability-accumulation paths and learning processes is examined. These issues have been addressed in both the LCL and the TFCL. However, neither provides an adequate framework to analyse the issues in the way this book seeks to do. Nevertheless, drawing on both bodies of literature, the book builds its own frameworks for technological capability-accumulation paths and the underlying learning processes.

Chapter 4 characterises the steel technology and the industry. The main processes and products and the key technological characteristics of the industry are briefly described. Chapter 5 clarifies the elements of the research design and methods. Its primary objective is to explain how this research was undertaken.

Part II (Chapters 6 to 8) describes the technological capability-accumulation paths and the underlying learning processes in the two case-study companies. The chapters are organised on the basis of a framework of phases. The introduction to Part II presents this framework, an overview of case-study companies, and the organisation of the empirical chapters. Part III analyses and draws conclusions from the preceding outlines of technological capability-accumulation paths and the underlying learning processes. Chapter 9 explores the influence of the learning processes on the paths of technological capability accumulation. Chapter 10 explores the implications of this relationship for cross-company differences in operational performance improvement. Chapter 11 outlines the conclusions of the book, the implications for management and policy, and suggestions for future research.

PART I

Background and Analytical Frameworks

2. Review of Empirical Studies

2.1 THE LATECOMER COMPANY LITERATURE

This section reviews some empirical studies from the technical perspective; the production organisation perspective and the wider perspective.

2.1.1 The Technical Perspective

From the early-1970s, research on technology in developing countries adopted a dynamic perspective. Leaving aside the static question of choice out of a given set of techniques, the new approach began to look at changes over time in technology and how companies achieved them (Stewart and James, 1982). This new approach influenced the emergence of a pioneer and revealing set of studies on indigenous technological capability generation in latecomer companies. These studies paid great attention to the changes over time in the technical dimensions of technological capability. In Latin America, in particular, much of the studies were undertaken under the ECLA/IBD/IDRC/UNDP Research Programme on Science and Technology, some of which are summarised in Katz (1987). Most of the Asian studies were undertaken under the World Bank's research project 'The Acquisition of Technological Capability', later summarised in *World Development* (1984). A few studies were undertaken in Africa (e.g. Mlawa, 1983).[1]

The issue of technological capability in Latin American steel plants received considerable attention in the 1970s. Indigenous efforts to design and build equipment for the initial plant are described in the study on the path of technological development of two plants of the Acindar company in Argentina from 1943 to 1978 (Maxwell, 1981). Systematic 'stretching' of nominal capacity, ranging from 60 per cent to 130 per cent, took place as a reaction to external pressures (e.g. raw material constraints). Acindar also produced a 'paradoxical' response to inconsistent government policy. This consisted of engaging in the improvement of an outdated plant rather than closing or radically updating it.

Impressive 'capacity-stretching' rates, above 200 per cent, can also be seen in the Brazilian Usinas Siderúrgicas de Minas Gerais SA (USIMINAS) from 1956 to 1976, in response to the company's financial crisis and delayed investments (Dahlman and Fonseca, 1978). The successful experience of upgrading

7

existing plant with the acquisition of new plant was also explored in the study of USIMINAS. This success was associated with a consistent, long-term, and aggressive technological strategy, personnel training, and development of a technological infrastructure (e.g. engineering, research, etc.). The case of the technological development in Altos Hornos de Mexico SA (AHMSA), from 1940 to 1977, indicates that 'capacity-stretching' may also arise as a 'defensive' response concerned more with output increase than cost reduction (Pérez and Pérez y Peniche 1987).[2] Unlike the USIMINAS case, AHMSA's path was characterised by a defensive and short-term reaction to inconsistent government policy, an excess demand situation, and obsolete equipment.

Successful 'capacity-stretching' efforts as a 'remedial' measure to cope with previous faulty investment activities, are illustrated by the study of Acerías Paz del Rio in Colombia from 1947 to 1956 (Puerta, 1979 in Maxwell, 1982). Successful 'remedial' plant improvements and 'stretching' efforts following faulty investment activities, resulting from inconsistent government policy and the firm's inexperience, were also found in the case of the Siderúrgica de Chimbote (later SIDERPERU) from 1956 to 1967 (Gianella, undated, in Maxwell, 1982). 'Capacity-stretching' was such a major finding in those studies that three individual cases (Acindar, AHMSA, and USIMINAS) were chosen in order to take a closer look at their characteristics (Maxwell and Teubal, 1980). The study argued that the Latin American 'capacity-stretching' efforts produced much higher rates than those reported in the existing literature.

However, these five Latin American studies focused on situations in which companies somehow responded actively to external constraints by engaging in successful 'capacity-stretching'. They did not consider situations in which a company did not respond, or responded passively, to changes in external conditions and continued, for decades, to operate the plant below design capacity. The problem of over and under-utilisation was investigated in Sercovich (1978) in a study of two petrochemical plants. Differences in capacity utilisation rates were associated with the initial design conditions. Perhaps because the study was confined to technical aspects only, other factors that might have influenced those rates (e.g. organisation and managerial arrangements, learning mechanisms) were ignored. Nevertheless, all these studies emphasised improvements in those plants as being associated with the presence of creative and intense endogenous efforts. However, in non-successful cases, the absence of improvements in process and equipment was associated with the absence of explicit in-house efforts to create change-generating resources, as in the case of a steel galvanising plant in Thailand (Bell et al., 1982).

Some of these studies also suggested that different intensities of in-house innovative efforts are associated with different patterns of performance. Evidence ranged from fast and positive rates of productivity change as in USIMINAS (Dahlman and Fonseca, 1978), to slow and negative rates as in five

Tanzanian textile mills (Mlawa, 1983). Intermediate performance increase (labour productivity) was found in an Argentinian rayon plant (Katz et al., 1978). In this plant, however, the in-house efforts (e.g. improving plant's operating standards, enhancing product quality) were not continuous (Bell et al., 1984). The problem in obtaining more reliable generalisations from these studies concerning the implications for performance lies in that none of them has made comparisons with established firms producing similar products (Bell, 1982; Bell et al., 1984). As a result, only crude comparisons can be made as between the Argentinian rayon plants in Katz et al. (1978) and the US rayon plants in Hollander (1965), as undertaken in Bell et al. (1984).

All these studies uncovered a number of learning mechanisms underlying the accumulation of technological capability in those plants. These were later analysed in Bell (1984) where a distinction between 'doing-based' and other mechanisms is made. Mechanisms such as 'learning from changing' were illustrated by Acindar's systematic experimentation, engineering and original inventive efforts. These created a spiral of greater understanding, confidence and psychological boost that fostered the plant's further improvements. 'System performance-feedback' mechanisms, 'learning through training' and 'by searching' through the constant acquisition of external knowledge adding to the companies' existing knowledge, were illustrated by the USIMINAS case. Similarly, the creation of specific stimuli by the Production Dept worked as a knowledge-creation mechanism at AHMSA. 'Learning by hiring', through the exchange of technical staff across the units of the plant and the hiring of new trained professionals, was illustrated by the Argentinian Du Pont plant.

The merit of that set of studies is to have uncovered the significance of in-house commitment to the process of technical knowledge-generation to create plants' own technological capabilities (Katz, 1987). They also illustrated that the accumulation of these capabilities is at least a necessary condition for significant technical change, especially in the long term (Bell, 1984). However, as those studies focused on individual plants, they did not provide a comparative analysis of the plant's path of technological capability-accumulation. Although those studies explored the underlying learning mechanisms, they focused on knowledge acquisition only. Therefore the processes by which individual learning is converted into organisational learning were not explored.

Additionally, that set of studies adopted a narrow perspective on the composition of technological capability. This perspective is more related to human resources, as pointed out in Tremblay (1994). Nevertheless, Maxwell looked at the way Acindar *organised* its technological activities by creating specialist technical divisions that were assigned systematic responsibilities. And Dahlman and Fonseca suggested that top management supported and enhanced USIMINAS in-house technical efforts. Later, as in Katz (1985), the issues of production organisation systems and learning mechanisms were pointed to as critical for the understanding of domestic technological capability development in developing countries. However, this study was not coupled with empirical studies any newer than those reviewed above. The importance of exploring changes in firms' industrial organisation to understand differences in improvements across plants, had been pointed out earlier in Hollander (1965). This study looked at the experience of a US company moving into a business on the basis of imported technology, a situation similar to that faced by late-industrialising companies.

Along these lines, Viana (1984) looked at the differences in performance between the plants using the HyL^3 and the Midrex reduction processes within the Venezuelan Siderúrgica Orinoco SA (SIDOR). These differences were associated with the implementation of the technology and, in particular, with differences in the plants' organisational arrangements, the 'suboptimal use' of learning mechanisms (e.g. 'learning from training', 'learning from changing') and with the acquisition of the Direct Reduction technology. Although Viana's study went a bit further than the previous studies, the way those knowledge-acquisition mechanisms were built and worked over time and the knowledge-conversion mechanisms were not explored. In addition, no cross-company analysis was developed. A relevant initiative to bring individual case-studies on Latin American steel plants together for a comparative analysis was taken in Maxwell (1982). However, this study was still limited by differences in the methodology used for each case-study and the lack of a comparative framework to contrast the firms' paths of capability-accumulation and learning processes.

2.1.2 The Production Organisation Perspective

The emergence of this set of studies was influenced by (1) the economic and industrial restructuring that took place in several developing countries during the 1980s, and (2) the emergence of a new set of studies in the TFCL by the mid-1980s, as reviewed later in Section 2.2.1. From the late-1980s, manufacturing companies in developing countries, particularly in Latin America, began to face up to the pressures from foreign competition. This derived from the opening-up of the previously protected market and the end of the import substitution (IS) policy, leading to an intense industrial restructuring. Changes in production organisation, based on those new concepts, were then studied as part of that restructuring process (Humphrey, 1995). It should be remembered that earlier studies (e.g. Fleury, 1977; Humphrey, 1982) had found that the way some companies organised their production systems prevented workers from becoming more skilled. This behaviour was associated with the market conditions during the 1960s and 1970s: a protected, expanding, and subsidised market that encouraged firms to be relatively inefficient (Fleury, 1985).

This new set of studies drew heavily on the principles of just-in-time (JIT) and total quality control or management (TQC/M) and continuous improvement – CI (Bessant and Kaplinsky, 1995; Humphrey, 1993, 1995; Kaplinsky, 1994). They explored how these principles were introduced within companies. At the operational level, practices included rapid die-change and pull systems of production; tracing of faults and the integration of production and quality checking functions; use of value engineering, statistical process control (SPC), stock reduction, cellular manufacturing, and quality circles, etc.[4] At the management level there were, for example, changes into a more decentralised system with less hierarchical levels, integration of production work with quality control and maintenance functions. Also workers were assigned more than one job and were able to move between jobs, an example of multitasking and team-working. Changes to the institutional context were based on shifts in pay and evaluation systems designed to involve, motivate and control workers.

Another sub-set of studies addressed the adoption of principles like JIT and TQC/M and techniques like materials requirement planning (MRP) around computer-aided design (CAD) and computer-aided manufacturing (CAM) technologies. They stressed the importance of changes in the organisational dimensions of production if the company were to achieve substantial gains (Hoffman, 1989) or even in wider dimensions like the 'social' (Meyer-Stamer et al., 1991). They also suggested that the focus on organisational practices (e.g. benchmarking) might be even more important than micro-electronics technologies (Mody et al., 1992) for late-industrialising companies. However, these studies did not go further to explore wider organisational dimensions. Therefore they made no substantial advance in relation to the previous sub-set of studies (e.g. Humphrey, 1993; Kaplinsky, 1994). In addition, because most studies focused on one point in time, they did not tackle the implementation process over time. Since latecomer companies are usually adopters of such techniques, a long-term approach to their implementation is critical to understand the success or failure of that adoption.

The merit of these studies, however, is that they explored the diffusion of those production organisation techniques in the latecomer context. They also found that some models cannot be simply replicated, but involve a dynamic adaptation to local circumstances (Humphrey, 1995). Some authors (e.g. Kaplinsky and Bessant, 1995) even suggested that those changes should be integrated into the strategic management of the company. Although not explicitly, they did contribute to exploring critical dimensions of technological capability in terms of production organisation practices (e.g. team-working and quality circles). These practices may also trigger knowledge-socialisation practices. Some of them (e.g. ISO 9000 certification for production standardisation) may even trigger knowledge-codification practices.

However, among the limitations of those mid-1980s studies were that: (1) they treated organisational practices in the manner of given 'techniques', rarely mentioning the word 'knowledge' or 'learning mechanism', unlike the studies from the 1970s reviewed earlier in Section 2.1.1; and (2) they did not develop a long-term description, again unlike the studies reviewed in Section 2.1.1. Nevertheless, they stimulated the development of studies to explore the organisational dimensions of technological capability; and the analysis of factors influencing the way those techniques were implemented.

2.1.3 Towards a Wider Perspective

From the early-1990s, the latecomer literature began to pay greater attention to the organisational and managerial dimensions of capabilities, learning mechanisms, corporate characteristics, and implications for performance, i.e. from a wider perspective compared to the studies from the 1970s and 1980s. In the electronics industry in Thailand, Tiralap (1990) found that in firms where the owner/management perceived technological change as essential for business growth, and workers' skills and knowledge as necessary for technological change generation, higher technical and business performance were achieved than in those that did not. In Piccinini (1993) the association between technological capability and energy performance in two steel companies in Brazil was investigated. The study found that the company that dynamically accumulated technological capability by making use of interactive knowledge flows had a better energy performance than the company that did not. However, the companies' paths of technological capability-accumulation were not constructed. Nor was the building and functioning of the mechanisms and processes explored, whereby those skills and knowledge were accumulated over time.

Other studies gave greater attention to some of the corporate characteristics influencing firms' technological behaviour. By studying a Jamaican metal company, Girvan and Marcelle (1990) found that the company's accumulation of technological capability was influenced by the company's leadership and commitment to knowledge acquisition practices (e.g. overseas training, foreign technical assistance). This study highlighted the role of leadership influencing technological capability-building. However, the fact that leadership might also constrain efforts on technological capability-building was not considered. Additionally, the path of technological capability-accumulation followed by that particular firm was not reconstructed and knowledge-conversion processes were not explored. Extending Girvan and Marcelle's study, Mukdapitak (1994) found differences in the technological strategies of twenty manufacturing firms in Thailand associated with factors like ownership, market orientation, size, and age. However, these findings were not associated with the firms' performance. In addition to these findings, the above set of studies revealed the importance of deeper investigation into factors influencing latecomer firms' technological capability-accumulation and performance.

Sharing that wide perspective and more concerned with knowledge-acquisition mechanisms, Scott-Kemmis (1988) found in the Brazilian pulp and paper industry, mechanisms ranging from training and technical improvement programmes, collaborative knowledge development, and also the management culture of openness to external knowledge. These were combined with the transfer of knowledge, skills, attitudes, and approaches from foreign to local workers. However, they were related to knowledge-acquisition mechanisms only. Therefore knowledge-conversion mechanisms were again not explored.

Tremblay (1994), commenting on Scott-Kemmis (1988), pointed out that this study did not systematically explore the organisational and managerial dimensions of those firms' technological capabilities. Neither did the study associate the findings with the firms' performance over time. Seeking to overcome those types of limitation, Tremblay (1994) developed a comparative analysis of the organisational dimensions of technological capability in a sample of Indian and Canadian pulp and paper mills associated with their performance over time. Among those dimensions were motivation and commitment to change; leadership; supportive relationships; decision-making process, control and channel of communication, information flow, interaction-influence, type of hierarchy; organisational slack and management attitude.

The study found no positive correlation between productivity growth and technological capability when the latter was narrowly defined, in other words, as embodied in a stock of individuals rather than in organisational systems. In contrast, an explicit association between the firms' technological capability embodied in organisational systems and their productivity growth was found. Tremblay's study contributed to overcoming the limitations of previous studies that had sought to explain international firm-level productivity differences within the latecomer context based on a narrow composition of technological capability (e.g. Pack, 1987). However, Tremblay's study did not reconstruct the technological capability-accumulation path followed by those firms and did not explore the underlying learning processes.

Drawing on the tradition of describing the path of technological capability development, initiated in the Latin American studies during the 1970s, Lall (1987) described the technological capability-development path in a set of industries in India (e.g. cement, steel, textiles). The study found diversity in the paths followed by those industries. Those paths illustrated a complex sequence of moving from basic into higher levels of technological capability acquisition. In addition, the study uncovered the influence of government policy in that process which, in the Indian case, was negative. However, because the study focused at the industry level, very little was said about technological capability-building inside firms. More specifically, inter-firm differences in the rates of technological capability-accumulation were not examined. As far as learning mechanisms were concerned, the study focused on knowledge acquisition only. Knowledge conversion was not explored.

In Hobday (1995) the paths followed by a set of electronics companies in East Asia were described. Although describing the paths in ways similar to those adopted in Katz (1987) and Lall (1987), this study was influenced by the latecomer marketing literature (e.g. Wortzel and Wortzel, 1981). The study specified the way the companies achieved their technological and market transition as they moved along the path evolving from simple to complex activities: from original equipment manufacturing (OEM), to own design manufacturing (ODM), and to own brand manufacturing (OBM). Hobday's study provided rich evidence of the characteristics of latecomer companies' paths of technological capability-accumulation. While the study supported Katz and Lall's conclusions, it argued against conventional innovation models based on technological-frontier companies.

Hobday's study paid great attention to the inter-firm mechanisms and organisational arrangements (e.g. subcontracting, joint-ventures, licensing, overseas training) that permitted the firms to acquire knowledge and engage in the adaptation of foreign technology. Much less attention was given to the intra-firm organisational and managerial dimensions of the technological capability-accumulation paths. Also, as in Lall (1987), differences in the rate of technological capability accumulation between the firms and knowledge-conversion mechanisms were not explored.

Focusing on the Korean electronics industry, Hwang (1998) examined the building-up of technological capability in the semiconductors (DRAM) and personal computer (PC) product areas in three *chaebols*. While DRAM firms caught up rapidly and performed well in export markets, the PC firms performed poorly in capability development and export markets. The study indicated that organisational capabilities in a particular industrial context can equate with organisational rigidities in a different technological and market environment. The study explored different aspects of organisational capability development in latecomer firms. However, rates of technological capability accumulation and operational performance improvement were not explored. In addition, the comparative analysis was limited to single product areas only.

Supporting Hobday's findings of non-linearity in latecomer firms' paths of technological capability-accumulation, Ariffin and Bell (1996) found diverse technological-accumulation paths associated with different types of subsidiary–parent linkages in electronics firms in Malaysia. Using a framework developed in Bell and Pavitt (1995), adapted from Lall (1992), the study gave great attention to the evolution of the intra-firm 'routine' and 'innovative' technological activities (e.g. process and production organisation, product-centred, and equipment). The study also looked at the role of learning mechanisms built in each company to acquire knowledge for capability-accumulation (e.g. external training, different types of 'learning-by-doing'). Although not explicitly, knowledge-conversion mechanisms (e.g. internal training) were also explored. In addition, the study has raised the importance of cumulative interaction between learning mechanisms for technological capability-accumulation paths. In doing so, this study stimulated an interest in deeper investigation into that issue.

As far as a deeper intra-firm perspective on technological capability-accumulation path is concerned, a revealing contribution was made in Kim (1995, 1997a). Drawing on individual case-studies, successful paths of technological capability-accumulation and the importance for those paths of the processes by which individual learning is converted into organisational learning were examined. These issues were explored in Hyundai Motors (Kim, 1995) and Samsung Electronics (Kim, 1997a). The studies also explored the positive role of leadership in creating a sense of crisis, as a contributor to those learning processes.

While Hwang (1998) argued that firms may adapt differently to their market and technological environments, Kim (1997b) seems to take it for granted that successful adaptation will occur if the external environment works effectively. As argued in Kim (1997b), successful technological learning requires an effective national innovation system, 'to force firms to expedite that learning' (p. 219). Therefore greater importance seems to be given to external conditions than to the intra-corporate learning processes as in Kim (1995, 1997a). However, what if companies operating within the same industry respond differently to the same government technological policy and follow different technological capability-accumulation paths associated with different learning processes and different performance improvement patterns? Indeed firms' paths of technological capability-accumulation can be influenced by external conditions. Nevertheless, it seems inadequate to conclude that those paths are totally, or perhaps even heavily, dependent on external conditions.

While Kim (1995, 1997a) focused on the positive aspects of the intra-firm learning process, Dutrénit (2000) focused on the constraints to creating a coherent knowledge basis to develop strategic technological capabilities in the long term. The study, which focused on a Mexican glass company, found that the company's uneven learning process was influenced by several factors central among them being: (1) the limited efforts to convert knowledge from the individual into the organisational level; (2) the different learning strategies pursued by the company and their limited co-ordination; (3) the limited integration of knowledge across organisational boundaries; and (4) the instability of the knowledge creation process. The study argued that the intra-firm learning processes played a major role in influencing that company's technological capability-accumulation path. This conclusion therefore differs from Kim's (1997b) argument.

The merit of Dutrénit's (2000) study is the long-term, detailed, and deep analysis of the intra-firm learning process. However, this study has not examined how the relationship between technological capability-accumulation and learning processes differs across companies. Indeed the study has underlined the need for further research to look more closely at factors influencing firms' learning process and distinguishing the specific from the general problems. An inter-firm comparative analysis therefore would contribute to overcoming that limitation and disentangling the 'debate' between Dutrénit (2000) and Kim (1997b).

Conclusions

The review of some empirical studies in the LCL in Sections 2.1.1 to 2.1.3 suggests that:

In general, most of the studies have focused on the description of paths of technological capability-accumulation, rather than on the relationship between the paths and the underlying learning processes. Indeed in-depth and long-term firm-level studies describing paths of technological capability-accumulation became quite scarce in the latecomer literature between the late 1970s and the early 1990s. It was not until the mid-1990s that new studies began to emerge: Kim (1995, 1997a) and Dutrénit (2000). Although these studies examined firms' paths and the underlying learning processes, they were based on single-case studies. As a result, in-depth comparative case-studies on those issues are still scarce.

Some studies have described the long-term path of technological capability-accumulation (e.g Dahlman and Fonseca, 1978; Katz et al., 1978; Maxwell, 1981; Lall, 1987; Hobday, 1995). However, they have not given adequate attention to: (1) the intra-firm organisational and managerial dimensions of technological capability; (2) the inter-firm differences in the rate of technological capability-accumulation; and (3) the functioning of the underlying knowledge-acquisition processes. In particular, knowledge-conversion processes have not been explored.

Other studies have tackled the relationship between capability building and the external environment (e.g. Hwang, 1998). However, the intra-firm knowledge-acquisition and knowledge-conversion processes were not analysed. There were studies in which some corporate characteristics (e.g. leadership) were explored (e.g. Girvan and Marcelle, 1990; Tiralap, 1990), but the long-term path of technological capability-accumulation was not reconstructed, and the learning processes were not explored. In studies in which comparative analysis of the organisational dimensions of technological capability and its implications for performance were undertaken (e.g. Tremblay, 1994;

Hwang, 1998), the influence of the learning processes on the accumulation of those capabilities was not explored.

2.2 THE TECHNOLOGICAL FRONTIER COMPANY LITERATURE

2.2.1 Technological Capability-accumulation Paths in the TFCL

By the mid 1980s, the attention in the industrial manufacturing studies had turned to changes in the organisation of production. These were seen as a major source of competitive advantage. From the investigation of the competitive success of Japanese companies, concepts like '*kaizen*' (Imai, 1987) and 'lean manufacturing' (Womack et al., 1990) were elaborated together with the renewal of 'continuous improvement': CI (Schroeder and Robinson, 1991; Bessant, 1992; Bessant et al., 1994). The approach to CI has been further defined as company-wide efforts on incremental and cumulative innovative activities (Bessant and Caffyn, 1997; Bessant, 1998). Application of CI (and TQM) has been extended to new product development processes (e.g. Caffyn, 1997), and CI practices underlying recent performance improvements in steel rolling lines (e.g. enforcement of 'plan', 'do', 'check' 'action': PDCA; Process Enhancement Groups: PEGs) have been highlighted in Collinson (1999).

Another subset of studies highlighted the importance of activities like inter-group, management culture, functional integration, and administrative practices to improve manufacturing activities (Ettlie, 1988; Bessant, 1991). In parallel, the 'mutual adaptation' approach (Leonard-Barton, 1988) focused on changes in both technology and organisation (e.g. sales) by tracing mutual changes as users implemented new technologies. However, this study did not explain how the resources to produce those changes had been created. Voss (1988) pointed out that one process technology (e.g. MRP) may succeed at one attempt and fail at another in the processes of implementation. This issue seems critical for latecomer companies. However, it was not tackled in the studies under the 'production organisation' perspective.

From the early-1990s, a new subset of studies has focused on how firms' competitive advantage can be strengthened by renewing and sustaining their 'capabilities' or 'competencies'. They have focused on both technical and organisational aspects. Sony's miniaturisation achievements are described in terms of its organisational capability or 'core competencies' (Prahalad and Hamel, 1990). Stalk et al. (1992) point out that large firms should be able to move competencies or capabilities from one business to another for technical achievements like Honda. Conversely, Patel and Pavitt (1994) argued that large firms' technological strategy is not based on 'core' competencies. Instead, to maintain and renew their competitive advantages they have to build a broad range of technological competencies.

Other studies have explored issues like the management of research and development (R&D) and technological collaboration, the integration between technology strategy, human resource management and organisational culture, combined with the accumulation of technological knowledge. This is the case with Dodgson's (1991) study on a biotechnology firm. While Dodgson examined these issues in a small company, Galimberti (1993) examined the way large chemical and pharmaceutical companies combined their accumulated capabilities with the assimilation of radical biotechnology. The merit of these two studies is how they explore the way companies coped with discontinuities in the technology they use.

Firms' competitive advantage was also associated with the building of technological competencies through R&D activities (e.g. Cohen and Levinthal, 1990; Mitchell and Hamilton, 1988; Miyazaki, 1993; Coombs, 1996), and the organisational capability to bring other capabilities together, like financial, marketing, and production (Miyazaki, 1993). These perspectives seem to be related to the conceptual approach to R&D capabilities as being the leading ones in defining the dynamic capabilities of a firm (e.g. Nelson, 1991). However, even in technological frontier companies, one should be wary of assuming that innovative activities are confined to R&D laboratories (Bessant, 1997; Tidd et al., 1997; Bessant and Caffyn, 1997). Firms' innovative activities are becoming a corporate-wide task, involving production, marketing, administration, purchasing and other functions (Bessant and Caffyn, 1997). In latecomer firms, in particular, a substantial part of their innovative activities is associated with incremental improvements. These take place in areas other than conventional R&D departments, for instance in operations, product and process engineering and design (Katz, 1985; Bell, 1997).

'Integration', as an organisational capability, has been associated with superior product development performance in a highly competitive environment. This can be operationalised in terms of small teams with broad responsibilities and cross-functional integration of teams for problem-solving within simplified and flatter organisations (Clark and Fujimoto, 1991). Cross-functional integration is also used as one of the foundations of process development performance (Pisano, 1997). The capability for integration has also been associated with the firm's overall competitive advantage, as in Tsekouras (1998). However, there seems to be no 'single best' way in which companies should organise their innovative activities. A key challenge for them is to achieve a balance between 'organic and mechanistic' and 'formal and informal' approaches to the organisation of their activities under particular technological and market conditions (Bessant, 1997; Tidd et al., 1997). Integration is also seen as a specific organisational capability to merge new knowledge with deep accumulated knowledge

(Iansiti and Clark, 1994). 'Technological integration' is viewed as consistently linked to both project and product performance (Iansiti, 1998).

In Leonard-Barton (1990, 1992a, 1995) several improvements in processes, products, and equipment in US manufacturing companies have been found. These have been associated with the firms' deepening and renewing of their core capabilities on the basis of four dimensions: (1) technical systems, (2) employee knowledge, (3) managerial systems, and (4) values and norms. However, most of these studies do not examine these improvements on a long-term basis. As a result, little is known about the rate at which the capabilities and the performance improvement have evolved over time and, in particular, how they differ across companies.

Conclusions

The review of empirical studies on technological capability in the TFCL suggests the following.

The TFCL framework is explicitly concerned with how firms can sustain, renew, routinise, and integrate technological capabilities that have already been accumulated. Therefore they are unconcerned with the creation and building-up of those capabilities in the first place, which is a critical issue for latecomer firms. In other words, the TFCL does not examine how companies have arrived at the level of innovative capability which they have today. This suggests that the TFCL falls short in explaining 'how', 'how fast', and 'why' those capabilities were built and accumulated over time.

As far as technological activities are concerned, most of the empirical studies in the TFCL have focused on process and, in particular, on product innovation activities. In several cases, those activities are addressed as being associated with R&D and 'breakthrough' activities. Little attention is given to (1) the way innovative activities starting from processes and products already available in the world are undertaken within companies (e.g. 'reverse engineering'); (2) different types of continuous improvement activities at the plant level (e.g. modifications in production processes, 'stretching' of plant capacity, etc.); and (3) project engineering and equipment incremental innovative activities. These three activities are critical in latecomer companies. This suggests that the TFCL framework has a narrow coverage of companies' technological activities.

Most of the empirical studies in the TFCL argue that 'competencies' and 'technological capabilities' are critical resources for any firms' competitive advantage in a changing environment. However, most of them do not describe the accumulation of those capabilities or explore the practical implications of this accumulation for firms' operational performance in the long term.

Nevertheless, the TFCL explores with an adequate level of detail the organisational and managerial dimensions of technological capability, and the importance of those dimensions in undertaking innovative process and product improvements and innovations to gain and maintain competitive advantage. In this respect, the TFCL provides good insights into the description of technological capability-accumulation paths in latecomer firms: (1) once they have accumulated innovative levels of technological capabilities they must be sustained, renewed, and routinised; (2) they begin to operate within open market competition; and (3) the firms are run, from start-up, in such a way that they are pushed to achieve competitive performance with attention to organisational and managerial dimensions of technological capability.

2.2.2　Learning Processes Underlying Technological Capability-accumulation Paths

The TFCL is rich in specifying several learning processes and practices underlying the building of firms' capabilities. In particular, the TFCL addresses processes for both knowledge acquisition and the conversion into the organisational level.

As far as external knowledge-acquisition processes are concerned, Cohen and Levinthal (1990) point out that incorporating outside knowledge into the firm is critical for innovative capabilities. This can be achieved by individuals through different 'internal mechanisms'. Other studies have pointed out the relevance of practices for importing and absorbing technological knowledge from outside the company for capability building: through vendors, national laboratories, customers, consultants (e.g. Leonard-Barton, 1990, 1995; Garvin, 1993). Knowledge may also be acquired from suppliers, competitors or through forming a technological alliance with a firm that possesses the knowledge (Huber, 1996a, 1996b). 'Integrating external knowledge' has been viewed as one of the practices underlying the building of capabilities in the successful near-net-shape project in Chaparral Steel (Leonard-Barton, 1992b).

Other processes involve pulling in expertise from outside by inviting experts to give talks to personnel, hiring in experts, hiring back retired employees, nurturing 'technological gatekeepers' and individuals who can search, interpret and disseminate external knowledge across the company, or fighting the 'not-invented-here' practices (Garvin, 1993; Huber, 1996a, 1996b; Leonard-Barton, 1992b, 1995; Leonard and Sensiper, 1998). Individuals may be hired to bring in expertise in 'problem-solving' and also in 'problem-finding' or 'framing' (Leonard and Sensiper, 1998). Users may be critical providers of knowledge for the firm through feedback and/or their involvement in development projects or their lead in new development projects (Leonard-Barton, 1995; Iansiti and Clark, 1994).

As far as internal knowledge-acquisition processes are concerned, the TFCL tends to give great attention to R&D as a mechanism for knowledge-building within firms, and also as a contributor to their increased competitive advantage. In the 1990s, some studies gave great attention to the way practices inside R&D units might contribute to capability or competency building (e.g. Miyazaki, 1993; Coombs, 1996; Iansiti and West, 1997). Practices involve the reorganisation of the internal processes inside R&D units to make them more effective (Coombs, 1996). They also involve the integration of research teams (Iansiti and West, 1997). However, these studies seem to consider R&D units as a given learning mechanism and examine how companies can use it better. They are unconcerned with how the mechanism has worked over time. Nevertheless, these studies provide useful insights into a description of the knowledge-acquisition and conversion processes in R&D units in latecomer companies.

A few studies have been concerned with different practices for knowledge acquisition through day-to-day manufacturing operations and/or experimentation at plant level (e.g. Leonard-Barton, 1990, 1992b, 1995; Garvin, 1993). Practices can be purposefully created (e.g. through 'implementation teams') which allow individuals to acquire knowledge during day-to-day work on 'how' and 'why' a technology is designed and operates in a given fashion (Leonard-Barton, 1990). Individuals might also be encouraged to engage in independent problem-solving in daily operations (Leonard-Barton, 1992b). Successful cases of experimentation on the basis of on-going improvement programmes (e.g. diversification of new materials) are associated with continuous knowledge flow from outside the company (Garvin, 1993). This suggests that interaction between learning mechanisms does matter. Companies may create a climate and mechanisms to encourage experimentation. The creation of such a climate will involve tolerance of individuals' failures and taking advantage of and highlighting the knowledge gained from such failures (Leonard-Barton, 1995). However, these studies did not explore (1) how these practices have worked over time; and (2) how companies differ in the way they build and use them.

The TFCL also specifies several processes by which individual learning is converted into organisational learning. For Garvin (1993) a variety of mechanisms may lead to the spread of knowledge throughout the organisation (e.g. written, oral, and visual reports, rotation of personnel, education and training, standardisation practices). Other practices, like shared experience, on-the-job training, 'brainstorming camps', and meetings may lead to knowledge-socialisation (Nonaka and Takeuchi, 1995). 'Internal knowledge integration' encompasses a collection of practices to facilitate the spread of knowledge across the company and the deepening of technological capabilities (Leonard-Barton, 1992b; Leonard-Barton et al., 1994; Leonard-Barton, 1995; Garvin, 1993). In other studies a more specific treatment was given to knowledge integration such as problem-solving activities (Iansiti and Clark, 1994) or the integration of groups of individuals for product development (Clark and Fujimoto, 1991). Standardisation of production practices and systematic documentation were specified as key practices for knowledge codification in Japanese companies (Nonaka and Takeuchi, 1995).

Conclusions

The review suggests that the empirical studies in the TFCL provide a wider treatment of learning processes, particularly the conversion of individual into organisational learning, than the LCL. However, although rich in the specification of different learning processes, the TFCL is much less clear about the way these processes work over time. Notable exceptions are Leonard-Barton (1990, 1992b, 1995) and Iansiti (1998). However, these studies have not traced the functioning of learning processes over the long term. Also scarce in the TFCL are studies on inter-firm differences in building learning process and using them over time and the implications for inter-firm differences in paths of technological capability-accumulation. Nevertheless, the TFCL suggests that the way these processes work has critical implications for capability building.

In sum, both LCL and TFCL have provided rich empirical contributions related to the issues of technological capability-accumulation paths and learning processes. However, they have left some issues unexplored. In particular, they have not explored how learning processes influence paths of technological capability-accumulation and in turn the rate of operational performance improvement. This study seeks to locate in between these two bodies of literature to address the relationship between these three issues in latecomer companies.

NOTES

1.   For a comprehensive, critical, and interesting review of this literature see Bell (1982). For a relatively recent literature review on some Latin American case-studies see Herbert-Copley (1990). For a review of some African case-studies see Herbert-Copley (1992).
2.   Originally published in 1978.
3.   Hojalata y Lamina.
4.   For a review of those studies see Humphrey (1995).

3. Conceptual and Analytical Frameworks

This chapter builds the frameworks within which the evidence on paths of technological capability-accumulation and the underlying learning processes are examined in the book. Both the LCL and TFCL provide useful frameworks for these issues. However, neither on its own provides an adequate framework to examine the issues which this book seeks to do. Nevertheless, it draws on them to build its own frameworks.

3.1 FIRMS' PATHS OF TECHNOLOGICAL CAPABILITY-ACCUMULATION

3.1.1 The Latecomer Company Literature

Technological capability has been defined in several ways in the LCL. Early definitions referred to an 'inventive activity' or systematic creative effort to achieve new knowledge at the production level (Katz, 1976). Another definition referred to 'technological capacity' including the skill and knowledge embodied in workers, facilities, and organisational systems, for both production and technical change activities (Bell, 1982; Scott-Kemmis, 1988).

Lall (1982, 1987) has defined the term as 'indigenous technological effort' in mastering new technologies, adapting them to local conditions, improving on them and even exporting them. Dahlman and Westphal (1982) developed the concept of 'technological mastery', operationalised through 'technological effort' to assimilate, adapt, and/or create technology. This is similar to Bell's (1982) and Scott-Kemmis's (1988) 'technological capacity'. Refining the concept, Westphal et al. (1984:5) have defined technological capability as 'the ability to make effective use of technological knowledge'. All these definitions are clearly associated with the development of indigenous efforts to undertake adaptations and improvements to imported technology. These in-house efforts are related to improvements in process and production organisation, products, equipment, and engineering projects.

23

From a narrower viewpoint, Pack (1987) has defined technological capability as being embodied in a stock of individuals. However, this becomes too narrow a definition and misses out the organisational context within which those resources are developed. Enos's (1991) definition of technological capability involves (1) technical knowledge (as residing in engineers, operators); (2) the institution; and (3) common purpose. However, like Pack (1987), Enos's definition also suggests that individuals are the locus where technological capability resides, and that institutions only bring them together but do not embody the technological capability.

A broader definition was developed in Bell and Pavitt (1993, 1995) in which technological capability was seen as embodying the resources needed to generate and manage technological change. These resources are accumulated and embodied in individuals (skills, knowledge and experience) and organisational systems. This definition seems to build on those developed earlier (e.g. Katz, 1976; Lall, 1982, 1987; Dahlman and Westphal, 1982; Bell 1982; Westphal, 1984; Scott-Kemmis, 1988). Adding to that, technological capability has a pervasive nature. Drawing on the resource-based approach (Penrose, 1959), and on empirical evidence, Bell (1982) has distinguished two kinds of resources: (1) those needed to 'operate' existing production systems; and (2) those needed to 'change' production systems. The latter should not be identified as a distinct set of specialised resources. Because of their 'pervasive nature', they are widely dispersed throughout the organisation.

Latecomer firms' technological capability-accumulation paths

Because the latecomer company starts from a condition of being uncompetitive in the world market ('industrial infancy'), the basic problem of industrial maturation is the accumulation of technological capability to become and continue to be competitive in the world market (Bell et al., 1984). This accumulation seems to involve an evolutionary sequence of in-house technological efforts (Katz, 1985). The accumulation of technological capability tends to reverse conventional sequences by following patterns like 'production-investment-innovation' (Dahlman et al, 1987). In this sense, it is possible to start only with the barest production capability and, on the basis of that, to build other technological capabilities to achieve high levels of technological development (Dahlman et al., 1987).

Drawing on Katz (1987), Dahlman et al. (1987) and Lall (1987), a framework was developed in Lall (1992, 1994) in which firm-level technological capabilities are categorised by functions. This framework indicates that the accumulation proceeds from simpler to more difficult categories. In addition, 'there is a basic core of functions in each major category that have to be internalised by the firm to ensure successful commercial operation. . . That basic core must grow over time as the firm undertakes more complex tasks' (Lall,

1994:267). At the more advanced stages, the firm would become a 'technologically mature' firm. Its key characteristics are 'the ability to identify a firm's scope for efficient specialisation in technological activities, to extend and deepen these with experience and effort, and to draw selectively on others to complement its own capabilities' (Lall, 1994:267 and 269).

This framework seems particularly useful to describe paths of technological capability accumulation followed by latecomer firms. However, it suggests three limitations. These seem to reflect a problem of the available frameworks in the LCL as a whole: (1) the framework points out that differences between firms in paths of technological accumulation are expected to occur. However, little is known about 'how' and particularly 'why' these differences occur; (2) although the framework points out that adaptations in organisational arrangements have to accompany technical ones in that path, the organisational aspects of technological capability have been left unexplored; and (3) although some key activities of the 'technologically mature' company are outlined, the framework does not go further to explain how latecomer companies differ in undertaking them.

Other frameworks have been used to describe technological capability-accumulation paths from different perspectives. The 'reversed product-cycle' in Hobday (1995) is more related to the accumulation of capabilities for export markets, while the 'acquisition-assimilation-improvement' framework in Kim (1997b) is more concerned with capability-accumulation for products than for other types of technological functions (e.g. equipment, investments, process and production organisation). Firms' paths of technological capability-accumulation would be influenced in the first place by the nature of their learning processes. This influence could be either positive (Kim, 1995) or negative (Dutrénit, 2000). This relationship can be influenced by the role of leadership in constructing crises and building coalitions and goal consensus, as explored in Kim (1995, 1997a). Paths also seem to be influenced by local industry and macroeconomic government policies, and industry trends (Lall, 1987; Bell, 1997).

In sum, the LCL provides fundamental concepts on technological capability in latecomer companies. Indeed the available frameworks in the LCL, particularly Lall (1992, 1994), seem useful for a description of firms' paths. However, these frameworks are not sufficient, if considered alone, to describe paths of technological capability-accumulation.

3.1.2 The Technological Frontier Company Literature

By the early-1980s, an evolutionary perspective on firms' technological activities had been developed (Rosenberg, 1982; Nelson and Winter, 1982; Winter, 1988; Nelson, 1991). This perspective drew on March and Simon's (1958) 'routinised behaviour' in organisations, Simon's (1959, 1961) 'bounded rationality' and Penrose's (1959) view on firm-specific capabilities as the main determinant of firm performance. From this perspective, firms are seen as dynamic organisations that know how to do things. The firm is a repository of productive knowledge that distinguishes it even from similar firms in the same line of business (Winter, 1988). This evolutionary perspective can explain the diversity that one would find when investigating firms' technological activities, even when they have evolved under the same economic conditions (Nelson, 1991).

Associated with this perspective is the argument related to the 'permanent existence of asymmetries between firms in terms of their process technologies and quality of output. That is, firms can be generally ranked as "better" or "worse" according to their distance from the technological frontier' (Dosi, 1985:19). Within this perspective, inter-firm differences in performance are interpreted as an implication of different accumulation of technological capabilities (Dosi, 1985, 1988). Those differences are also associated with the main characteristics of the innovative process within firms which is 'uncertain' (Nelson, 1991) and 'path-dependent' (Dosi, 1988; Pavitt, 1988; Teece, 1988). In addition, the knowledge on which a firm draws, in the form of competencies, is 'firm-specific', that is, it is stored in its 'routines'. These 'routines' it is argued, underlie the effective performance of the organisation (Nelson and Winter, 1982). However, this framework does not say how effective routines are created and how they differ from ineffective ones (Bessant, 1997).

In the early-1990s, the framework of firms' capabilities was refined in Teece et al. (1990). Drawing on the resource-based approach, but moving beyond it, the framework focused on the mechanisms by which firms accumulate new capabilities. Within this framework, competencies, capabilities, or strategic assets are seen as the source of sustainable competitive advantage for the firm (Prahalad and Hamel, 1990; Pavitt, 1991; Dodgson, 1993; Malerba and Orsenigo, 1993). Great attention has been given to the firm's stock of resources and the firm's ability to develop specific competencies (e.g. Prahalad and Hamel, 1990). Pavitt (1991) argues that firm-specific competencies explain why firms are different, how they change over time, and whether or not they are able to continue to be competitive. Later, the 'dynamic capabilities' framework (Teece and Pisano, 1994) emerged to explain differences in firms' competitive advantage. Building on some of these approaches, the 'core capability' framework was developed in Leonard-Barton (1992a, 1995). These frameworks are concerned with how firms become more competitive by strengthening the capabilities that already exist. They seem unconcerned with how these capabilities were created in the first place.

In sum, the TFCL is rich in concepts (e.g. 'routines', 'path-dependency') and other conceptual arguments (e.g. inter-firm differences in technological capability-accumulation and performance). It seems useful in helping to explain inter-firm differences in technological capability-accumulation paths. However, less clear in that body of literature is (1) how inter-firm differences in technological capability-accumulation paths and performance emerge over time and (2) the empirical applications of the concepts to explaining inter-firm differences.

3.1.3 Building a Framework to Describe Technological Capability-Accumulation Paths

The book draws particularly on the frameworks available in the LCL and also on the conceptual approaches in the TFCL to interpret the empirical evidence. It makes use of the term 'technological capability' in the sense defined in Bell and Pavitt (1993, 1995). This sense is in line with earlier definitions of the term (e.g. Katz, 1976; Lall, 1982, 1987, 1992; Dahlman and Westphal, 1982; Bell, 1982; Westphal et al., 1984; Scott-Kemmis, 1988). As a result, technological capability is defined here as the resources needed to generate and manage technological change, including skills, knowledge and experience, and organisational systems. Specifically, technological capability refers to the firm's abilities to undertake in-house improvements across different technological functions, such as process and production organisation, products, equipment, and investments. The reasons for using this definition are that: (1) its sense is embedded in the characteristics of the latecomer company therefore it is more adequate than the sense available in the TFCL; and (2) its sense is broad enough to fulfil the book's objective of describing paths including both technical and organisational dimensions of technological capability.

In addition, the book uses a disaggregation of different types of technological capability as provided in the framework developed in Bell and Pavitt (1995), adapted from Lall (1992), to describe paths. This framework distinguishes between 'routine' technological capabilities and 'innovative' technological capabilities across different technological functions. 'Routine' capabilities are the capabilities to carry out technological activities at given levels of efficiency and given input requirements; they may be described as technology-using skills, knowledge and organisational arrangements. 'Innovative' capabilities are those to create, change or improve products and processes; they consist of technology-changing skills, knowledge, experience and organisational arrangements.

One reason for choosing this framework is that it has been used successfully by other researchers to trace paths of technological capability-accumulation (e.g. Lall, 1987; Ariffin and Bell, 1996). Another reason is that the framework has proved feasible in tracing paths of technological capability-accumulation in steel companies. To modify the original framework for steel, this book has drawn on empirical information on technological capability provided by steel companies. Details about the modification process are outlined in Chapter 5.

The modified framework is indicated in Table 3.1. The columns set out the technological capabilities by function; the rows, by level of difficulty. They are

Table 3.1 Technological capabilities in the latecomer steel company: an illustrative framework

Levels of Technological Capabilities	Technological Functions and Related Activities				
	Investments		Process and Production Organisation	Product-centred	Equipment
	Facility User's Decision-making and Control	Project Preparation and Implementation			
ROUTINE					
(1) Basic	Engaging prime contractor. Deciding on plant location. Securing and disbursing finance. Terms of reference.	Preparation of initial project outline. Synchronising civil constructions with installations works.	Routine production co-ordination across plant. Absorbing plant designed capacity. Basic PPC and QC.	Replicating steels efficiently following widely accepted specifications. Routine QC. Supplying export markets.	Routine replacement of equipment components. Participating in installations and tests of performance.
(2) Renewed	Active routine monitoring of existing plant units and infrastructure.	Routine engineering services in new and/or existing plant. Simple ancillaries engineering.	Stability of BFs and Steel Shop. Improved plant coordination. Obtaining certification for routine process QC (e.g. ISO 9002, QS 9000).	Improved replication of given and/or own steels specifications. Routine product QC awarded international certification (e.g. ISO 9002, QS 9000)	Routine manufacturing and replacement of components (e.g. cylinders) under international certification (e.g. ISO 9002).
(3) Extra Basic	Active involvement in technically assisted technology sourcing and project scheduling.	Broad outline of project planning. Technically assisted feasibility studies for major expansions. Standard equipment procurement.			
(4) Pre-intermediate	Partial monitoring and control of: expansion feasibility studies, search, evaluation, and selection of technology/suppliers.	Installations E (civil and electricity, piping, mechanical, metallic, refractories structures and architecture). Technically assisted expansions. Detailed E.			

Table 3.1 (Continued)

Levels of Technological Capabilities	Investments		Technological Functions and Related Activities		
	Facility User's Decision-making and Control	Project Preparation and Implementation	Process and Production Organisation	Product-centred	Equipment
INNOVATIVE					
(3) Extra basic			Minor and intermittent adaptations in process, de-bottlenecking, and 'capacity-stretching'. Systematic studies of new process control systems.	Minor adaptations in given specifications. Creating own standards for steels' dimension, shape, surface quality, and mechanical properties. Systematic study of new steels' characteristics.	Minor adaptations in equipment to adjust it to local raw materials production organisation. Own breakdown maintenance.
(4) Pre-intermediate			Systematic 'capacity-stretching'. Manipulating key process parameters (eg. reduction). Designing models for static automated systems. New organisational techniques (e.g TQC/M, ZD, JIT).	Systematic improvements in given specifications. Systematic 'reverse engineering'. Technically assisted design and development of non-original steels. Developing own product specifications. Licensing new product technology.	Large equipment revamping (e.g. BF) without technical assistance. Detailed and basic reverse E. Large equipment manufacturing.
(5) Intermediate	Full monitoring, control, and execution of: feasibility studies, search, evaluation, and selection, and funding activities.	Basic E of individual facilities. Expanding plant facilities. Procurement E without technical assistance. Procurement E (specifications, project analysis). Plant commissioning. Intermittent provision of technical assistance.	Continuous process improvements. Designing models for dynamic automated systems. Integrating automated process control and PPC. Routinised 'capacity-stretching'. Logistics systems for JIT delivery	Continuous improvements in own specification. Non-original design, development, manufacturing, and commercialisation of complex and high-valued steels without technical assistance. Product development certification (e.g. ISO 9001). Involvement in world projects (e.g. ULSAB).	Continuous basic and detailed equipment E and manufacturing of individual facilities (e.g. BF, Sinter). Preventive maintenance.

29

Table 3.1 (Continued)

Levels of Technological Capabilities	Technological Functions and Related Activities				
	Investments		Process and Production Organisation	Product-centred	Equipment
	Facility User's Decision-making and Control	Project Preparation and Implementation			
INNOVATIVE					
(6) High-intermediate	Own overall project outline and execution. Providing technical assistance in expansion decisions and negotiations.	Basic E of whole plant. Systematic provision of technical assistance in: feasibility studies, basic, detailed, procurement E, and plant start-up. Working with suppliers in new facilities projects.	Integrated automated operations systems with corporate control systems. Engaging in process innovation based on research and engineering.	Adding value to steels developed in-house. Non-original design and development of extra-complex and higher-valued steels (e.g. ultra high/low carbon, coated, sandwich sheets). Engaging in users' product design and development projects (e.g. car makers). Complex JIT distribution systems.	Continuous basic and detailed equipment E and manufacturing of whole steelworks and facilities and/or components for other industries. Continuous technical assistance (e.g. BF revamping) for other steel companies.
(7) Advanced	World-class project management. Developing new production systems via R&D.	World-class engineering. New process design and related R&D.	World-class production. New process design and development via E and R&D.	World-class new steels design and development. Original product design via E, R & D.	World-class design & manufacture of equipment. R&D for new equipment and components.

Notes: Based on integrated works. BF = Blast Furnace; E = Engineering; JIT = Just-in-Time; PPC = Production planning and control; QC = Quality Control; TQC/M = Total Quality Control and Management; ULSAB = Ultra Light Steel Auto Body; ZD = Zero Defect.

Sources: Adapted from Lall (1992) and Bell and Pavitt (1995); own elaboration based on the research.

measured by the type of activity expressing the levels of technological capability, in other words, the type of activity the company is able to do on its own at different points in time. The framework consists of seven levels of capability across five technological functions: (1) facility user's decision-making and control; (2) project engineering; (3) process and production organisation; (4) product-centred; and (5) equipment. Functions (1) and (2) will be examined together under the heading of 'investments'.

In addition, the framework disaggregates 'routine' capability into Levels 1 and 2 for process and production organisation, product-centred, and equipment activities: (1) the capability to operate steel facilities on the basis of minimum accepted standards of efficiency in the industry, hereafter 'routine basic capability'; and (2) the capability to operate steel facilities on the basis of international standards, or recognised international certification, hereafter 'routine renewed capability'. This latter draws on the definition of 'enabling capability' (Leonard-Barton, 1995). As far as routine capabilities for investments are concerned, they are disaggregated into Levels 1 to 4. 'Innovative' capabilities are disaggregated into Levels 3 to 7 for process and production organisation, product-centred, and equipment activities. Innovative capabilities for investments are disaggregated into Levels 5 to 7.

To facilitate the visualisation of the inter-firm differences in paths, these will be represented similarly to the 'ladder' framework in Bell (1997), as indicated in Figure 3.1. This book seeks to trace the paths over as long a period as possible throughout the companies' lifetime. This will allow the rate of accumulation to be tackled, in other words, the number of years needed to attain each level and type of technological capability for different technological functions. The accumulation of a level of capability is identified when a company has achieved the ability to do a technological activity that it had not been able to do before. In addition, the book will take into account the building, accumulation, sustaining (or weakening) of technological capability for different technological functions, in other words, the consistency of the paths.

3.2   THE UNDERLYING LEARNING PROCESSES

Learning in this book is understood as the various *processes* by which technical skills and knowledge are acquired by individuals and through them by organisations (Bell, 1984). In other words, the processes by which individual learning is converted into organisational learning. Learning processes permit the company to accumulate technological capability over time. In the light of the LCL and the TFCL this book breaks down learning into two distinct processes: (1) the knowledge-acquisition process and (2) the knowledge-conversion process. The former is more related to learning at the individual level. The latter is re-

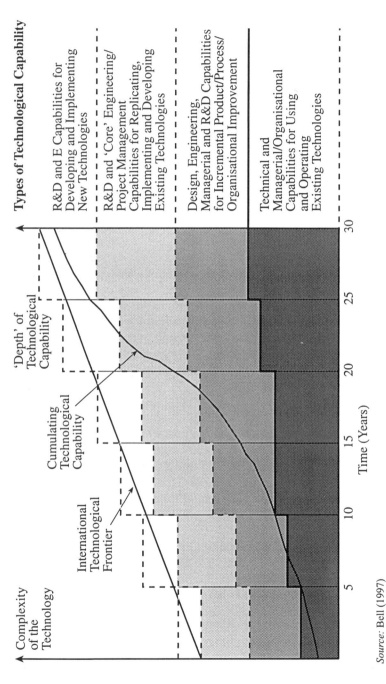

Source: Bell (1997)

Figure 3.1 Path of technological capability-accumulation in latecomer firms: an illustrative framework

32

lated to learning at the organisational level. One reason for this decomposition is that learning first takes place at the individual level, as discussed later in this section. Another reason is that latecomer companies lack even basic capability at their start-up. Therefore knowledge-acquisition processes are critical to understanding their technological capability-accumulation paths.

As demonstrated in Chapter 2, knowledge-acquisition processes have been addressed in both the LCL (e.g. Katz et al., 1978; Lall, 1987; Hobday, 1995; Kim, 1995; Dutrénit, 2000) and the TFCL (e.g. Cohen and Levinthal, 1990; Garvin, 1993; Leonard-Barton, 1990; 1992b; 1995; Huber, 1996a, among others). Although the knowledge-acquisition processes, as addressed in the TFCL, are predominantly for the sustaining of existing capability they can be adapted for latecomer companies. Examples of external knowledge-acquisition processes addressed in both bodies of literature are the pulling in of expertise from outside, overseas training, use of technical assistance, and participation in conferences. Individuals may also acquire tacit knowledge within the company by doing routine and/or innovative technological activities. These activities may take place on the production lines and/or within R&D laboratories.

As far as the knowledge-conversion processes are concerned, the LCL has little to contribute from either empirical or conceptual approaches. Conversely, the TFCL provides relatively rich empirical contributions, as reviewed in Chapter 2. Also, the TFCL provides conceptual approaches that seem useful in building a framework for learning processes in this book, as reviewed below.

3.2.1 Conversion of Individual into Organisational Learning[1]

The 'organisational learning' perspective
The wide use of the term 'organisational learning' has made the concept ubiquitous (Dodgson, 1993). In one of the early studies on that issue Cyert and March (1963) suggested that organisations are entities that act and learn. However, the role of individuals in organisational learning was not given adequate attention in that study (Hedberg, 1981). Argyris and Schön (1978) introduced a framework whereby organisational learning takes place through individual learning. Indeed: 'Although organisational learning occurs through individuals, it would be a mistake to conclude that organisational learning is nothing but the cumulative result of their members' learning' (Hedberg 1981:6).

As for Simon (1996), an important component of organisational learning is the transmission of knowledge from one organisational member to another. This implies the relevance of knowledge socialisation as one of the processes by which individual learning is converted into organisational learning. Organisational learning is also viewed as being associated with: (1) the organisation's absorptive capacity, consisting of prior knowledge; and (2) the intensity of its members' efforts on interactive problem solving to facilitate knowledge creation at the organisational level (Cohen and Levinthal, 1990). Therefore any analysis of learning processes needs to consider learning at the individual level in the first place and then at the organisational level.

Organisations also learn through different modes: (1) through challenging existing practices ('double-loop' or 'generative' learning); or (2) through correcting situations without changing existing practices ('single-loop' or 'adaptive' learning) (Argyris and Schön, 1978; Senge, 1990). Rather than being discrete, these modes are part of a continuum (Argyris and Schön, 1978). The authors focused on those learning modes as organisations and their members respond to a changing environment. These modes of learning are analogous to the five-level learning model developed in Bessant (1998). In this model continuous organisational learning, following a reorganisation, may lead to the consolidation and routinisaton of the learning process. This suggests that 'one-off' mechanisms for knowledge conversion would be unlikely to lead to a long-term effective organisational learning process. The organisation that learns has been defined as that which is skilled at creating, acquiring, and transferring knowledge, thus permitting continuous improvements to take place. However, organisations may be effective at creating or acquiring new knowledge but less successful in applying that knowledge to their own activities (Garvin, 1993).

Nevis et al. (1995) argued that all organisations develop learning systems. Drawing on Huber (1996b)[2] they developed a three-stage model to examine companies' learning systems: (1) the knowledge acquisition; (2) the knowledge sharing; (3) the knowledge utilisation. The framework identifies a variety of 'learning orientations' related to 'facilitating factors'. However, it seems to give greater attention to the facilitating factors than the learning process. This differs from the perspective in this book. In addition, that framework, drawing on evidence from successful technological frontier companies, is concerned with how 'learning organisations' may become better learning systems. However, this book is concerned with how latecomer companies can build up their 'learning systems' in the first place. The framework suggests that some learning systems may be dysfunctional while effective learning systems are hard to achieve. In this respect, some of its elements can be useful in the investigation of learning processes in latecomer companies.

Drawing on Argyris and Schön (1978) and Huber (1996b) among others, Kim (1993) developed a framework focusing on the link between individual and organisational learning. However, it seems to give little attention to knowledge-acquisition processes. In addition, it is concerned with how existing learning processes are improved rather than how they were created in the first place. Therefore such a framework is not adequate for this book. Nevertheless, it draws attention to the complexities involved in learning processes, particularly at the organisational level.

The 'knowledge-building firm' perspective

Drawing on Polanyi's (1966) distinction between two knowledge dimensions (tacit and codified), another group of authors has been concerned with firms' knowledge base as a source of their competitive advantage. Some are more concerned with the tacit dimension of knowledge, although in an organisational setting (e.g. Nelson and Winter, 1982; Dosi and Marengo, 1993; Winter, 1988; Teece et al., 1990). Others are concerned with the interaction between those two dimensions to create organisational knowledge (Nonaka, 1994; Nonaka and Takeuchi, 1995; Leonard-Barton, 1995; Spender, 1996; Leonard and Sensiper, 1998). The latter have also taken into account the interplay between individual and organisational learning. This is in line with the 'organisational learning perspective' (e.g. Argyris and Schön, 1978; Hedberg, 1981; Simon, 1996).

Some authors (e.g. Leonard and Sensiper, 1998) argue that certain dimensions of knowledge are unlikely ever to be wholly codified. Spender (1996) points out that tacit knowledge does not mean knowledge that cannot be codified, and Dutrénit (2000) points out that the possibility of tacit knowledge being codifiable changes over time. This book considers both tacit and codified dimensions of knowledge. However, it is more concerned with the processes by which individual learning is converted into organisational learning than with the interplay between those two knowledge dimensions.

Nonaka (1994) and Nonaka and Takeuchi (1995) argued that the accumulation of the tacit and codified dimensions of knowledge, in a separate way, is unlikely to lead to the creation of a knowledge base within the firm. They introduced a framework whereby the tacit knowledge of individuals becomes organisational learning. This consists of four 'modes of knowledge conversion' between the tacit and codified dimensions: (1) from tacit to tacit (socialisation), whereby tacit knowledge is shared among individuals; (2) from tacit to codified (externalisation), whereby individuals articulate the foundations of their tacit knowledge; (3) from codified to codified (combination), whereby discrete pieces of explicit knowledge are combined into a new whole; and (4) from codified to tacit (internalisation), whereby explicit knowledge is embodied in individuals.

Another framework was developed in Leonard-Barton (1995) consisting of four 'key knowledge-building activities'. This is in line with the processes of organisational knowledge creation in Nonaka and Takeuchi (1995). Authors like Leonard-Barton et al. (1994), Leonard and Sensiper (1998), Pavitt (1998) and Bessant (1998) suggest that the organisation of certain mechanisms plays a key role in overcoming the difficulties encountered in bringing different expertise together. Their insights are useful to build a framework for examining learning processes in latecomer companies.

The above review suggests that the TFCL is rich in conceptual approaches to learning processes. However, with a few exceptions like Garvin (1993), Leonard-Barton (1990, 1992a, 1995), and Nonaka and Takeuchi (1995), most of them do not provide a framework to describe how learning processes work within companies. These are, however, more concerned with improvements to learning processes that already exist than with the building up of these processes. Nevertheless, the TFCL provides fundamental concepts and frameworks for learning processes that can be combined into a framework to tackle this issue in latecomer companies.

3.2.2 Building a Framework for the Description of Learning Processes

The key concern here is how learning processes work in latecomer companies. This practical approach to learning processes has been adopted in other studies, for example, the 'learning mechanisms', in the LCL, and the 'knowledge-building activities' (e.g. Garvin, 1993; Leonard-Barton, 1995). Coombs and Hull (1998) have suggested that the focus on 'knowledge management practices' has some benefits: (1) they can be empirically observed; and (2) they have common features, which can be introduced in different companies, but can be given different degrees of importance or implemented in different ways. The framework for learning processes used in this book is outlined in Table 3.2.

Defining the four learning processes
In the light of the LCL and the TFCL the framework disaggregates learning into knowledge-acquisition and knowledge-conversion processes. Knowledge-acquisition processes are further disaggregated into external and internal. Drawing on two components of the framework developed in Nonaka and Takeuchi (1995), this book disaggregates the knowledge-conversion process into knowledge socialisation and knowledge codification. As a result, the framework consists of four learning processes. Each of these processes contains different sub-processes, mechanisms or practices. The learning processes (the rows in Table 3.2) are defined as follows.

1. External knowledge-acquisition processes   These are the processes by which individuals acquire tacit and/or codified knowledge from outside the company. This can be done through pulling in expertise from outside, drawing on technical assistance, and overseas training. Other processes may involve systematic channelling of external codified knowledge, inviting experts for talks, etc.

2. Internal knowledge-acquisition processes   These are the processes through which individuals acquire tacit knowledge by doing different activities inside the company, for example, through daily routines and/or by engaging in

Table 3.2 Key features of the underlying learning processes in the latecomer company: an illustrative framework

LEARNING PROCESSES	Variety	Intensity	Functioning	Interaction
	Absent–Present (Limited–Moderate – Diverse)	One-off – Intermittent – Continuous	Poor – Moderate – Good – Excellent	Weak – Moderate – Strong
Knowledge-acquisition Processes and Mechanisms				
External Knowledge Acquisition	Absence/presence of processes for acquiring knowledge locally and/or overseas (e.g. pulling in expertise from outside, overseas training). Diverse variety may bring different expertise into the company.	The way the company makes use of this process may be continuous (e.g. annual overseas training for engineers and operators), intermittent or even one-off (e.g. interrupting overseas training).	The way a process is created (e.g. criteria to send individuals to overseas training) and the way it works over time may strengthen or mitigate variety and intensity. Timing: 'learning-before-doing'.	The way a process influences other external or internal knowledge-acquisition processes (overseas training, 'learning-by-doing') and/or other knowledge-conversion processes.
Internal Knowledge Acquisition	Absence/presence of processes for acquiring knowledge by doing in-house activities (e.g. 'capacity-stretching', plant experimentation). These may be routine operational and/or innovative activities (e.g. product development)	The way the company makes use of different processes for internal knowledge acquisition. This may influence the individuals' understanding of the principles involved in the technology.	The way a process is created (e.g. research centres), and the way it works over time have practical implications for variety and intensity. Timing: 'learning-before-doing'.	Internal knowledge acquisition may be triggered by external knowledge acquisition process (e.g. plant improvements triggered by overseas training). This may trigger knowledge-conversion processes.
Knowledge-conversion Processes and Mechanisms				
Knowledge Socialisation	Absence/presence of different processes whereby individuals share their tacit knowledge (e.g. meetings, shared problem-solving, OJT).	The way processes like supervised OJT are continued over the years. Continuous intensity of knowledge socialisation may trigger knowledge codification.	The way knowledge-socialisation mechanisms are created (e.g. in-house training) and work over time. This has implications for variety and intensity of the knowledge-conversion process.	Bringing different tacit knowledge into a workable system (e.g. creating knowledge links). Socialisation may be influenced by external or internal knowledge-acquisition processes.
Knowledge Codification	Absence/presence of different processes and mechanisms to codify tacit knowledge (e.g. systematic documentation, internal seminars, etc.)	The way processes like operations' standardisation are repeatedly done. Absent or intermittent codification may limit organisational learning.	The way knowledge codification is created and works over time has implications for the functioning of the whole knowledge-conversion process. This also influences variety and intensity of the process.	The way knowledge codification is influenced by knowledge-acquisition processes (e.g. overseas training) or by other knowledge-socialisation processes (e.g. team-building).

Source: Drawn mainly from frameworks available in the TFCL.

37

improvements to existing processes and production organisation, equipment, and products. The process may also take place through research activities within formally organised R&D centres, plant laboratories and/or systematic experimentation across operational units.

3. Knowledge-socialisation processes   These are the processes by which individuals share their tacit knowledge (mental models and technical skills). In other words, any formal and informal processes by which tacit knowledge is transmitted from one individual or group of individuals to another. They may involve observation, meetings, shared problem-solving, and job rotation. Training may also work as a knowledge-socialisation process. For instance, during training programmes individuals with different background and experience may socialise their tacit knowledge with trainees and instructors. This framework takes into account different types of training like in-house training (course-based), on-the-job training (OJT), and the provision of training to other companies. These types of training may be preceded by basic training (e.g. to improve numeracy and literacy skills).

4. Knowledge-codification processes   These are the processes by which individuals' tacit knowledge (or part of it) becomes explicit. In other words, the process whereby tacit knowledge is articulated into explicit concepts, in organised and accessible formats, and procedures and becomes more easy to understand. As a consequence, the process facilitates the spread of knowledge across the company. This may involve the standardisation of production procedures, documentation, and internal seminars. The elaboration of training modules by in-house personnel may involve both knowledge-socialisation and knowledge-codification processes. Therefore processes (3) and (4) are critical for the conversion of individual into organisational learning.

Defining the key features of the learning processes
The key features of the learning processes (the columns in Table 3.2) involve variety, intensity, functioning and interaction, defined as follows.

1. Variety   Both the LCL and the TFCL suggest that a diversity of processes is necessary to guarantee adequate knowledge acquisition by individuals and its conversion into the organisational level. 'Variety' is defined here as the presence of different learning processes within the company. Variety is assessed here in terms of the presence/absence of a whole process (e.g. the knowledge-codification process) and other sub-processes it may involve (e.g. the standardisation process). The latter may involve different mechanisms (e.g. the updating of basic operating standards, project design codification). Therefore variety is assessed both between the four learning processes and within them.

2. Intensity   The TFCL suggests that 'one-off' learning processes are unlikely to lead to effective knowledge-acquisition and its conversion into organisational learning. Over time, some practices may be routinised and form part of the company's daily routine (Garvin, 1993; Bessant, 1998). 'Intensity' is defined here as the repeatability over time in creating, upgrading, making use of, improving, and/or strengthening the learning process. Intensity is important because: (1) it may secure a constant flow of external knowledge into a company; (2) it may lead to greater understanding of the technology acquired and the principles involved through internal knowledge-acquisition processes; and (3) it may secure the constant conversion of individual into organisational learning therefore routinising it.

3. Functioning   The way companies organise their learning processes is critical to capability building (e.g. Leonard-Barton et al., 1994; Leonard and Sensiper, 1998; Pavitt, 1998). Some learning processes may be dysfunctional as pointed out in both the TFCL (e.g. Nevis et al., 1995) and the LCL (e.g. Dutrénit, 2000). Companies may organise their learning processes differently (Bessant, 1998). 'Functioning' is defined here as the way learning processes work over time. Although intensity may be continuous, the functioning of the processes may be poor. Learning processes may start out functioning well but deteriorate over time. Functioning may contribute to strengthening and/or mitigating 'variety' and 'intensity'.

4. Interaction   The TFCL suggests an 'organic system' perspective on learning processes (e.g. Senge, 1990; Garvin, 1993; Leonard-Barton 1990, 1992, 1995; Teece and Pisano, 1994). More specifically, interaction between knowledge-acquisition and knowledge-conversion processes matter for capability building (e.g. Leonard-Barton, 1992b, 1995; Garvin, 1993). The LCL also suggests the importance of cumulative interaction between learning mechanisms for technological capability-accumulation (e.g. Ariffin and Bell, 1996). 'Interaction' is the way different learning processes influence each other. For example, a knowledge-socialisation process (e.g. in-house training programme) may be influenced by an external knowledge-acquisition process (e.g. overseas training).

The framework in this book draws on the 'organic system' perspective (Senge, 1990; Garvin, 1993; Leonard-Barton, 1990, 1992b, 1995) to examine how learning processes work as a 'whole'. On the basis of the four features outlined above, different learning processes may give rise to effective or ineffective 'learning systems' within the company. The way the 'learning system' works may have practical implications for the path of technological capability-accumulation and, in turn, the rate of operational performance improvement over time. This relationship is the basic analytical framework of this book. This is represented in Figure 3.2.

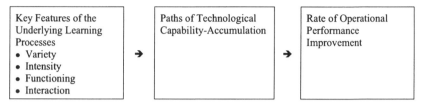

Figure 3.2   The book's basic analytical framework

This book recognises the fact that different factors within the company may influence the learning processes and, in turn, the paths of technological capability-accumulation. For instance, Senge (1990) pointed out that leaders of learning organisations play a key role in fostering learning processes. This is in line with the role of leadership explored in Kim (1995, 1997a). Learning processes can be influenced by organisational beliefs (Schein, 1985; Dodgson, 1993). Argyris and Schön (1978) pointed out that organisations and their members may engage in different modes of learning as they respond to external and internal environments. Also, paths can be affected by external factors (Lall, 1987; Bell, 1997). However, this book does not intend to address these issues. The present study recognises the presence of leadership behaviour and external conditions in the case-study companies. However, these two issues, which are outside its scope, are addressed only very superficially. The analytical framework of the book is represented in more detail in Chapter 5.

NOTES

1.   This structuring builds on Dutrénit (2000).
2.   Originally published in 1991.

4. The Steel Technology and the Industry

This chapter provides a background to the book by briefly characterising the steel technology and the steel industry. It is organised in two sections. The first section describes the main processes and products of a coke-based integrated steelworks where the research for this book is concentrated; the second outlines the key technological characteristics of the steel industry.

4.1 INTEGRATED STEELWORKS: MAIN PROCESSES AND PRODUCTS[1]

The making of steel in an integrated works consists of four major phases: the raw materials preparation; the conversion of the iron ore into pig iron (the reduction process); the conversion of pig iron into molten steel (the refining or steelmaking process); the solidification of the molten steel into desired forms (the shaping process). These phases take place in different types of steelworks which have been defined on the basis of their production processes:

1. Integrated. The term 'integrated' means that all four phases are undertaken within the plant. An integrated works may be coke- or charcoal-integrated, depending on the reducing agent (coke or charcoal) used.
2. Semi-integrated (or minimills). The term 'semi'-integrated means that only two phases are undertaken in the plant: refining and shaping.
3. Non-integrated. The term 'non'-integrated means that only one of the four phases is undertaken in the plant.

Whatever the type of steelworks, they run twenty-four hours a day, seven days a week. A plant normally has four crews of operators. An integrated works usually occupies an area of about 7–10 km². The flow of the main production processes is represented in Figure 4.1.

41

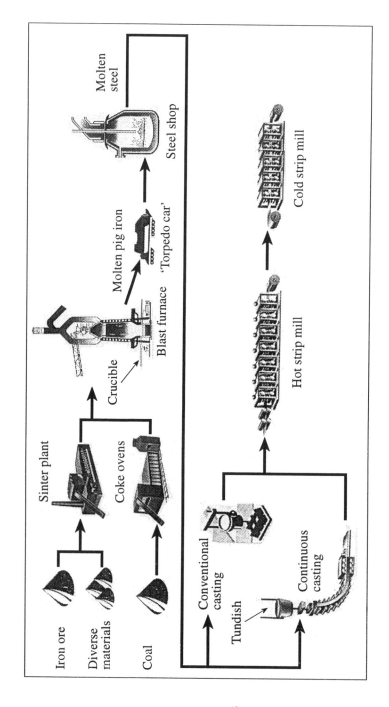

Source: Adapted from printed material from the researched steel companies.

Figure 4.1   The main production processes of an integrated steelworks

4.1.1   The Raw Materials Preparation Processes

The basic raw materials are iron ore, coal, scrap, ferro-alloys, and fluxes (limestone). Iron ore, the basis for the production process, is metal-containing earth (ore) consisting of iron, oxygen and silica (as sand). The fine ores are converted into pieces of strong roundish shaped material ranging from 6 to 25mm in size (agglomerates). Sintering has been the most commonly used agglomerating process and takes place in the sinter plant. This process takes place on a travelling grate that carries a bed of fine iron ores mixed with fuel. The mixture is burnt at a temperature of about 1,400°C to produce porous pieces called sinter.

A reducing agent – carbon – is indispensable to extract the iron from the unwanted elements of those materials (e.g. silica and oxygen). Carbon is extracted from coal, which also contains impurities like ash, benzene, and pitch. As a result, before it reaches the blast furnace (BF) the coal goes into the coke ovens where it is heated to a temperature of about 1,450°C to remove these impurities in the form of gas – the coke oven gas (COG). The product of the coke ovens is the metallurgical coke.

4.1.2   The Reduction or Ironmaking Process

In this process, which takes place in the BF, those iron-bearing materials (iron ore, sinter, or pellets, etc.) are reduced to molten iron. Coke is used as reducing agent. The coke-based reduction process consists of charging iron ore, coke and fluxes into the top of the BF. In parallel, a large amount of heated air is blown – injected in the sides at the bottom – up through the descending materials.

This reaction takes place at a temperature of around 1,500°C. The molten pig iron is tapped, at the bottom of the furnace, into a ladle (the 'torpedo-car') which travels on rails to the steelmaking phase. The use of agglomerates (sinter or pellet) and the injection of oil or oxygen contributes to lowering the coke rate. Coke rate is the consumption of coke per tonne of pig iron produced. It is a key performance indicator of a blast furnace. The reduction area consumes about 70 per cent of the energy consumed in the whole steelplant. Thus the development of capabilities for continuous performance improvements (e.g. reducing the coke rate and increase of productivity of the reduction process) is a critical task for any steelplant.

The period in which the BF is in operation is called the 'campaign'. A campaign may be terminated to revamp the furnace, which normally takes place every 8–10 years. A campaign may also be interrupted as a result of operational problems (e.g. cracks in the crucible) or strikes. Effective steelplants achieve long BF campaigns associated with their capabilities for routine operation and continuous improvements in the reduction process parameters and

the capabilities for equipment planned revamping and preventive maintenance.

4.1.3   The Refining or Oxygen-steelmaking Process

In this process, which takes place in the lined converters in the basic oxygen
furnace (BOF) shop molten pig iron and scrap are converted into molten
steel.[2] Hereafter the BOF shop will be referred to simply as steel shop. The
process seeks to 'adjust' the pig iron and scrap to make steel of the desired
carbon content. The basic oxygen process is characterised by the use of gas-
eous oxygen as the sole refining agent; a metallic charge composed largely of
molten iron and scrap; and chemical reactions that proceed rapidly in a metal
bath involving the elements carbon, silicon, manganese, phosphorus, sulphur,
nitrogen, and oxygen. This combination results in a process that needs no external fuel.

Since the 1960s, the steelmaking process has been controlled by automated
systems. Although minimum expected weights of both scrap and molten iron
are prepared for charging into the furnace in advance, the final weight determination is made by computer. The computer calculates the raw materials composition for a given heat on the basis of the specifications for the finished steel,
which are indicated in the heat order. These are fed into the computer by the
plant's production and planning department. On the basis of those data, the
computer calculates the required weights of molten iron, and scrap, and deter-
mines the amount of oxygen that will be blown.

However, even more than the availability of an on-site process computer
programmed with a charge control, good control of the high-purity oxygen
process depends on the information with which it is supplied. Effective control is associated with accuracy of the inputs into the computer; consistency
of practices and quality of materials in use; reliability of the computer system and measuring devices; convenience of methods to obtain the calculations and adherence to the computer recommendations and to the steel shop
daily practices. To increase the reliability of the computer system, some
steel companies build their own mathematical models. Thus the plant's capabilities for hitting the steel composition desired in the heat order, for accurate and fast tapping and after-tapping procedures, and for consistent daily
organisational practices, are critical for high-utilisation rates and overall
performance of the steelmaking process. These capabilities contribute to reduce the reblow rate – the proportion of heats that needs to be reblown by
oxygen for correction in the steel composition and/or temperature – and the
addition of costly minutes to the heat time – the elapsed time between the
heats in the oxygen converter.

4.1.4   Casting, Rolling Processes and the Main Products

As far as the casting processes are concerned, steelplants may use the conventional process, the continuous process, or both. In conventional casting, molten steel is poured into ingot moulds of various shapes and sizes. These cast iron moulds are usually made in the foundry of the steel mill. After the steel has been cooled and hardened, the ingots are removed from their moulds by a crane. They are placed into soaking pits, heated by oil or gas, to achieve a uniform temperature. From these soaking pits the ingots are again lifted by a crane and move into the rolling phase.

Currently, most steel companies have been replacing conventional ingot casting by continuous casting. In the continuous casting shop, the ladles of molten steel are drained into a receptacle, called a tundish, at the top of the machine. Steel from the tundish flows into a long series of moulds that, at the same time, contain it, cool it, and move it forward. The formed steel, called strand, solidifies from the outside in and leaves the series of moulds as a fully solid shape. This process is not really continuous, since it works with batches of molten steel. The crew seeks to achieve several batches in succession so that the casting machine achieves a long run. This is why the continuous casting needs very accurate control and the plant's capabilities are critical to achieving this. Following this process, the steel becomes ready for use – the so-called 'net-shaped steel' – or needs further adjustments – the 'near-net-shaped steel'. These initial shapes are known as billets or slabs.[3] Then they move into the rolling process. The slabs first go through a hot rolling process, in the hot strip mill, where they are first heated in furnaces to reach the rolling temperature of 800°C.

Then they are passed through a series of rolls in line until they are reduced to a very long and thin plate, the 'coils' of which then move into the cold strip mill, where they are pickled in acid to clean the surface. They are passed through a further series of cold rolls to become finished coils of thin, wide strip, hence the term 'cold-rolled steel'. The thickness is usually measured in millimetres or decimals of millimetres. These coils can be delivered to the user immediately or passed through an automatic tinning line, where they are given a microscopically thin coating to become tinplate. Some coils are coated with zinc (galvanised). Shaped steel products can be classified into flat and long products. Far from being bulky and rough, today's steel mill products are quite thin, diverse, and have a wide range of applications in different industries, as indicated in Table 4.1.

In sum, the multi-stage, multi-product, and large-scale character constitute the key features of most steelplants. These develop their specific characteristics

Table 4.1   Main steel products and their applications

PRODUCTS	MAIN APPLICATIONS
FLAT STEELS[a]	
Slabs	Intermediate product for the manufacturing of other flat products or for the basis of machines.
Uncoated sheets and coils	
Plates and coiled plates	Shipbuilding; railroad rolling stock.
Hot-rolled sheets and coils	Motor vehicle parts and accessories; heavy mechanical industry.
Cold-rolled sheets and coils	Cycles, motorcycles, and car and truck bodies; cookers, refrigerators, washing machines and other domestic and commercial appliances.
Coated sheets and coils	
Terne plates	Roofs of buildings; agricultural and highway machinery; light mechanical industry.
Galvanised sheets	Civil construction; agricultural machinery; domestic appliances.
Chrome-plated sheets	Electric, electronics and mechanical industry.
Tinplates	Tins; packages and containers.
Special sheets	
High-carbon steel	Agricultural machines and domestic appliances.
Stainless steel sheets	Medical equipment; cutlery.
Engineering and 'super-clean' sheets	Manufacturing of components for stressed applications: aircraft industry; components of steam and gas engines, and turbines (rotors, disks, blades); watchmaking industry; surgical implant materials.
LONG STEELS	
Blooms and billets	Locomotive wagon axles; auto parts.
Bars (tool and die steel)	Tools in general; domestic appliances; autoparts.
Structural shapes (light, medium, and heavy)	Civil construction; agricultural machinery; heavy mechanical industry; shipbuilding; automobile industry.
Wire rod	Screw, bolts and rivets; domestic appliances.
Concrete reinforcing bars	Civil construction (bridges, roads).
Rails and track accessories	Rails and tracks.
Seamless tubes and welded pipes	Water and gas pipes; furnaces for nuclear power plants; Oil pipes and off-shore platforms; petrochemical, automobile and aircraft industry.
DRAWN PRODUCTS	
Wire	Automotive industry; telecommunication cables.
Forging	Forging for vehicles (tractors)

Notes: (a) As far as flat steels are concerned, a steel 'plate' is distinguished from a steel 'sheet' by its thickness. Normally, a thickness in excess of 6.3mm is called plate, while thinner products, which may be coiled, are called sheet or strip. Plates and hot-rolled sheets are normally 2,000-6,000mm long. Hot-rolled coils are normally 1.5-12.7mm thick, cold-rolled sheets and coils are 0.4-1.9mm thick. Galvanised sheets are 0.3-2.7mm thick and coated sheets and 'tin-plate' (very thin) are 0.18-0.45mm thick and may be double-reduced to 0.15-0.28mm thick.

Sources: Adapted from IBS (1999) and Steel Technology International (1997/8).

so that the simple copying of techniques used in other plants is not viable in this industry. As a result, efforts to develop capabilities for incremental improvements in existing facilities (e.g. BF, steel shop), processes, production organisation, and products are quite critical for performance improvement in any steelplant (Maxwell, 1982).

4.2 KEY TECHNOLOGICAL CHARACTERISTICS OF THE STEEL INDUSTRY

The key characteristics of innovative activities in the steel industry can be summarised as follows:

1. The industry has been involved in major innovations in process, equipment, and products on the basis of R&D and engineering (e.g. the Corex process; the compact strip mill). Although suppliers are involved in these activities, a large part has been led by the steel industry across different countries.
2. In parallel, incremental innovations in processes, production organisation, equipment, and products constitute a key technological feature in the development of the industry world-wide. Steel producers play a key role in undertaking these incremental innovative activities.
3. Therefore there is an interplay between the role of suppliers of processes and equipment and steel producers in the development of the steel industry and technology. For the steel producer the development of a firm's technological capability is critical to the undertaking of incremental innovative activities and to the building and sustaining of competitiveness in the industry. In doing so they also contribute to the innovative activities in (1).

4.2.1 The Relevance of Technological Capability Development in the Steel Producer

The characteristics outlined above are in line with the categorisation of 'modern' steelmaking as a 'heavy process industry'. Its distinctive characteristic is the demand for the project owner/operator's technological capability development, particularly in late-industrialising countries (Bell et al., 1995). In addition, innovative activities within steel companies should not be seen as being polarised in terms of basic or advanced (e.g. R&D based). Indeed, in the light of Table 3.1, a major part of their innovative activities would be within Level 5 (intermediate) and Level 6 (high-intermediate). Therefore to undertake innovative activities steelmakers need to develop technological capability at these levels. In the light of these characteristics, one can outline at least four types of technological capability whose development is critical for the steel company:

1. *Investment capabilities*   Since the setting-up of steel production facilities is associated with costly investment, the development of capabilities for decision-making and control of plant expansion and project engineering (basic, detailed, installations, procurement, and technical and feasibility studies) is crucial. These activities may well be provided by equipment or process suppliers, by other steel companies, or by specialised project engineering firms. However, the more the steel producer delegates those activities to third parties, the less is its control over project costs, risks, and overall management. It becomes even more crucial for the steel producer to respond effectively to the emergence of recent innovations in the industry (e.g. compact strip production, Corex process). Therefore to undertake in-house innovative investment activities, steel companies need to accumulate technological capability up to Levels 5 and 6.

2. *Process and production organisation capabilities*   Process activities are a major part of the industry. The steel producers' capability to adapt the process technology to local conditions, improve processes parameters and the production organisation practices continuously is crucial for performance improvement over time and for building up and sustaining competitiveness. For these activities, steelmakers need to accumulate technological capability to at least Level 4. In addition, today's steel shops based on the oxygen process are supplied with automated process control systems. However, the steel plant plays a major role in creating effective mathematical models and effective daily production organisation practices and systems to increase the yield of the steelmaking process. These activities are associated with the development of capabilities for process automation/integration at Level 5. In parallel, for a stable operation of large-scale processes and equipment, the steelmaker needs also to accumulate adequate routine operating capabilities (Levels 1 and 2).

3. *Product capabilities*   Steel products have been utilised in different industries and the demand for higher-valued steels for specific uses has increased from the early-1980s. During the 1990s, following a world industrial trend, steelmakers' capabilities to improve product quality on the basis international standards (e.g. ISO or QS 9000) and to meet users' specific delivery needs have become critical to compete. In particular, the steel industry has been pressured by the automobile and household appliance industries world-wide to supply lighter and more high-resistant steels. This has put additional pressure on the development and/or deepening of the steelmaker's capability to accelerate the rate of continuous improvement in existing products and/or new product development. These activities call for the accumulation of innovative capability at Levels 5 to 6. However, it seems unlikely that steel companies could achieve successful product development without the routine capability to manufacture the new products adequately. In other

words, they also need to develop routine capability for products and process and production organisation (Levels 1 and 2).

4. *Equipment capabilities*   Steel plants operate large and complex equipment (e.g. blast furnaces and steel shops) where improvement in daily performance is also associated with effective maintenance and planned revamping. This calls for the development of Levels 4 to 5 capability. These activities may be provided by suppliers or specialised engineering firms. However, this leaves the steelmaker with limited knowledge on the equipment functioning and its intricacies. In the long term, it may mitigate against overall process performance. In addition, the capability to design and manufacture the plant's own components and large equipment is also critical for the steelmaker. The accumulation of this capability may also increase the knowledge on the principles underlying the equipment, hence contributing to the achievement of rapid process improvement performance.

NOTES

1. This subsection draws on the researcher's direct-site observations and interviews in six Brazilian steel companies during the pilot study and the main fieldwork, McGannon (1985) and Araujo (1997).
2. This shop is also referred to as LD Steel Shop, as a reference to the name of the two Austrian towns – Linz and Donavitz – where this process was first used.
3. A 'billet', normally square or rectangular, is used for long products. A 'slab', normally wider and thinner than a billet, is used for flat products.

5. Research Design and Methods

5.1 KEY ELEMENTS OF THE RESEARCH DESIGN

5.1.1 The Research Questions and Related Assumptions

This study has been structured to answer three questions:

1. How different were the paths of technological capability-accumulation followed by two large steel companies in Brazil, over time?
2. To what extent can those differences be explained by the key features of the various processes by which knowledge is acquired by individuals and converted into the organisational level – the underlying learning processes?
3. What are the implications of the technological capability-accumulation paths for operational performance improvement in these companies?

In answering these questions this study will explain that:

1. The paths of technological capability-accumulation followed by those two companies are diverse and have proceeded at differing rates.
2. The differences in their paths are associated with the key features of their knowledge-acquisition and knowledge-conversion processes.
3. Differences in the rate of operational performance improvement are associated with differences in the way and rates at which technological capabilities have been accumulated over time.

5.1.2 The Research Analytical Framework (elaborated)

In the light of Figure 3.2, the research analytical framework is represented in more detail in Figure 5.1 (adapted from Bell, 1997). Central to this book is how the key features of the learning processes influence the technological capability-accumulation paths. This relationship is represented by the two central rings in the figure. The implications of this relationship for the rate of operational performance improvement is represented by the horizontal dotted line running in parallel with the company's lifetime. Other rings represent the influence coming from other factors which, although recognised, are outside the focus of the study.

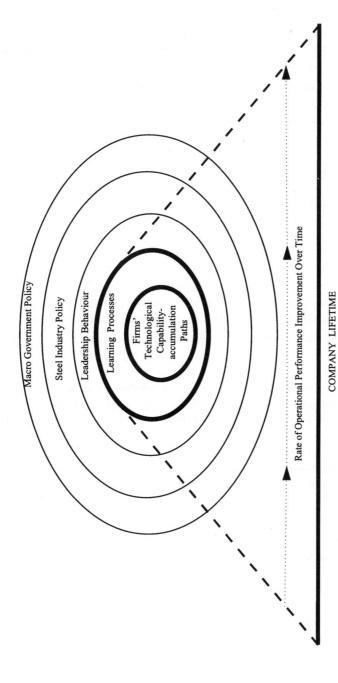

Macro Government Policy

Steel Industry Policy

Leadership Behaviour

Learning Processes

Firms' Technological Capability-accumulation Paths

Rate of Operational Performance Improvement Over Time

COMPANY LIFETIME

Figure 5.1   The book's analytical framework (elaborated)

5.1.3 Unit of Analysis

There are different ways of defining the unit of analysis. The term may express the problem focused on in the study (Yin, 1994) or it can be related to the explanation aimed at by the end of the study (Patton, 1990). In this study the unit of analysis is the issue the study seeks to describe and then explain, in other words, the paths of technological capability-accumulation followed by the two case-study companies. In particular, it examines how and why they differ.

5.1.4 Modifying the Framework for Technological Capability in Steel

The original framework for Table 3.1 was first presented to key informants in the steel companies and institutions in the early stages of the pilot work. As the feasibility to adapt it was confirmed, the adaptation process proceeded through the pilot work. However, this was more painstaking than had been expected. Through the pilot interviews, information on several types and levels of technological activities was collected. This was supposed to match the logic of the original framework. However, it was provided in a rather scattered way. The main challenge was to group and classify those activities accurately and coherently.

The original idea was to disaggregate the framework into different areas of an integrated steelworks (e.g. reduction, steelmaking, rolling mills, etc.). However, following a consultation with two key informants in different companies, this idea was ruled out from the outset. Although the disaggregation itself was feasible, analysing the capability-building in one large company and a cross-company analysis would be extremely complicated. As a result, it was decided to build the framework on a global basis, following the logic of the original one. Following the pilot work, different aspects of a steel company's activities were brought together to build a new framework. 'Routine' technological capabilities were disaggregated into two levels and 'innovative' capabilities into five levels. This then was used to analyse the pilot-study evidence. Although it worked quite well, it needed further refining.

During the early stages of the main fieldwork, the adapted framework was again presented to key informants in the companies. Taking their comments into consideration, the shaping process proceeded during the interviews. Following the main fieldwork, the framework was improved on the basis of the new information collected. The outcome was then mailed to three key informants for a check. A further check took place six months after the fieldwork, during a four-hour meeting in one of the case-study companies. During that meeting, a detailed and interactive review of the adapted framework was carried out.

5.1.5 Combining Qualitative with Quantitative Elements

Although this research is primarily qualitative, it involves elements of quantitative research. Indeed quantitative and qualitative approaches are not mutually exclusive strategies for research. Both qualitative and quantitative data can be collected in the same study (Patton, 1990). The qualitative approach will permit the tackling of the issues of paths and learning processes, through histories of the companies. The quantitative approach would strengthen the comparative analysis of those issues and, particularly, the cross-company differences in operational performance improvement.

5.1.6 In-depth and Comparative Case-study Methodology

The choice of a case study is particularly conditioned by the research questions and unit of analysis. The case-study methodology is more appropriate when 'how' and 'why' questions are being asked about a past or current phenomenon drawing on multiple sources of evidence (Yin, 1994). This is the case with this research. Therefore the case-study method was chosen as more appropriate for this research. A wholly historical method would focus on the past events and sources. This would not permit tracing out the problem by linking distant and recent events. In addition, this study is interested in how learning processes have worked over time rather than only in their incidence or frequency. Therefore a wholly survey methodology would also not be appropriate. While the results of surveys are generalised to populations – statistical generalisations – the results of case studies are generalised to theoretical propositions – analytical generalisation (Yin, 1994). This latter is the primary concern in this research.

5.1.7 Criteria and Process to Select the Cases

Understanding the critical phenomenon may depend on choosing the cases well (Patton, 1990; Yin, 1994). As opposed to probability sampling, the logic and power of purposeful 'sampling' is to select information-rich cases to study in depth (Patton, 1990). 'Information-rich cases are those from which we can learn a great deal about issues of central importance to the purpose of the research' (Patton, 1990: 69). The process undertaken to identify and select the cases went through four stages.

During the exploratory work: selecting steel

The primary function of the exploratory work was to decide on the type of industry within which the research would be developed. By meeting researchers and analysts in industry associations and research institutions in Brazil, information on the main characteristics of three industries – engineering, petrochemical, and steel – was gathered. Following some comparisons, steel was selected. In addition to the reasons outlined in Chapter 1, other reasons influenced the choice: the characteristics of steel technology, the industry, and companies fulfilled the research topic, whose design was then being formed; and I had recently participated in a study, as a research assistant, which included a few steel companies.[1] By conducting the field interviews for this study in Brazil, this researcher became particularly interested in the steel industry.

During the pilot study
On the basis of the material collected from the exploratory work, six steel companies were selected for the pilot study. The early stages of the pilot work consisted of interviewing specialists in steel-related institutions. They provided an overview of the technological activities in the main steel companies in Brazil. They also nominated a few key informants in those companies. The pilot study met its objectives successfully: differences in the companies' paths of technological capability-accumulation (and rates) were found; the feasibility of tackling the learning mechanisms was confirmed; companies that could illustrate different paths associated with different learning mechanisms were identified; therefore the practicability of the research topic, the primary objective of the pilot work, was confirmed; and the pilot work paved the way for the main fieldwork.

During the main fieldwork
On the basis of the material from the pilot study, four steel companies were selected to illustrate variety in the paths of technological capability-accumulation associated with different learning mechanisms and operational performance. From these, two companies with the same process and product technologies were selected: Companhia Siderúrgica Nacional (CSN) and Usinas Siderúrgicas de Minas Gerais S.A. (USIMINAS). Two other companies with similar process technology, but different product technology were also selected. It took two months to select these companies. However, as the fieldwork proceeded, it was realised that: (1) the scale of information that had been gathered from the three initial companies was quite large; (2) those three companies had fulfilled the objective of information richness and variety. Therefore the fourth company was dropped.

During the writing of the case-studies
Although three companies had been fully researched, the third company was dropped when nearly 70 per cent of the case-study had been written in first draft. This decision sought to cope with scale of information in the study; and to keep an adequate level of detail in the analysis of the two companies in the study. As a result, the study ended up by examining two companies: USIMINAS and CSN. Key details about them are provided in the introduction to Part II.

5.2   OPERATIONALISING THE RESEARCH STRATEGY

5.2.1   Sources of Information

This study is primarily based on empirical information gathered in different areas of two steel companies. The sources of information were conditioned by the research methodology. As a result, information-gathering based essentially on closed-ended questionnaires was ruled out from the outset. That was not appropriate for the kind of information needed to illuminate the research questions. Complementary information was also gathered from steel-related institutions in Brazil.

In the steel companies

There were four sources of information within the steel companies, as indicated in Table 5.1. As far as the stay in the steel companies is concerned, during the pilot study, the activities concentrated on some interviews and casual meetings. These were held in head offices with informants who had worked in the plant. Some were held in the plant followed by tours around the operations units with engineers or technicians. Institutional publications were also gathered. A former chief engineer who had retired from one of the companies was located and interviewed. He provided invaluable information to confirm the practicability of the research.

Table 5.1   Sources of information in the steel companies (pilot and main fieldwork)

Sources of information	Details
(1) Open-ended interviews. The informants are organised in three groups:	*Group 1* General plant managers, managers, superintendents, engineers, technicians, foremen and operators of the coke, sinter, blast furnace, steelmaking and continuous casting, and the hot and cold strip mills. Managers, engineers, researchers and technicians from the research units and laboratories, engineering, metallurgical, PPC and quality control, and maintenance units *Group 2* Industrial and development directors. Vice-president and president-directors. Managers (and analysts) of the human resources, corporate planning and marketing areas. Advisers to corporate directors *Group 3* Former company members: presidents, plant and corporate managers, and engineers
(2) Direct-site observations	These involved the observation of individuals at work (e.g. engineers, operators), meetings, and their presentations in the companies' events. They have also contributed to capturing aspects of current features of the learning processes and companies' behaviour
(3) Casual meetings	These involved relatively unplanned meetings with individuals in the sites
(4) Companies' documentary and archival records	This material was gathered in the form of annual reports, internal bulletins and newspapers, organisational charts, institutional CD-ROMs and videos, commemorative and historic institutional publications, technical papers published by the companies' individuals, and copies of overheads of their presentations in or outside the company

5.2.2   Preparation for the Field Activities

The preparation for the pilot study consisted of sending a letter of introduction to six companies, four months in advance. This was followed by the elaboration of the financial and operational plan. A key activity in that phase, was the elaboration of the interview guide. During the pilot work the informants were told they would be interviewed again within six months.

The preparation for the main fieldwork began two months in advance. The logistical planning activities involved an introductory letter to four companies detailing dates, estimated duration of stay, description of interviews, the areas and individuals to be interviewed; a detailed financial and operational plan. The substantive activities involved the elaboration of 'research intermediate categories'. They were 'intermediate' because their level of disaggregation was between the main research questions and the interview questions. They were built to clarify the 'kinds of information' needed to illuminate the research questions.

The building-up of those categories consisted of breaking the components of the research questions down into simpler and more comprehensible terms. Only when those categories had been well elaborated, did the preparation move to the elaboration of the interview guide. Following the structure of the pilot study interview guide, the questions were grouped in two levels: generic and specific. The specific level was further disaggregated into types of technological activity, learning mechanisms, and types of informants.

5.2.3   In the Field: Operationalising the Research Strategy

Carrying out the interviews

Interview strategy   In the pilot study the strategy had been to conduct the interviews in as open-ended a way as possible. As a standard procedure to begin each interview, about 3 minutes were spent presenting the research topic. To save time and to make the message clear, this presentation was aided by Figure 3.1. This procedure sought: to generate interest in the research; to build confidence in the research by expressing the researcher's commitment to the high-quality outcome; to make the informant feel more involved in the research and aware of the type of information needed. Drawing on the interview guide, the informants were asked for 'stories' about their technological activities. More than 100 in-depth interviews were carried out during the fieldwork.

It was decided, before the pilot work, not to tape-record the interviews. Following this strategy, each interview was conducted as a 'structured conversation'. This sought to create as informal and comfortable an atmosphere as possible. Hence 20 × 12cm cards were used to take notes. At the beginning of each interview, the date, the informant's name, position, and the start time of the interview were entered on the card; each card was numbered as the interview proceeded and the finish time recorded. Having a clear idea about the duration of each interview was important in planning and controlling subsequent interviews.

The interview strategy proved quite successful but extremely exhausting. While the informants were telling their stories, the contents had to be assimilated and recorded quickly. In parallel, the researcher had to be in control of the interview to make interventions, ask for clarification and, particularly, to keep the interview focused. This experience demonstrated the difficulty involved in conducting the interview according to the sequence in the guide. Sometimes the informant made a point related to a later stage in the guide. This made it necessary to skip to different parts to avoid missing the chance of exploring a certain point. Therefore the researcher had to be very familiar with the interview guide. Another standard procedure was to finish each interview with the question: 'Could it be better?'. This question proved quite powerful in gathering information about the constraints (and stimuli) to the activities in the informants' areas. On several occasions, this opened up the opportunity for more detailed questions.

Taking notes   After the experience of the pilot interviews, it was realised that the note-taking had to be improved. This meant being more active during the interviews. A new procedure, aided by a kit consisting of the cards and two pens (one red and one blue), was adopted. The blue pen was used to take normal notes, using key words or the informant's verbatim quotations. The red pen was used to note down any critical points raised during the talk: differences and similarities between companies, conflicting or complementary pieces of stories from earlier interviews, insights relating to the way that information could be used in the study. After each critical interview, more notes were made to include what could not be written down during the interview. The key words were expanded.

Adjusting the interview guide   From the mid-point of the fieldwork, the interview guide began to be adjusted on a daily basis. As the interviews proceeded, more details were gathered about the problem, triggering the need to deepen the level of information required. Early each morning, at the hotel, specific questions were formulated for the interviews of that day. Some of these interviews had been arranged only the day before. In the adjusted guide, questions about a specific project or event, which had taken place in that area, were asked. The questions also sought to confirm the accuracy of a particular story, to complete or to sort out conflicting stories.

Interview snowballing   Some informants suggested other individuals for interview. If more information was needed, the suggested interview was arranged. On other occasions, the informant was asked to nominate another informant, when the need for more information became clear. Interview snowballing was extremely useful to locate new interesting informants. However, it was quite time-consuming to keep arranging or re-arranging the interview schedule.

Off-company interviews   These were interviews with members who had retired from the company. They were identified through the snowballing process. The interviews took place at their home or current workplace, in the evenings or during the weekends. They had all been involved in critical positions and on projects for many years in the companies. Six former members of the companies were interviewed, among them two former presidents. They provided invaluable information for the research.

Collective interviews   On three occasions three or four informants were interviewed together. One such occasion was in the pilot study but the result was not satisfactory. The other two were during the main fieldwork and were successful. In one interview – in an automation unit – the interview followed the whole guide without the need for additional information, each informant helping to cover different periods and pieces of the history. In another interview – in a steelmaking unit – the interview also went through the guide, but additional information was needed. At the end of that particular interview, one engineer was asked for a further interview. He had more stories to tell about the company's product development activities.

Asking for the interview papers   On several occasions, the informants made some drawings on pieces of paper as they spoke (e.g. about changes in process or equipment, in organisational charts and flows, etc.). Those drawings were normally accompanied by qualitative and quantitative notes. By the end of each interview, the informants were asked to hand over these notes, which generally looked quite messy. They were 'translated' immediately after the interviews in order to keep them meaningful. They were stapled to the respective interview cards. They provided a great deal of additional detailed information and were particularly useful during the analysis of the fieldwork material and in the writing-up process.

Casual meetings
Casual conversations were useful to complete stories, check for their accuracy, and even to sort out conflicts between them. On some occasions, they snowballed more interviews or revealed the existence of projects or events that needed to be investigated. They took place at lunch time or in the evenings. Some took place during tours around the plant and laboratories. Some were not literally 'casual', since they were deliberately provoked by the researcher (e.g.

starting a conversation with an operator in a lunch queue leading to a fruitful meeting). Obviously, notes were not taken during such conversations and had to be made afterwards. Casual conversations were also helpful in learning more about steel technology. In the tours around the mills, wearing a hard hat and special glasses, interactions took place with operators and foremen. They talked enthusiastically about the way their units worked. Sometimes, changes in process and production organisation and equipment, that had been mentioned in the interviews, could be verified in detail. Such moments provided an education on steel technology.

Making direct-site observations

These observations were useful to gather information about the firms' present phase. In particular, they helped to gather information about the way some learning mechanisms worked (e.g. presentation of QCC groups) and about the role of leadership. While some of these observations confirmed the information that had been gathered in the interviews, others revealed discrepancies. The observations took place in different areas in the companies. Notes were seldom taken while observing, but written up later in a different place.

Decision to leave each company

The decision about when to stop the information-gathering process within each company was made on the basis of four conditions: if key individuals involved in critical technological activities had been met with and essential archival records had been consulted and/or copied; if different individuals in different areas had been met with and interviewed about the same issues and their stories confirmed; if evidence on the key relationships between the research design issues had been gathered, confirmed and clarified, in other words, if sufficient evidence for internal validity had been achieved; and if the interviews and casual meetings were no longer adding new information. Although these criteria caused some delay to engaging with a new company, meeting these conditions was crucial for the fieldwork effectiveness. Several practical problems arose in the field during the operationalisation of this strategy. Coping with them was critical for the successful implementation of the fieldwork. This proved to be an enormous learning experience for the researcher.

5.3   ANALYSING THE FIELDWORK MATERIAL AND WRITING THE CASE-STUDIES

The process for analysing the collected material started during the fieldwork. A 'memo-book' was used where 'notes for analysis' were written on the basis of the interview cards produced during the day, particularly the 'red notes'; and the notes on casual conversations and site-observations (e.g. differences between companies, implications between interviews, implications of some findings for the research questions, some insights for the study conclusions). The first activity after the main fieldwork was the basic organisation of the material collected. The preliminary analysis of the fieldwork material went through three main stages as follows.

1. This consisted of building a set of analytical tables drawing on the organised empirical evidence. Each table focused on one research issue and covered three companies across time. In addition, the field analytical notes were organised in one document: 'reflections from the fieldwork'. The building of those tables permitted: (1) the initial selection and systematic distinction and grouping of the scattered pieces of empirical evidence into organised categories (e.g. types of technological activities, knowledge-acquisition, and knowledge-conversion mechanisms); and (2) an overview of the different patterns followed by each company. However, the information in these tables was too condensed and the variables needed further distilling. In addition, the cross-company differences in operational performance were not quite clear at that stage.
2. This consisted of building a new set of analytical tables with a greater level of detail and disaggregation of variables. As a result, not only the number but the size of the tables increased. In the tables, the case-study companies were normally displayed in the rows and the variables in the columns across time. The tactic was to identify whether and how each variable (and its components) differed across companies. To enrich the analysis process, an analytical text, of about three pages, was elaborated for each table. This exercise permitted (1) the identification of a different evolution of the variables in each company; (2) making relationships between variables with a greater level of accuracy; (3) the influence of intervening variables (e.g. external conditions; leadership); and (4) making plausible interpretations and drawing conclusions from the empirical evidence. Although this stage proved more effective than Stage 1, there were some discrepancies in the vocabulary used across the tables and analytical texts. Further homogenisation of terms was necessary.
3. Before building a new set of tables, the analytical frameworks were improved. The main tactic was to improve the research 'yardsticks', Tables 3.1 and 3.2, to examine the empirical evidence. In addition, a framework of phases was built to improve the cross-company comparison (see introduction to Part II for details). This exercise facilitated: the homogenisation of the vocabulary across the new set of analytical tables and texts; a greater connection of the empirical evidence with the conceptual framework; an even more reliable interpretation of the cross-company differences and more ac-

curate relationships between the variables. Unlike the previous stages, a deeper analysis of the indicators of operational performance was undertaken consisting of: re-grouping the indicators into categories and contrasting them across companies to highlight differences; and combining the quantitative with the qualitative information to interpret those differences.

Only when the analytical tables had been consolidated, did the work move to the writing-up of the case-studies. The final analytical tables operated as a guide for the writing process. The strategy was to write up the cases chronologically in the light of: (1) Tables 3.1 and 3.2; and (2) the comparative framework of phases. To cope with long time-periods and large amount of information, short sketches were initially written for each issue in each company within a limited time period. The first draft of those sketches was hand-written and then gradually extended in word processing, as part of the writing tactic.

NOTE

1. See Saravia (1996). In Brazil, the study involved the steel and petrochemical industries. In Argentina, it involved the steel, gas, oil, and railways industries.

PART II

Technological Capability-accumulation Paths
and the Underlying Learning Processes in the
Case-study Companies

Introduction

Part I of the book has set out the background and framework for studying technological capability-accumulation paths and the underlying learning processes. In the light of this framework, Part II investigates these two issues in the case-study companies. In particular, the empirical chapters in Part II will describe the paths of technological capability-accumulation followed by two large integrated steel companies in Brazil over their lifetime. In parallel, the chapters will describe the learning processes underlying those paths. As the start dates and ages of the case-study companies are different, common phases have been created. Section II.1 provides a brief overview of the case-study companies and their common phases, leading to the organisation of the empirical chapters.

II. 1   THE CASE-STUDY COMPANIES AND THE COMMON PHASES

USIMINAS was established in 1956. Since the start-up of the plant, in 1962, the company has been producing flat steels using the oxygen-steelmaking process technology. Following its start-up, USIMINAS went through three major expansions. The company was state-owned until October 1991 when it was privatised. Table II.1 summarises the main phases over USIMINAS's lifetime.

USIMINAS is located in the state of Minas Gerais. The 'Intendent Câmara' steelworks is located in Ipatinga, 320 km north-east of the city Belo Horizonte. USIMINAS ranks second in Brazil in steel production volume. The turnover in 1997 was US$2.1 billions.[1]

CSN was established in 1941. Since the start-up of the plant, in 1946, the company has been producing flat steels. It started by using the Siemens-Martin steelmaking process technology.[2] In 1977 it moved to the oxygen-steelmaking process. Following its start-up, CSN went through three major expansions. The company was state-owned until April 1993 when it was privatised. Table II.1 summarises the main phases over CSN's lifetime. CSN is located in the state of Rio de Janeiro. The 'President Vargas' steelworks is located in Volta Redonda, 180 km north-east of the city of Rio de Janeiro. CSN ranks first in Brazil and

Table II.1 The main phases of the case-study companies

Companies	Project planning and plant installation	Period between start-up and major expansions[a]	Major expansions and plant upgrading	Privatisation and plant upgrading
USIMINAS	• Project preparation: 1956–61 • Creation: 1958 • Plant start-up: 1962 • Completion of the installation of first planned integrated works: 1965 • Plant initial designed capacity: 0.5million tpy	• 1966–72	• Expansion I (1973–76): 1.4million tpy • Expansion II (1974–78): 2.4million tpy[b] • Expansion III (1979–86): 3.5million tpy	• Privatisation: October 1991 • Plant updating and increase in the scale of activities: 1990s

CSN	• Project preparation: 1938–41
	• Creation: 1941
	• Plant start-up: 1946
	• Completion of the installation of the first planned integrated works ('Plan A'): 1948
	• Plant initial designed capacity: 0.27–0.36 million tpy
		• 1949–53	• Plan 'B': (1950–53): 0.75million tpy
			• Plan 'C': (1956–60): million tpy
			• Intermediate plan: (1961–63): 1.4million tpy
			• Plan 'D':
			  • Stage I: (1969–74): 1.7million tpy
			  • Stage II: (1975–78): 2.5million tpy
			  • Stage III: (1979–86): 4.6million tpy
				• Privatisation: April 1993
				• Plant updating and increase in the scale of activities: 1990s

Notes: (a) Within that period although the first major expansion had, to some extent, been planned it was not executed. As a result, the companies had to operate with the original facilities.

(b) The overlap in the years means that Expansion II began to be planned while Expansion I was being implemented.

Sources: USIMINAS and CSN.

Table II.2 Time boundaries between the common phases and the related chapters

Case-study companies	Start-up and initial absorption phase	Conventional expansion phase	Liberalisation and privatisation phase
USIMINAS	1956–72	1973–89	1990s
CSN	1938–53	1954–89	1990s
Related Chapters	**Chapter 6**	**Chapter 7**	**Chapter 8**

Source: Own elaboration.

Latin America in terms of steel production volume. In addition, CSN is the world's largest tinplate producer within one plant. The turnover in 1997 was US$2 billion.[3]

To make the cross-company comparisons more meaningful the case-study companies are investigated under common phases. The time boundaries between these phases and the related chapters addressing them are outlined in Table II.2. The 'start-up and initial absorption phase' covers the activities from the years of project definition and operational start-up to the year before the planned installations of the new large facilities began to take place. The 'conventional expansion phase' refers to the major planned expansions of the plant's capacity.[4] This phase covers the years from the start of installation of the new facilities to the late-1980s, when new economic conditions were set in Brazil and in its steel industry. These two phases are associated with the dominance of the import substitution (IS) policy, the state-ownership of the steel industry, and the protectionism of the economy. The 'liberalisation and privatisation phase' refers to the new set of economic conditions that emerged in Brazil and covers the activities in the early-1990s up to 1997, when the fieldwork for this research took place.

NOTES

1. USIMINAS, Annual Report 1997.
2. This process was also known as the 'open-hearth' process because the charge was exposed to flames moving over its surface.
3. CSN, Annual Report 1997.
4. The terms 'initial absorption' and 'conventional expansion' are adapted from Maxwell (1982).

6. The Start-up and Initial Absorption Phase

This chapter describes the paths of technological capability-accumulation followed by the two companies during the start-up and initial absorption phase. In parallel, the chapter describes the learning processes underlying those paths. The chapter covers the 1956–72 period for USIMINAS and the 1938–53 period for CSN.[1]

6.1 TECHNOLOGICAL CAPABILITY-ACCUMULATION PATHS

In the light of the framework in Table 3.1, this section traces the technological capability-accumulation paths followed by USIMINAS (Section 6.1.1) and CSN (Section 6.1.2) during the start-up and initial absorption phase. The cross-company differences in their technological capability accumulation paths are outlined in Section 6.1.3.

6.1.1 USIMINAS: 1956–72

Within ten years, USIMINAS had accumulated Levels 1 and 2 routine capabilities (basic and renewed, respectively) across all five technological functions: investments (facility user's decision-making and control, and project preparation and implementation), process and production organisation, product-centred, and equipment. In parallel, USIMINAS had also accumulated Levels 3 to 4 innovative capabilities (extra basic to pre-intermediate, respectively) across three technological functions, as indicated by the evidence below.

Investment activities
As this subsection indicates, USIMINAS engaged actively in the accumulation of both routine and innovative investment capabilities between 1956 and 1972. By the end of the start-up and initial absorption phase, within ten years, USIMINAS had accumulated Levels 1 up to 4 routine (basic to pre-intermediate) capability for investment activities.

In April 1956, USIMINAS was established as a pilot company. In parallel, a working group was created to conduct the initial negotiations with foreign investors, particularly the Japanese. Drawing on a strong engineering background, members of the working group engaged actively in some of the project engineering activities. This group formed the initial basis for the accumulation of USIMINAS's investment capabilities. In May 1957, following the signing of the 'Horikoshi–Lanari Agreement', USIMINAS was transformed from a pilot into a definitive company. That agreement was the basis of an eight-year technical assistance agreement with Yawata Iron and Steel Co (Yawata), to assist USIMINAS in initial plant operation. The agreement also detailed the project funding. Following the signing of that agreement, most members of USIMINAS's initial board pulled out from the project.[2] This event triggered the appointment of the metallurgical engineer, Mr Lanari, as USIMINAS's president in early 1958.[3] He and his directors led the dominant group in USIMINAS. This leadership championed several audacious projects, contributing to the development of USIMINAS's initial key technological capabilities, as described later.

Taking the initiative of deciding 'who' should do 'what', USIMINAS's leadership assigned the technical area of the company to the Japanese until the Brazilians had the capability to operate the plant efficiently. When the 'Preparatory Mission' came to USIMINAS in May 1958 to develop further project details, USIMINAS had completed detailed technical studies on the plant location in the Ipatinga region and the availability of raw materials.[4] This initiative contributed to the gradual building of its own investment capabilities.

In 1959, through its purchasing office in Tokyo, USIMINAS engaged in project design, approval of equipment specifications and bids for equipment acquisition, contracts with suppliers and inspection of equipment manufacturing activities with the Japanese teams. Taking the lead in the construction activities, despite the financial difficulties that began in 1962, USIMINAS was able to synchronise construction with installation activities, keeping them on schedule. Instead of waiting for its partner's final word, USIMINAS engaged in independent strategic decision-making. This reflected, as early as 1959, top management's concern with USIMINAS's 'technological independence'[5] from the Japanese. By 1965, USIMINAS had completed the installation of the initial integrated works. By that time, USIMINAS had been able to carry out key investment activities independently. These included defining plant location, preparing terms of reference, and synchronising civil construction with installation activities.

On the basis of Table 3.1, this indicates that USIMINAS had accumulated Level 1 routine investment capabilities, in other words, the basic level of capabilities necessary for routine investment activities. As early as 1965, as a result of the financial crisis, USIMINAS's president made it clear that break-even point would only be reached with the production of one million tpy of ingots, pushing the company into an immediate expansion decision. Two consecutive expansion plans were prepared by USIMINAS, assisted by Nippon-USIMINAS. They were submitted to the group formed by the National Bank of Economic Development (BNDE) and Booz-Allen & Hamilton Intl. (BAHINT),[6] but not approved.[7]

It should be remembered that, from 1964 to 1967, Brazil, and its steel industry in particular, were passing through an economic and financial crisis. As a result of that crisis and the BNDE/BAHINT's report, in 1967, the Consulting Group for the Steel Industry (GCIS) was set up by the government to prepare the National Steel Plan. Following the GCIS's recommendation to increase USIMINAS's capacity to 1.4 million tpy, the engineering nucleus was created in the company to co-ordinate its expansion activities. However, it was not until 1973 that those facilities were fully installed.[8] The severe delays in obtaining the Brazilian government finance for the expansion were associated with USIMINAS's high debt to sales ratio.[9] Constrained by its financial problems and its urgent need to break even, USIMINAS engaged in the rapid absorption of its initial designed capacity. The company had to achieve cost reduction and production expansion without any investment in new facilities. This factor had critical implications for USIMINAS's accumulation of production capabilities, as described below.

Process and production organisation activities

Within ten years, USIMINAS had accumulated Levels 1 and 2 routine capability in association with Level 3 to Level 4 innovative capability for process and production organisation activities. As mentioned in different interviews, USIMINAS seemed strongly concerned with the accumulation of capabilities to operate the plant efficiently. Following the start-up in 1962 USIMINAS sought to take over gradual and effective plant control from the Japanese. In addition, USIMINAS engaged actively in innovative process and production organisation activities.

As early as 1958, USIMINAS set up the system of industrial costs and appropriation and the auditing department to control construction costs. Later they were adapted to control production costs. In that same year, the technical department, consisting of five Brazilian engineers and headed by one Japanese engineer, was created to study the raw materials to be used in production. One year later, that unit was undertaking studies on the search, storage, and supply of raw materials and on the properties of the local refractory materials for the pig iron and steelmaking processes. In 1961, the units activities were strengthened by the creation of a laboratory which, supported by the raw materials department, engaged in the analysis of properties of the agglomerates for the reduction process.[10]

The building of this initial capability permitted USIMINAS to respond successfully to the controversial directions of the National Coal Plan. The way USIMINAS responded to this external factor contributed to the accumulation of a higher level of that capability. Initially, USIMINAS was opposed to the compulsory use of at least 40 per cent of local coal in the reduction process, because of the negative effects on the coke rate and production costs.[11] However, the company moved beyond the use of the sintering process, as a conventional way to bring the coke rate down, engaging in the manipulation of key parameters in the reduction process.

Drawing on his strong technical background, USIMINAS's president knew that the coke rate could be brought down to 550kg if the sinter properties and the metallic charge composition of the BF, combined with the high quality of local iron ore, were manipulated.[12] From 1960 the technical dept and the raw materials dept engaged, under Japanese supervision, on systematic studies on the quality and chemical composition and granulometric properties of the iron fines for the sintering process.[13] In February 1963, when the sinter machine started up, USIMINAS was able to charge the BF with 85 per cent of sinter. In 1964, that 'sinter ratio' increased to 98 per cent. The coke rate was brought down from 731kg in 1962 to the competitive rate of 584kg in 1965.[14] This indicates that USIMINAS was about to meet its own target of 550kg thus approaching international standards.

As far as the company's management was concerned, by 1964 all the administrative positions were occupied by Brazilians. By late-1965, they had 100 per cent control of the technical positions. As early as 1964, USIMINAS took its own initiative to restructure the corporation to 'make the company more professional', to improve the decision-making process, clarify the definition of authority and responsibility and increase training of the Brazilians in the managerial area.[15] Although the management consulting firm BAHINT had been contracted to propose and execute that restructuring in 1964, USIMINAS seemed to be in control of the proposed changes and their outcomes: 'Booz-Allen's people used to come up with several proposals, usually far away from what we were expecting. Their proposals were refused until they came up with a reasonable operationalisation of what we had envisaged for USIMINAS's organisational system.'[16]

The restructuring involved two major projects: the corporate management project (organisational and managerial structures and flows in the plant, corporate office, and regional offices) and the industrial engineering project (expansion of the industrial engineering department).[17] As a result, USIMINAS was organised on a line-staff basis.[18] The metallurgical control and inspection dept was involved in the absorption and application of new metallurgical techniques, quality and operations control, process and product improvements, and the analysis of product performance by the steel users. The industrial engineering dept engaged in the implementation of the standard cost systems, economic analysis of improvements in processes, operational assessment and applied mathematics, control of inventory levels, and time and motion studies.[19] This suggests that USIMINAS was building its own organisational system, in other words, its own technological capabilities to engage in innovative process, product-centred and equipment activities.

As far as the evolution of production in the start-up and initial absorption phase was concerned, the production of molten steel in 1964 was four times that of 1963. This was associated with high utilisation rates which, in the third year of production, were above 55 per cent in the main operational units. By 1965 USIMINAS had achieved a balanced production growth which may be a reflection of its capability for routine production co-ordination across the operational units. Therefore, on the basis of Table 3.1, USIMINAS had accumulated, within 3 years, Level 1 routine (basic) capability for process and production organisation activities.

As a result of its financial difficulties, USIMINAS was forced to try to increase scale to break even, but without new facilities. The company had the operational and support units engage in intense efforts to absorb and 'stretch' the capacity of the original plant. Those efforts contributed to the development of the 'capacity-stretching capability', from 1965, a key part of USIMINAS's history. The increases over design capacity ranged from 20.8 per cent (coke ovens) to 135.8 per cent (steel shop), more than twice the design capacity, particularly between 1968 and 1972. Those rates were associated with several improvements in equipment, process and production organisation. A large proportion of these improvements were done by USIMINAS on its own, reflecting that the 'capacity-stretching capability' was being accumulated, year by year, as higher capacity utilisation rates were achieved. Because of the long-term implications of that capability for USIMINAS's path of technological capability-accumulation, a closer look into the characteristics of its development is particularly relevant.[20]

Implementation of known practices   Capacity 'stretching' derived from practices well described in the steel literature and implemented in steel companies world-wide. It should be remembered that it was not until 1971 that the research centre was set up. Therefore the capacity-stretching activities were undertaken by the operational and support units only. The BF dept and the then technical dept had been engaged, since 1963, in the manipulation of the metallic charge of the BF 1 to bring down its coke rate. Building on that experience, further improvements were undertaken like the screening of the raw materials characteristics, control, preparation, and classification of the charging (raw material). These led to higher yields per tonne of input and fewer processing difficulties in pig iron production.[21] In addition, these improvements had a positive influence on the operations performance of the steel shop, as analysed later in Chapter 10.

In April 1970, when USIMINAS had reached a utilisation rate of 170 per cent of its designed capacity, the company took the initiative of introducing its zero defect (ZD) campaign at the plant. Strongly supported by top management, it started in the hot strip mill and, by June, it had extended to other units in the plant. This campaign sought to eliminate production defects and bottlenecks, improve quality of process and product, increase productivity and reduce accidents, and to increase the operators' involvement in production decision-making. The campaign gave rise to a fruitful competition between the production units for improved utilisation rates and productivity.[22] At the corporate level, the technique of management by objectives (MBO) was introduced in 1970. In association with the ZD campaign, it contributed to the setting of challenging goals for units at both plant and corporate levels.[23]

Progressive systematisation   The capacity 'stretching' evolved from an intermittent activity, in the initial years (1965–67), into a systematic and continuous activity (1968–72). An example of this evolution was the improvement efforts within the sinter plant involving the reduction area, the foundry, and the technical department.[24]

Relatively unplanned outcome   Because the 'stretching' was not a primary goal in itself, its outcome could mistakenly be interpreted as wholly unplanned or unexpected over time. Indeed, in 1969, USIMINAS formally saw annual production of 790,000 tonnes as the 'maximum probable production' that could be achieved with the original plant. By 1972 over 1.2 million tonnes annually had been achieved.[25] The evidence suggests that, month by month, the activity became more planned. Hypotheses were built and preparatory tests and systematic comparison with other plants were done to establish new 'stretching' possibilities (e.g. the steel shop). These continuous efforts seemed to have been triggered by the performance benefits being generated. The case of the steel shop, where the highest utilisation rates were achieved, is illustrative of these characteristics of the accumulation of that capability, as described in Box 6.1.

Relatively low investment needed   The costs of the 'capacity-stretching' expansion, from 0.5 million tpy to 1.2 million tpy, were roughly estimated to be only about US$40 million, compared to US$261 million in the original plant. And 'what little investment occurred was in minor peripheral equipment such as sintering screens, roll crushers, minor modification in major equipment units and technical assistance'.[26]

Pervasiveness   As individual production units engaged in the 'stretching' of their capacity, they generated imbalances in the subsequent units (e.g. the in-

Box 6.1   USIMINAS's steel shop: accumulating 'capacity-stretching' capability

The accumulation of this capability went through four phases as described below. Until 1966 USIMINAS drew on Yawata's assistance. From 1966, most of the activities were carried out by USIMINAS on its own.

Phase 1 (June 1965 – June 1967). It involved the increase of production capacity from 500,000 to 600,000 tpy. It first consisted of increasing tonnes per heat: from 56 to 58tonnes/heat, by changing the charging procedure. It also consisted of reducing the heat time by increasing the oxygen flow (from 7–8,000 to 9,000 Nm³/h).

Phase 2 (July 1967–April 1969). To meet the challenging target of 36 heats/day the following activities were undertaken:

- *Overcoming the converter capacity*   Before undertaking any change in the converter, a careful comparison was made between USIMINAS' converter and 16 other converters world-wide involving 8 variables. This was to assess whether the planned increase was within acceptable metallurgical standards. Then the lining of the converter was increased from 42 to 47.5m³.
- *Changes in the equipment limits*   To increase production, some electrical and mechanical conditions had to be changed (e.g. the tilting mechanism, charging measures in the tilting motors, brake capacity, and lining lifetime). These changes were made by the engineering dept and the maintenance division of the steel shop.
- *Increase in the casting speed*   This was to synchronise the heat time of the converter with the cycle of the casting bridge. This sought to increase the capacity from 600,000 to 700,000 tpy. Indeed the actual production was 790,914 tpy

Phase 3 (from May 1969). By early-1969, it was realised that that capacity 'stretching' was not sufficient. The increase in the sinter production associated with other changes in the BFs led to a production of pig iron above 60,000tonnes/month. Therefore a new plan was set up to increase production capacity to 850,000 tpy. Building on the capability accumulated in the two previous phases, several hypotheses were studied relating to the implications of the new increase of the charge volume for the subsequent units. The resulting modifications involved: (1) the increase in the tonnes per heat (62 to 67tonnes/heat); (2) greater flow of oxygen (9,000 to 11,000 Nm³/h); (3) reduction in breaks time; (4) introduction of a larger scrap gutter; (5) reduction in time spent for repair, improvement in the cleaning underneath the converter, through a system developed by USIMINAS; and (6) re-arrangement of role of supervision and lead operators.

> *Phase 4 (from June 1970).* Psychologically boosted by the successful re-
> sults and building on its own capability, USIMINAS then engaged in new
> improvements to increase the capacity to 950,000 tpy. Among the main im-
> provements were the increase in tonnes per heat (67 to 70tonnes), re-sizing
> the ingot moulds, increase in the oxygen flow (11,000 to 12,500 Nm^3/h),
> increase in oxygen storage, improved co-ordination between BFs and the
> steel shop.
>
> *Note*: Nm^3 = Newton cubic metres.
> *Sources:* Interview with a manager of the steel shop technical division. Plant tour in the steel
> shop with a technician. See also paper by two USIMINAS's engineers: Oliveira, A.A. de
> Andrade & J.A.M. Caldeira (1970), 'Etapas de desenvolvimento da capacidade de produção
> da aciaria da USIMINAS', *USIMINAS Revista,* vol. 1, no. 1, pp. S/14-20.

crease in production capacity of the BF triggered the production increase in the
steel shop which, in turn, triggered the stretching of capacity of the rolling mills,
under the pressure to reduce stocks of intermediate products). As a result, this activity rapidly spread across the whole plant. This contributed to the pervasive accumulation of the 'capacity-stretching' capability in USIMINAS from that time
on.[27]

USIMINAS's capacity-stretching capability, accumulated during the
start-up and initial absorption phase (1962–72), can be seen as a 'platform capability' for the company. This seems to have had implications for its long-term
trajectory of technological capability-accumulation. In addition, it was not only
associated with the improvements, utilisation rates and other performance indicators within the original vintage of plant. It seems to have worked as a platform capability for the company to undertake further innovative activities in
equipment and process and production organisation in further vintages of plant.

By 1972 USIMINAS had undertaken key technological activities like systematic 'capacity-stretching', the building of its own organisational basis and
system (through the corporate restructuring), the introduction of new organisational techniques (e.g. ZD and MBO). In the light of Table 3.1, this indicates
that, within 10 years, USIMINAS had accumulated Level 3 to Level 4 innovative capability for process and production organisation.

Product-centred activities
During the start-up and initial absorption phase (1956–72) USIMINAS was
engaged in the accumulation of capabilities to do routine and innovative
product-centred activities.

In 1965, the oversupply in the steel industry in Brazil and USIMINAS's financial difficulties forced the company to face up to its main domestic competitor, CSN, and to seek export markets. To do this, USIMINAS sought to increase
product quality and diversification by focusing on the production of

high-quality steels for the petrochemical and automobile industries, initially drawing on the Japanese for assistance. USIMINAS's top management made it clear to the plant that 'product quality, rather than quantity' mattered.[28] By 1962, one year before the plate mill's start-up, USIMINAS had established its own technical department, a laboratory and the plate mill to study its plate quality. From June 1963 to May 1964, about 40 different types of steels were produced efficiently to given specifications.[29]

In late 1963, following its entry into the plates segment, USIMINAS was accredited by the 'Lloyds Register of Shipping of London'. This permitted the company to supply the Brazilian shipbuilding industry, as well as Petroleo Brasileiro SA (PETROBRAS), and export markets.[30] From August 1964, USIMINAS slabs began to be shipped to Argentina, Uruguay and the US. The company was awarded the 'Veritas Bureau of Brussels', as a result of the quality of its hot sheets.[31] In the light of Table 3.1, USIMINAS had accumulated, within 3 years, Level 1 routine (basic) and some Level 2 (renewed) capability for product-centred activities. However, this was not sufficient for USIMINAS in 1966 to move into the manufacture of more complex and higher-quality steels for domestic appliances, automobiles, and the petrochemical industries. When the technical assistance agreement (TA-1) was signed with Yawata in 1966, assistance in product quality improvement was particularly emphasised.[32]

As a result, USIMINAS engaged in the manufacture of higher-valued steels associated with higher-quality processing – the product improvement strategy. By August 1966, four months after the signing of TA-1, killed steels with aluminium for extra-deep stamping, and ageing-resistance, for the automobile industry, began to be manufactured in USIMINAS. These activities involved intense efforts on minor adaptations of existing specifications by the metallurgical dept, the steel shop and the hot strip mill, under Yawata's supervision, to meet local demand. This also involved the creation of USIMINAS's own standards for dimension, shape, surface quality, and mechanical properties of steels. In parallel, systematic studies of the chemical characteristics and mechanical properties of new steels were being undertaken by the metallurgical dept as a preparation for engaging in new product developments.[33]

In September 1966, USIMINAS engaged in the introduction of plates and hot sheets for high pressure pipes, following the American Petroleum Institute (API) specification, with some adaptations to meet PETROBRAS's needs. In parallel, the company engaged in the manufacture of high-resistant plates for traction and weldability. A further 65 products (plates and hot and cold sheets) were introduced between 1966 and 1972 including the F-36. This steel was further applied to the manufacture of flotilla by the German firm Hapag AG.[34]

By 1969, USIMINAS was supplying 45 per cent of the steel used by the car and truck industry in Brazil. One year later, 93 per cent of the steel used by the shipbuilding industry and 52 per cent used by automobile industry in Brazil were being supplied by USIMINAS.[35] This suggests that the product improvement strategy had been critical for USIMINAS to achieve greater financial returns to overcome its financial problems.[36] Despite Yawata's technical assistance during 1966–72 period, the evidence gathered on those years suggests that USIMINAS was engaging in product quality improvements on an increasingly independent basis. This reflects the gradual accumulation of its own capabilities.

In addition to the product improvement strategy, the implementation of the ZD campaign in USIMINAS from 1970, as referred to earlier, led to the setting of rigorous limits of tolerance for product defects in the rolling mills (e.g. scrapping rates in plates, hot sheets, and coils) on a progressive basis. As these goals were met, lower limits of defect tolerance were continually set, improving product quality. By 1972, USIMINAS was able to manufacture steels efficiently to given specifications, to supply export markets. In parallel, it improved products by adapting their key properties therefore adding value to them. Thus, on the basis of Table 3.1, USIMINAS had accumulated, within 10 years, Levels 1 and 2 routine capability in association with Levels 3 to Level 4 innovative capability (extra-basic to pre-intermediate levels, respectively) for innovative product-centred activities.

Equipment activities
During the 1962–72 period USIMINAS had moved into the accumulation of innovative capabilities for equipment. The Foundry started in 1963. This unit engaged in the routine manufacture of small components for plant equipment and minor adaptations to equipment to adjust it to the local raw materials and to production organisation. In association with other operational units, the foundry also engaged in 'capacity-stretching' efforts.[37] Evidence related to that period helps to describe the fast rate at which USIMINAS accumulated innovative equipment capability.

In March 1971, the BF 1 was blown out for its first revamping. The revamping was to wholly replace the refractory lining and balance and/or replace some existing auxiliary equipment. In 1967, four years before the revamping, the revamping commission of BF 1 was created, involving the engineering, the coke ovens, and blast furnace. Supported by the then technical dept, these units began to study for the revamping. The commission co-ordinated, supervised and prepared the final report for the revamping.[38] This evidence indicates the preparation that used to take place in USIMINAS before they engaged in any new technical activity.

As USIMINAS did not have the capability to do the revamping on its own, it drew on the assistance of Nippon Steel Corp. (NSC) and Ishikawajima Heavy Industries Corp., for the planning and co-ordination activities. It also drew on

Montreal Engenharia SA, a local engineering firm, for execution activities. After the revamping, BF 1 had an increase of 30 per cent in production over the course of a month. The four-year preparation and the close contact with NSC's assistance contributed not only to a revamping within an international record time of 45 days, but also to USIMINAS learning the details of BF revamping.[39] In addition, while the revamping of BF 1 was being prepared for, in March 1970, BF 2 came up with irregularities in operation. The furnace did not respond to the actions taken by the BF dept to increase its production. The company detected that these irregularities were associated with the development of a hard crust on the furnace internal walls. As the scheduled revamping of BF 1 was approaching, leading to a production fall, USIMINAS decided to take quick action to try to eliminate this crust.

Although the BF dept had been able to identify the hard crust characteristics, USIMINAS still did not have the capability to eliminate it on its own. As a result, a technical assistance contract was signed with the Japanese firm, Nishin Kogyo KK. The evidence indicates that USIMINAS worked closely with Nishin Kogyo KK to learn how to eliminate the problem. BF 2 was successfully blasted with dynamite five times between March and April 1971, with the furnace in operation. After the first dynamiting, the volume of blast air increased by 20 per cent. The furnace's productivity increased by 32 per cent (1970–71).[40]

By June 1972, it was BF 1 that was displaying operation irregularities. The cause was diagnosed by the BF dept as being associated with the development of a hard crust on the furnace internal walls, even though it had just been relined. This crust was different in volume, size, type and location from that developed in BF 2. Although a more complex problem, USIMINAS engaged in the removal of this crust drawing on its own capabilities. It involved various prolonged and painstaking stoppages from February to March 1972, but the revamping was successful.[41] This indicates that USIMINAS seems to have learnt about the dynamiting process. In the light of Table 3.1, the evidence indicates that in addition to the accumulation of Level 1 and 2 routine capability, USIMINAS had accumulated, within 10 years, Level 4 innovative capability for large equipment revamping.

6.1.2 CSN: 1939–53

During this period, CSN failed to accumulate adequate Levels 1 to 3 capability across four technological functions: investments (facility user's decision-making and control, and project preparation and implementation), process and production organisation, product-centred, although Level 1 (routine) basic equipment capability was accumulated. Indeed the evidence gathered about that phase in CSN reveals a low level of concern of the company with the long-term development of in-house technological capability.

Investment activities

During the 1938–53 period CSN did not accumulate adequate Level 1 capability for investments. Project engineering activities during that period were characterised by inadequate preparation and intense delegation of activities to consulting firms. This was associated with decision-making processes that seemed to pay very little attention to in-house technological capability-accumulation.[42] These characteristics may have been influenced by the strong political aspects permeating CSN's project during the first Getulio Vargas government (1930–45). Despite the merits of their genuine efforts to build the first large-scale all-coke steel company in Brazil from the late 1930s, evidence reveals very little commitment, if any, to the long-term development of local technological capability.[43] As a result, by the end of the start-up and initial absorption phase, CSN had not accumulated the capabilities to carry out investment activities independently.

In June 1939, following an invitation by the Brazilian government, a project outline for an all-coke steel plant was prepared by the United States Steel Corp. (USS). This company was expected to engage in the project. However, nationalist groups in Brazil, demonstrated their opposition to the project. They managed to pass the Mining Code in January 1940, which prohibited any foreign capital in mining and metallurgy in the country. This led USS to pull out of the project.[44] This nationalist behaviour had further implications for the way the learning processes worked in CSN and the development, in turn, of production technological capabilities (1940s–1950s), as described later in this chapter.

In 1940 a new project was proposed by the consulting firm, Arthur McKee & Co. This was submitted to the US government, which added some political and military conditions to the project. Another condition was that CSN's steelworks should be led by US engineers and managers, particularly from USS, until the Brazilians had learnt how to operate it efficiently.[45] However, as raised in different interviews in CSN, USS seemed only weakly committed to the success of the project. Added to this, the evidence suggests that technological capability-accumulation was not a key issue for the Brazilian project leaders in their relationships with foreign suppliers. Also, influenced by their nationalist values, the Brazilians did not seek to learn systematically from the Americans how to operate the plant efficiently.[46] These factors had further implications for how the accumulation of technological capability occurred in CSN.

By 1940, the project design and equipment acquisition activities had been concluded by the firm Arthur McKee in CSN's purchase office in Cleveland.[47] In that same year, CSN's technical office was established in Rio de Janeiro. Its decision-making became dependent on the consulting firm. The delegation of construction activities to contractors, the distant involvement of the technical office with the sites, and the delegation of the project design, installations, testing and commissioning to Arthur McKee, all suggest CSN's low level of concern with the development of its own investment capability. The initial integrated works started in June 1946 under the supervision of USS and Arthur McKee.

The creation of CSN in April 1941 was followed by the appointment of its first president, a banker, Mr Guinle,[48] and technical director, a metallurgical engineer.[49] Under this presidency (1941–45) there was little concern with in-house accumulation of technological capability. This might reflect the president's low level of interest in this issue, probably associated with the lack of a technical background. Later, leadership more involved with technological activities was developed by the technical director, during the 1950s, when he became president. However, this presidency contributed to opening up the way for the military engineers to take over key technical positions until the 1970s. Their military background in itself was not the problem, but the way they steered CSN had some negative implications for the functioning of the learning processes and, in turn, the technological capability-accumulation path in CSN, as described later in this chapter.

Process and production organisation activities

During the start-up and initial absorption phase (1938–53), CSN failed to complete the accumulation of Level 1 routine (basic) capability for process and production organisation activities. In other words, particularly during the 1946–53 period, CSN was not able to operate the plant efficiently. In addition, CSN did not move into the accumulation of innovative level of that capability.

CSN's first corporate organisation took place in 1945. It was not until the late 1950s that it changed. During the start-up and initial absorption phase, CSN's organisation was marked by an intense centralisation of activities in the technical directorate and an absence of organisational units to engage in preparatory activities for the operations and further support to the production lines.[50] In March 1946, the appointment of a new president for CSN, former head of the CSN's Purchase Office in the US,[51] contributed to intensifying the presence of military managers and engineers in CSN.

In 1945, following the US government's recommendation, key technical positions were given to Americans until CSN had the capability to operate the plant efficiently. However, at the plant level, things evolved in a rather different way. This marked the beginning of a series of incoherences between the corporate management and the plant as far as technological activities and learning processes were concerned. Most of CSN's engineers refused to work under American supervision. They sought to take over the technical positions within the shortest possible time, probably influenced by their nationalism.[52] By late-1947, less than two years after start-up, the Brazilians had taken full control of the plant. However, the capability to operate it efficiently had not been accumulated.[53]

From 1947 the plant began to be pressed by top management to increase production volume despite cost or process quality. As a result, there was a lack of co-ordination within and across the reduction area and the SM steel shop. Each unit began to work in isolation.[54] The evolution of production in the first three years was characterised by a great instability of the operations of BF 1 associated with frequent stoppages.[55] In principle, the absorption of the original capacity should have taken place during the 1949–53 period. However, very small efforts were made to reach higher utilisation rates and to absorb that capacity. Even so, increased targets for production volume continued to be set by top management.

The average capacity utilisation rates of the coke ovens, BF 1, and the SM steel shop during that period were 72.7 per cent, 81.3 per cent, and 91.9 per cent, respectively.[56] It was not until the mid-1950s that the steel shop reached higher capacity utilisation rates on a more continuous basis.[57] These rates of capacity utilisation may be associated with inadequate routine production co-ordination across the operational units and the absence of continuous efforts to achieve higher utilisation rates. This absence of efforts on incremental innovative activities is also illustrated by evidence on the reduction process.

In 1942, forced by the National Coal Plan to use 40 per cent of local coal in the reduction process, CSN built a facility for coal washing. However, this was not sufficient to minimise the bad effects of the low-quality local coal on the coke rate.[58] Instead of creating organisational units to engage in continuous efforts to improve the reduction process and bring the coke rate down, CSN relied on the acquisition of large facilities: the coal washing mill and the sinter plant. The sintering process, however, was not installed until 1960. This fact suggests that CSN began to rely on the acquisition of the latest large equipment technology rather than developing its own technological capabilities to continually improve its production activities. This initial behaviour had serious implications for CSN's trajectory of accumulation of that capability and for operational performance improvement, as analysed later in the book. In the light of Table 3.1 CSN had failed to complete the accumulation of Level 1 routine (basic) capability for process and production organisation activities.

Product-centred activities

During the start-up and initial absorption phase (1938–53), CSN failed to complete the accumulation of Level 1 routine (basic) capability for product-centred activities. In addition, CSN did not move to the development of the innovative level of that capability. Being the largest flat steel producer in Brazil, CSN began to control the price and the steel supply in the country. Until the late 1960s, the prices of CSN's products were determined by the government in response to its economic policy rather than the company's needs. This indicates the government's low level of concern with the implications of such a policy for CSN.[59]

The absence of any product quality improvement and diversification strategy in that period can be associated with two factors. On the one hand, until the early 1960s, CSN was the only local flat steel supplier for the construction and railroad industries. They usually demanded a narrow range of simple and low-value steels. In 1956, CSN's production represented 55 per cent of Brazilian steel industry output, while 40 per cent of flat steel consumption in Brazil was supplied by CSN.[60] In other words, local competitive pressure was practically absent.

On the other hand, despite having the competition from imported steels, CSN did not seek to develop the capabilities even to replicate the production of simple given specifications efficiently. Neither did the company seek to introduce higher-value and higher-quality steels. Indeed CSN became notorious for failing to replicate simple product specifications efficiently.[61] On the basis of Table 3.1, this indicates that Level 1 routine (basic) product-centred capability had not been accumulated. The absence of efforts to develop in-house product-centred capabilities, particularly innovative capabilities, went through the 1960s. This contributed to mitigating CSN's capacity to face up to competitors (e.g. USIMINAS), as described in the next chapter.

Equipment activities
CSN created its foundry in 1947. During the start-up and initial absorption phase (1940s–1950s), this unit engaged in the manufacturing of simple small components (e.g. cast iron parts) for routine replacement in the plant equipment, stimulated by the IS policy. By the end of this phase (1953), CSN had accumulated Level 1 routine (basic) capabilities for the manufacture of simple parts and components. However, CSN did not move into the accumulation of innovative level of capabilities to carry out more sophisticated equipment activities (e.g. revamping of large equipment).[62]

6.1.3 Summary of the Cross-company Differences

The evidence outlined in Sections 6.1.1 and 6.1.2 indicates that both USIMINAS and CSN started as state-owned flat steel companies with inadequate technological capabilities. However, their technological capabilities were accumulated in different ways. During the start-up and initial absorption phase, USIMINAS accumulated Levels 1 and 2 routine capability across all five technological functions (facility user's decision-making and control, project engineering, process and production organisation, product-centred, and equipment). In addition, USIMINAS moved into the accumulation of Level 3 to Level 4 capability for process and production organisation, products, and equipment. In contrast, CSN passed through this phase without having completed the accumulation of Level 1 capability for those functions (except for equipment). In addition, CSN did not move into the accumulation of levels of innovative capability. More specifically, the differences are as follows.

Investment capability

In the early years of the project, USIMINAS's project leaders had the long-term accumulation of in-house technological capability as a central objective in the project. In contrast, CSN's paid little attention to this issue. The perspective of those project leaders seemed to have conditioned the different ways by which each company engaged in project implementation activities. USIMINAS's engaged actively in project design activities concerned with its long-term 'technological independence' from the Japanese. In contrast, CSN engaged passively in project engineering activities and delegated most to consulting firms (e.g. Arthur McKee). While USIMINAS accumulated routine capability for investments from Levels 1 up to 4, CSN continued to have inadequate investment capabilities in the start-up and initial absorption phase.

Process and production organisation capability

Even before the plant start-up, USIMINAS began to build organisational units to support its operational units. However, the building of such units was absent in CSN in the start-up and initial absorption phase. Constrained by financial difficulties and pushed by leadership to break even, USIMINAS engaged in intense efforts to absorb and even 'stretch' plant capacity, leading to the accumulation of its own 'capacity-stretching' capability. In contrast, the evidence indicates that CSN passed through that phase with little engagement in process quality improvement, operating the plant inefficiently, and at low capacity utilisation rates compared to USIMINAS.

It is striking to see how both companies responded differently to the same government policy (e.g. the National Coal Plan) with different implications for their capability-accumulation. USIMINAS rapidly engaged in several incremental activities to manipulate the metallic charge to bring down the coke rate, producing an active response to that policy. In contrast, CSN did not engage in continuous improvement efforts to reduce the coke rate. Indeed it waited for nearly fifteen years for the sinter plant, producing a passive response to that policy. Therefore while USIMINAS accumulated Levels 1 and 2 routine capability and moved into the accumulation of Level 3 to Level 4 innovative capability (extra-basic to pre-intermediate) for process and production organisation activities, CSN failed to accumulate either level of capability during the start-up and initial absorption phase.

Product-centred capability
USIMINAS rapidly accumulated Levels 1 and 2 routine capability for product-centred activities, with consistent product quality and diversification strategy. In parallel, the company moved into the accumulation of innovative levels of that capability (Levels 3 to 4) by engaging in innovative product activities on an independent basis. In contrast, CSN made no effort on efficient routine product manufacture or product improvement activities. As a result, the company failed to accumulate Level 1 routine capability and Level 3 innovative product-centred capability during the start-up and initial absorption phase.

Equipment capability
USIMINAS not only accumulated Levels 1 and 2 routine capabilities for equipment activities, but rapidly engaged in the revamping of large equipment (e.g. BFs), initially technically assisted and later independently. This led to the accumulation of innovative capability at Levels 3 to 4 for equipment activities. In contrast, CSN did not move beyond Level 1 routine equipment capability during the start-up and initial absorption phase.

6.2   KNOWLEDGE-ACQUISITION PROCESSES

In the light of the framework in Table 3.2, this section describes the knowledge-acquisition processes during the start-up and initial absorption phase in USIMINAS (1956–72) and CSN (1938–53). Sections 6.2.1 and 6.2.2 focus on the external and internal knowledge-acquisition processes, respectively.

6.2.1   External Knowledge-acquisition Processes

Pulling in expertise from outside

USIMINAS   In 1958, following the decision to give the top technical and administrative positions to the Japanese, several Japanese technicians were sent to USIMINAS, bringing new tacit knowledge into the company. In 1958, USIMINAS hired an experienced Brazilian engineer to run the newly created general construction unit.[63] Other engineers were hired for the Purchase Offices in Tokyo, Dusseldorf, and Paris. In 1958, USIMINAS began to recruit newly graduated engineers from the Ouro Preto School of Mines and Metallurgy, from the Federal Engineering School of Itajubá, and from the Minas Gerais Federal University (UFMG). While some of them were appointed to the newly created technical department and laboratory, others were sent to other sites to work on the construction and installation of the plant.[64]

In addition, in 1968, an experienced engineer and director of the mining firm Companhia Vale do Rio Doce (CVRD) was hired as a key adviser to

USIMINAS's president. Continuing this practice, in late-1960, USIMINAS hired Yawata's former director as 'chief adviser' (1960–63) and, in July 1961, a Japanese engineer was hired as operations director.[65] Having experience as engineers in large Japanese steel companies, these individuals brought substantial and updated tacit knowledge into USIMINAS. The evidence suggests that USIMINAS sought to pull in expertise ranging from local newly graduated to international and experienced engineers, i.e., a moderate to diverse expertise in terms of quality and quantity.

CSN In 1941, CSN recruited about 130 Brazilian engineers from the Federal School of Engineering of Itajubá and the Army Engineering School for construction activities. A group of about 60 designers was recruited from technical schools in Rio de Janeiro and São Paulo. In 1942, a group of 37 American engineers and technicians was sent to Volta Redonda to work in the installations and start-up of the plant. When operations began, about 150 additional engineers were recruited from the Ouro Preto School of Mines and Metallurgy and from the São Paulo Polytechnic School. Technicians were recruited from the Federal Railways Company (RFFSA), the Light & Power Company or from the steel firm Companhia Siderúrgica Belgo-Mineira (CSBM), established in 1921, then the first charcoal-integrated steelworks in Latin America. In this way CSN acquired external tacit knowledge for operational activities. The army engineers were given priority in the company's hierarchy, and CSN did not hire steel experts.[66] Although CSN hired a number of local engineers, most of them seemed not to be sufficiently experienced. In other words, the variety of this expertise was limited as far as the quality of the tacit knowledge being acquired was concerned. The evidence suggests that CSN did not engage in efforts to bring more robust expertise into the company.

Recruiting operational individuals

USIMINAS As the company needed a large number of operators and technicians for its operations, it engaged, from 1960, in a nation-wide call for operational individuals, whom it attracted from different parts of the country. To be recruited as technicians they were not expected to have previous experience, as long as they had the complete or even incomplete secondary level schooling. 'There weren't enough individuals with technical level in the country. We had to train our own technicians.[67] The new recruits underwent a series of interviews and tests through which their skills and aptitudes were assessed. Another practice consisted of recruiting workers beyond actual needs. Over the months, the company chose those who would stay while others were let go. This tactic, however, had negative social implications for the region and gradually it ceased.[68] Following their recruitment, some individuals were sent to the schools of the National Service for Industrial Apprenticeship (SENAI). In

1964, 160 out of 1,323 individuals that had applied to work in the plant were recruited, suggesting that USIMINAS was becoming more selective. By the late 1960s, reflecting the good functioning of its recruitment procedures, USIMINAS was called on by other organisations to assist them in recruitment.[69] This has been the result of improvements in these practices over the years. It also seem to have had further positive implications for knowledge internalisation by operators.

CSN   From 1940, individuals from all over the country had been recruited to work in the construction activities (from 1,500 in 1941 to 8,054 in 1946)[70] and, later, in the operations. Most of them were illiterate, with a rural background and no previous contact with industrial work, but there were plumbers, bricklayers, and electricians. Following a medical examination, they were registered and, in some cases, renamed by CSN, and sent to the sites.[71] Managers and engineers interviewed in CSN recalled this practice as being inadequate for the effective operation of a steelworks. This poor recruiting procedure had implications for the knowledge internalisation of CSN's operators.

Importing expertise to lead training

USIMINAS   In 1958, an agreement was signed with Nippon–USIMINAS to regulate the dispatch of the Japanese technicians to work in the plant. It was decided that highly-qualified engineers and technicians experienced in different areas of steel production would be sent to train USIMINAS's operators and technicians. They were required to learn basic Portuguese to facilitate communication with the Brazilians.[72] In late-1960, 400 Japanese technicians arrived in the plant, and rapidly became critical for tacit knowledge acquisition by USIMINAS's engineers, technicians, and operators during the installations, start-up, and initial operations activities. Their numbers decreased to 223, in 1963, as the Brazilians began to take over the control of the plant. In 1965, the group was only eight Japanese, and they began to be replaced every six months to meet USIMINAS's new knowledge-acquisition needs.[73] The diverse experience these Japanese technicians brought to USIMINAS contributed to bringing different sorts of tacit knowledge on steelmaking into the plant. In addition, the functioning of this process of external knowledge-acquisition seemed to have had positive implications for other types of training across the plant, as described later in this chapter.

CSN   In 1941, it was announced by CSN's top management that 'qualified engineers and technicians from the US will work together and assist the locally recruited personnel until they have become able to operate the plant themselves'.[74] However, at the plant level, things evolved in a rather different way. In 1944, a group of 55 American engineers and technicians arrived in CSN. However:

> Apart from some engineers, the Americans sent to CSN were poorly-qualified. Some of them had worked as second or third-rate operators of furnaces in US Steel. Once they came to CSN, they became foremen and even managers. . . but made no effort to learn Portuguese and. . . to upgrade their technical knowledge.[75]

On the one hand, this may reflect the low level of concern of the US steel companies with the success of the project. On the other, it suggests the low level of CSN's top management with the functioning of the knowledge-acquisition process at the sites. Although present in CSN, the functioning of the process was poor. In addition, practically no action was taken by top management to correct the situation. As a result, the expected tacit knowledge-acquisition by the Brazilians was mitigated. This led to further problems of internalisation of procedures by the CSN's operators, in other words, there was a weak interaction between the learning processes.

Channelling of external codified knowledge

USIMINAS   As early as 1966, USIMINAS took the initiative of building the technical information centre (TIC) to channel external codified knowledge into the company. This initiative was championed by its president and his adviser, hired from CVRD. The idea was not the building of a conventional library. The company seemed to be in search of a more active mechanism. USIMINAS undertook a pioneering and very expensive investment and seemed to be aware of its importance: 'In 1966, we were in the middle of nowhere with an ambitious goal of becoming an internationally competitive steel company. Bringing new knowledge into the company, quickly and effectively, was crucial to meet that goal.'[76]

To organise and run the TIC, USIMINAS relied on the strong technical background of TIC's first manager. This permitted an accurate scanning and interpretation of the codified knowledge on the steel industry and technology. [77] From 1966 to 1972 TIC consisted of a team of 12 individuals.[78] During the start-up and initial absorption phase, particularly from 1966 to 1972, TIC was a key provider of codified knowledge to the plant's workers. The channelling of that codified knowledge seems to have played a critical role in enhancing the 'capacity-stretching' efforts across the plant (e.g. detailed description and interpretation of practices from other steel plants and from the technical literature).[79] Interviews in USIMINAS revealed that, initially, the operationalisation of TIC was not welcomed by all managers. The reason was that TIC sought to centralise the subscriptions and the organisation of the technical publications. However, some groups of individuals wanted to have control over and exclusive access to publications. In addition, some groups were opposed to the idea of disseminating published knowledge across units to reach as large a number of individuals as possible.[80] However, the explicit support given to TIC by USIMINAS's presidency played a critical role in overcoming this.[81]

A second type of resistance came from individuals from the sites. Initially, papers sent to them by TIC used to be ignored. Indeed some individuals argued that the constant flow of reading material disturbed their daily routines.[82] However, as TIC continued their efforts, individuals gradually began to recognise the benefits of this mechanism. The next chapter describes some of the practices TIC used to encourage the site individuals to absorb this codified knowledge.

CSN   By the 1950s CSN had built a conventional library. In other words, the mechanism described above was absent from CSN.

Overseas training

USIMINAS   In September 1958, a trip by USIMINAS's president to Japan triggered the practice of sending small groups of engineers to do short-term practical training in Japan.[83] Groups of five USIMINAS engineers began to be sent regularly to Tokyo to provide detailed information to the USIMINAS' purchase office. This led to an intense acquisition of tacit knowledge by the Brazilians as they interacted with the Japanese and made technical visits to steel companies (e.g. Yawata) and equipment manufacturers (e.g. Hitachi). As a result, they brought 'fresh' tacit knowledge into USIMINAS, and codified knowledge in the form of projects, companies' published documents and technical literature.[84] In the light of Table 3.2, this illustrates a good functioning of that mechanism.

Following the president's trip, a longer training programme for engineers was designed. In September 1958, four years before start-up, USIMINAS sent a group of 10 engineers, known as the 'Samurai', to Japan for six to 12-months' training. Through supervised OJT in Iawa steel works' units, each engineer acquired tacit and codified knowledge by asking questions, observing, imitating the host engineers and also by visiting and observing other steelworks, and attending conferences. In order to prepare their Monday morning reports for the USIMINAS's Tokyo office, they worked together through the whole weekend, at the hotel, giving rise to a series of knowledge-conversion activities from the individual to the group level. These weekly reports were sent to USIMINAS in Brazil contributing further to the spread of that codified knowledge to other individuals.[85] On the basis of Table 3.2, this suggests not only a good functioning of, but also the beginning of a strong interaction between learning mechanisms.

In July 1969, three years before the launch of the research centre, USIMINAS sent a group of 32 newly recruited engineers to undertake eight-months' training in four different countries (Japan, US, France, and UK). The tacit and codified knowledge acquired through this training was related to research techniques and strategies on high-resistant plates development.[86] This is another example of continuous intensity of the overseas training process. There was therefore a mix of short and long-term overseas training taking place from the late 1950s to the early-1970s.

CSN　The initial training took place in the early-1940s. During the project design a group of nine engineers worked with McKee in New York and Cleveland. This training was limited to project activities only.[87] In 1946, when the installations and start-up activities were taking place, a short technical training for about 50 of CSN's engineers was arranged in the US. However, this was one-off. It was not until the mid-1970s that overseas training of engineers was resumed in CSN.[88] The evidence suggests that, although present, the overseas training process in CSN was limited in terms of variety of mechanisms and intermittent intensity.

Participation in conferences and related events

USIMINAS　The practice of encouraging individuals to participate in those events, locally or overseas, began in USIMINAS as early as 1960. A group of engineers participated in the Osaka International Fair where they acquired the latest knowledge on steel production and commercialisation. During the start-up and initial absorption phase, technicians and engineers were encouraged to participate in more events (e.g. meetings of the Latin American Iron and Steel Institute (ILAFA), II International Conference of the Oxygen Steelmaking Process, Training within Industry (TWI)[89] among others). As a result of these events individuals brought different knowledge into the company.[90]

CSN　As mentioned in different interviews in CSN, it was not until the 1960s that individuals began to be encouraged to participate in such events.

Using technical assistance for knowledge acquisition

USIMINAS　By 1966, when the initial technical assistance agreement had expired, a new five-year agreement, TA-1, was signed with Yawata. This agreement was renewed until 1976. An issue that came up in different interviews in USIMINAS was that individuals soon developed the practice of working as closely as possible with the technical assistance teams. In other words, foreign technical assistance was seen as providing the knowledge to permit the company to engage independently in further innovative activities.[91]

CSN　During the 1941–53 period CSN received foreign technical assistance, for instance, from Arthur McKee and suppliers. However, as mentioned by the managers interviewed, the systematic commitment to learning from technical assistance as in USIMINAS, was absent, and this situation continued over the following decades.[92]

Providing scholarships for research and teaching

USIMINAS   In 1960, USIMINAS began to provide, on a continuous basis, scholarships for research (e.g. the Gorceix Foundation) and for individuals to do engineering courses (e.g. the Ouro Preto School of Mines and Metallurgy). This seemed to work as a strategic mechanism for USIMINAS: it would later provide tacit knowledge (e.g. trained engineers) and codified knowledge to the company.[93]

CSN   It was not until the 1970s that this mechanism was developed in CSN.[94]

Educational infrastructure in the community

USIMINAS   In early-1959, USIMINAS's president championed the building of an education infrastructure in Ipatinga (e.g. 'Grupo Escolar Almirante Toyoda', 'Intendente Câmara Educational Centre' and others), despite initial Japanese opposition. The infrastructure was designed to provide basic literacy and numeracy skills up to the primary, vocational, secondary and technical levels. In the long term, it sought to educate USIMINAS's future operators and technicians for the future phases of the company.[95]

CSN   In 1943, the 'Grupo Escolar Trajano de Medeiros' was the only school built in Volta Redonda, where primary and secondary levels were taught, but this was at the initiative of Rio de Janeiro's government rather than CSN.[96] CSN's low engagement in building a sustainable education infrastructure in Volta Redonda seems to reflect low level of concern with the long-term development of in-house technological capability.

6.2.2   Internal Knowledge-acquisition Processes

USIMINAS   During the start-up and initial absorption phase (1956–72), tacit knowledge was acquired internally by USIMINAS's individuals, as they actively began to engage in project engineering activities during those years. The initial studies undertaken by individuals in the technical department and the laboratory led to the acquisition of substantial tacit knowledge of production process and product quality, before plant start-up. By 1963, individuals had been able to acquire key knowledge on the intricacies of the reduction process. This was achieved by engaging in the manipulation of different parameters of the metallic charge quality. Therefore those practices played a substantial role in the accumulation, within three years, of Level 1 routine (basic) capability for reduction processes.

Rather than waiting for Japanese instructions, USIMINAS's individuals were encouraged to engage in more complex and challenging activities. This engagement rapidly became systematic (e.g. 'capacity-stretching' efforts), suggesting their intensity over time. The intensity of 'learning-by-doing' more complex activities independently contributed to increased confidence and greater understanding of the principles involved in those activities and the newly-acquired technology. The way those practices worked seemed to have been influenced by the functioning of other practices like the systematic preparations and previous external knowledge-acquisition, or 'learning-before-doing' (e.g. the revamping commission and the 'Samurai'). The evidence suggests the development of a strong interaction between the learning mechanisms.

In addition, by the end of the start-up and initial absorption phase (1971) USIMINAS set up its research centre. However, it was during the 1960s that the idea grew. In 1961, following a visit of USIMINAS's top management members to Yawata's research facilities, the audacious decision of building a research unit in USIMINAS was taken. The unit started, informally, in 1964 with a group of four researchers studying new steel characteristics. As in the TIC, leadership played a key role in championing this project and overcoming resistance or scepticism of certain groups.[97] A key issue, however, has been that since its creation this unit has been treated as one of the key mechanisms for internal knowledge acquisition and for USIMINAS's long-term 'technological autonomy'. This had long-term implications for the functioning of this mechanism.

CSN During the start-up and initial absorption phase (1939–53), as referred to in the previous subsections, CSN's individuals used to delegate project engineering activities to consulting firms (e.g. Arthur McKee). This factor, in association with the resistance to work under the Americans' supervision, connected with nationalism, contributed to mitigating the internal knowledge-acquisition process. As a result, individuals began to turn to consulting firms to solve their basic production problems. This suggests that the acquisition of knowledge, skills and experience process was limited to doing routine simple plant operations – mostly inefficiently – or even by delegating them to consulting firms.[98] This pattern of very basic 'learning-by-doing' (and the absence of any 'learning-before-doing') contributed to constraining the accumulation of basic levels of routine and innovative technological capabilities in the 1946–53 period.

6.2.3 Summary of the Cross-company Differences

In the light of Table 3.2 and the above empirical evidence, this subsection summarises the cross-company differences in the features of the external and internal knowledge-acquisition processes during the start-up and initial absorption phase in USIMINAS (1956–72) and CSN (1938–53).

As far as external knowledge-acquisition processes are concerned, the differences are summarised in Table 6.1. Nine processes are summarised in the table. All nine were present in USIMINAS but only six in CSN. Although six processes were present in both companies, they differed in terms of the variety of practices they involved. For example, USIMINAS sought to bring in expertise with different backgrounds to the company. In contrast, CSN limited their intake to local and newly graduated engineers and technicians. Overseas training in USIMINAS involved a mix of short- and long-term courses, routine plant operations, innovative projects (e.g. research techniques for product development), technical visits and observation. These activities covered a relatively large number of individuals. In contrast, in CSN these activities were limited to a group of engineers who were trained in routine plant operations.

In addition, although the six processes were present in both USIMINAS and CSN, the evidence suggests that most of them also differed in terms of intensity, functioning, and interaction. For instance, both companies pulled in expertise continually during that phase. However, USIMINAS engaged in the deliberate searching for and hiring of more robust expertise. Experienced engineers were given strategic positions in the company. Some of them influenced the building up of other learning processes (e.g. channelling external codified knowledge) which, in turn, influenced the internal knowledge-acquisition processes. Although CSN tried to acquire qualified technicians (e.g. from CSBM and RFFSA) this expertise seemed not to be robust enough to bring strategic tacit knowledge and new ideas into the company. In addition, the evidence does not suggest that it influenced the building of other learning processes. For these reasons, the evidence suggests that the process of pulling in outside expertise functioned better in USIMINAS than in CSN. In addition, while this process in USIMINAS interacted in a moderate/strong way with other processes, in CSN this interaction was weak.

Another example is the overseas training process. In USIMINAS it operated on a continuous basis, pushed by an explicit concern with in-house capability-building. In contrast, in CSN it worked on a one-off basis disconnected from any concern with long-term capability-building. In addition, the evidence suggests that USIMINAS set up different schemes for overseas training in different countries. In addition, individuals were sent to the types of training associated with the company's objectives to build capabilities for different functions (e.g. project design, routine plant operation and product improvement/development). This suggests that the functioning of this process, was better in USIMINAS than in CSN.

For instance, in USIMINAS overseas training, influenced internal learning by 'capacity-stretching'. This meant that in USIMINAS the processes and

Table 6.1   Summary of the differences in key features of the external knowledge-acquisition processes during the start-up and initial absorption phase between USIMINAS (1956–72) and CSN (1938–53)

EXTERNAL KNOWLEDGE-ACQUISITION PROCESSES	VARIETY		INTENSITY		FUNCTIONING		INTERACTION	
	USIMINAS	CSN	USIMINAS	CSN	USIMINAS	CSN	USIMINAS	CSN
Pulling in expertise from outside								
• Newly graduated engineers	Present (diverse)	Present (limited)	Continuous	Continuous	Good	Moderate	Moderate	Weak
• Experienced engineers and managers								
• Steel experts as top technical advisers								
Recruiting operations individuals	Present (diverse)	Present (diverse)	Continuous	Continuous	Good	Poor	Moderate	Weak
Importing expertise to lead training	Present (diverse)	Present (limited)	Continuous	Intermittent	Good	Poor	Moderate	Weak
Channelling of external codified knowledge	Present (moderate)	Absent	Continuous	–	Moderate	–	Moderate	–
Overseas training								
• Routine plant operations	Present (diverse)	Present (limited)	Continuous	Intermittent	Moderate	Poor	Moderate	Weak
• Product development, process, and equipment improvement								
• Short-term and/or long-term courses								
• Technical visits and observation tours								
Participation in conferences and related events	Present (diverse)	Absent	Continuous	–	Moderate	–	Moderate	–
Using technical assistance for knowledge acquisition	Present (moderate)	Present (limited)	Continuous	–	Moderate/Good	–	Moderate	–
Providing scholarships for research and teaching	Present (moderate)	Absent	Continuous	–	Good	–	Weak	–
Educational infrastructure in the community	Present (limited)	Present (limited)	Continuous	Intermittent	Good	Good	Weak	Weak

Source: Own elaboration based on the research.

mechanisms began to work interactively (e.g. overseas training coupled with learning by 'stretching' plant capacity). In CSN this type of interaction was absent. While USIMINAS's effective knowledge-acquisition process contributed to enhancing its technological capability-accumulation path, CSN's ineffective process contributed to constraining the accumulation of its own capabilities.

As far as internal knowledge-acquisition processes are concerned, the differences are summarised in Table 6.2. Evidence relates to five processes which are outlined in the table. All five were present in USIMINAS and three in CSN. Although both companies were involved in project design and plant installation, USIMINAS had taken the lead in different project engineering activities, avoiding delegation to the Japanese. The presence of CSN's individuals in the McKee office, however, was limited. In addition, in the 1940s and early-1950s CSN was delegating critical activities to consultants rather than engaging independently in them. USIMINAS engaged in systematic improvements in existing facilities through 'capacity-stretching' and deliberate manipulation of key process parameters across different operational units. These practices were absent in CSN. As a result, individuals in CSN had a smaller variety of mechanisms with which to increase their understanding of the principles involved in the technology compared with USIMINAS's.

Although both companies engaged in routine plant operations, the evidence in this chapter suggests that USIMINAS worked together with the Japanese until it had learnt to operate the plant. Evidence here also suggests that CSN took over the plant control from the Americans before having learnt how to operate it efficiently. In addition, there was insufficient involvement by CSN's individuals in continuous process and equipment improvement. Additionally, USIMINAS had more varied criteria on which to base the selection of operators and also had a more diverse, continuous, and better functioning knowledge-acquisition processes (e.g. outside expertise, overseas training). This suggests that the process of acquiring greater understanding of the technology through daily plant operation functioned better and interacted more strongly with other processes in USIMINAS than in CSN. This different functioning and interaction contributed more to the building of routine capability for effective plant operation in USIMINAS than in CSN.

6.3   KNOWLEDGE-CONVERSION PROCESSES

6.3.1   Knowledge-socialisation Processes

Three processes of in-house training are distinguished: (1) course-based: basic; (2) course-based; and (3) on the job. Training modes (2) and (3) are related to

Table 6.2 Summary of the differences in key features of the internal knowledge-acquisition processes during the start-up and initial absorption phase between USIMINAS (1956–72) and CSN (1938–53)

INTERNAL KNOWLEDGE-ACQUISITION PROCESSES	VARIETY		INTENSITY		FUNCTIONING		INTERACTION	
	USIMINAS	CSN	USIMINAS	CSN	USIMINAS	CSN	USIMINAS	CSN
Involvement in project design	Present (diverse)	Present (limited)	Continuous	Continuous	Good	Moderate	Good	Moderate
Involvement in plant installations	Present (diverse)	Present (limited)	Continuous	Continuous	Good	Poor	Moderate	Weak
Routine plant operation	Present (diverse)	Present (diverse)	Continuous	Continuous	Good	Poor	Moderate	Weak
Doing systematic studies in laboratories and manipulation of production process parameters	Present (moderate)	Absent	Continuous	–	Good	–	Moderate	–
Systematic 'capacity-stretching' efforts	Present (diverse)	Absent	Continuous	–	Good	–	Moderate	–

Source: Own elaboration based on the research.

knowledge socialisation. The distinction is important because training mode (1) may have practical implications for knowledge socialisation and the building of basic capabilities.

In-house training (course-based): basic

USIMINAS   USIMINAS began by training the individuals working in the plant construction, in basic Portuguese and arithmetic. In addition, following the recruitment of individuals for operations, the illiterate workers were sent for basic training to learn to read and write. In August 1961, USIMINAS set the target of qualifying an operator within six months through a training programme consisting of two parts. USIMINAS was responsible for providing the course consisting of Portuguese language, mathematics, physics, and electricity.[99] Improving the literacy and numeracy skills of the trainees probably contributed to improving their assimilation of other training programmes and benefiting the routine site operations.

CSN   Following the limited recruitment procedure, as described earlier, operators were sent straight to the operations sites. In other words, basic training to improve literacy and numeracy skills was absent.[100]

In-house training: course-based

USIMINAS   Following this basic training, the operators went through a more specific training. This was provided by the Japanese and consisted of equipment and machinery operation. The individuals initially received training on steel production processes. Then they were grouped on the basis of the unit in which they would work and were trained in the principles, characteristics, functioning, and safety issues of the technology. To facilitate the trainees' assimilation, the Japanese trainers made use of wooden models which replicated some of the details of the operational units.[101]

Such disciplined training led to the socialisation of the instructors' tacit knowledge with the trainees via written, oral, and body language, observation and imitation of practices and other interaction. This in turn led to fruitful knowledge-socialisation between the trainees. These interactions permitted an expansion of the trainees' mental models, know-how and skills, triggering a socialisation and spread of basic tacit knowledge on steel production across the group and plant levels. One manager interviewed in USIMINAS even stressed that the training contributed to providing the trainees with a good notion of features of the equipment and processes they would operate and how to operate them.[102] By December 1961, around 400 individuals had undertaken this type of training and a plan had been prepared to train 2,000 more.[103] In the early 1960s, as its engineers acquired tacit and codified knowledge, particularly through overseas training, USIMINAS began to design and implement in-house training programmes. One example is the case of the steel shop, as described below.

In 1961, a group of USIMINAS's overseas trained engineers, including two 'Samurai', created a three-phase training for the steel shop's operators. This facilitated the socialisation and the articulation of the tacit knowledge acquired overseas, indicating a strong interaction between the learning mechanisms from the outset. The first phase started in July 1962, 11 months before start-up, and consisted of practical training in arithmetic and industrial safety followed by tests. Fifty-nine operators were approved for the second phase, involving steel production, safety, and operation of the oxygen furnace. In February 1963, the operators were re-grouped and went through the third phase, consisting of a near-normal operating condition training. Some operators had specialised training with Japanese technicians. By April 1963, three months before the start-up, the first two crews had been trained and their tacit knowledge on steelmaking had been effectively spread into the steel shop leading to efficient operation and further improvements.[104]

In the light of Table 3.2 the evidence in this section suggests that the functioning of the training processes was good. The company seemed to have a deliberate strategy to prepare operators effectively for their duties. Indeed, key training processes took place before start-up of operations (e.g. the steel shop). The evidence also suggests a moderate to strong interaction between the processes. For example, the specific training of operators was led by Japanese trainers, and the in-house training programme in the steel shop was designed and implemented by USIMINAS's own engineers, most of whom had been trained overseas.

CSN In the early-1940s, as the plant start-up approached, CSN took action to train individuals to work at the sites. In 1943, about 50 young individuals, recruited to work in the electrical equipment installations, were trained by a retired but experienced electric engineer. Following plant start-up, however, most of these individuals left CSN, taking their substantial tacit knowledge with them.[105] In 1944, a 'school for welders' was improvised and a retired American welder was hired to lead the training. Experienced and non-experienced workers undertook training which contributed to the sharing of tacit knowledge between these individuals (instructor and trainees). A few of them were engaged in the foundry. However, most of them left CSN to work elsewhere,[106] leading to another loss of tacit knowledge. This points to the beginning of the problem of retaining tacit knowledge in CSN.

In late-1944, another programme was created to train individuals for operations. This became known as the 'emergency course'. It consisted of short training given by the Americans to the individuals to work in the production units.

Some sessions were improvised on site. However, 'communication and assimilation were very poor. Most of the Brazilian workers were illiterate and could hardly understand technical Portuguese. . . The Americans could barely speak any Portuguese. . . and made no additional efforts to teach the Brazilians properly.'[107]

The functioning of this training may reflect the poor functioning of the recruitment process and the import of expertise for training. This also seems to reflect the absence of basic in-house training for the operators. As a result, the operators engaged in the units were poorly trained. This probably led to lack of confidence, misinterpretation of instructions, breakdown of equipment, and high accident rates.[108] Although information on the rate of accidents in the 1940s–1950s period is not available, interviews suggested it was high. In addition, the company formally used to express its concern with the frequency of accidents.[109] It should be noted that accident rates have negative implications for capacity utilisation rates (e.g. because of stoppages) and high maintenance costs.

In 1944, another training programme was established called the 'adaptation course for the job in the mill'. It was implemented through the creation of the CSN's Pandiá Calógeras Industrial School. CSN drew on assistance from the Americans and an education specialist. In 1944, CSN had about 100 individuals, from 13 to 17 years old, admitted to its industrial school. The objective was to form obedient, disciplined, and responsible professionals through a rigorous training routine. These trainees came from the state of São Paulo and Niterói (in the state of Rio de Janeiro). Every day, they had to get up at 6 a.m., undertake a gymnastic session and, dressed as soldiers, head for the sites where they worked from 7 a.m. to 2 p.m. supervised by engineers. In the evenings, they attended courses in the Portuguese language, mechanics, and materials resistance. However, most of them ran away from the school. By 1947, only 37 had completed the programme and CSN had abandoned this training. In 1948, the school was restructured but continued to have problems in the long term, as described later in Chapters 7 and 8.[110]

The idea of building the school certainly contributed to the socialisation of tacit knowledge between the instructors, site engineers, and the trainees. Indeed, evidence from archival records reveals that the trainees who completed the programme had acquired a respectable technical level. However, these achievements were not sufficient for the number of operators that CSN needed. Although present in the company, the intensity of an in-house training process was intermittent. The evidence suggests that the training activities in the 1943–53 period were rather improvised. Most were executed on a short-term basis. This may reflect the poor functioning of the process of importing expertise to lead in-house training and intermittent overseas training.

On-the-job training

USIMINAS On-the-job training (OJT) was created in USIMINAS during the construction phase: each foreman supervised ten workers and each engineer supervised six foremen. Several operators, who had worked in the plant construction and followed one training course, were recruited for the production units. The OJT for engineers, technicians and operators was supervised by the Japanese. In some cases, by the side of each Japanese, there was one Brazilian who observed, imitated, asked questions, and took notes. In other cases, behind each three Brazilians, there was one Japanese looking over their shoulders and giving instructions.[111]

The daily and disciplined OJT helped to socialise the Japanese's tacit knowledge with the Brazilians. This helped them to develop a greater understanding of how the Japanese prepared the raw materials and coped with the intricacies of grading and timing in the refining process to achieve high-quality plates. When the Brazilians showed they had learnt the job, the Japanese gave them full responsibility. Some took six months to learn, others took up to a year.[112] USIMINAS's top management expressed its commitment to the effective functioning and intensity of OJT, by encouraging the engineers to work under the Japanese supervision: 'We used to tell the engineers: The closer to the duty, the more you learn about it. . . The more the Brazilians work under the Japanese supervision, the more they will learn from them. . . We have learned a lot from that practice.'[113]

Engineers who had undergone supervised OJT soon began to fill technical positions. They then took over the roles of the Japanese in OJT supervision thereby sharing their accumulated tacit knowledge with different individuals. This gave rise to more interactions between the learning processes.

CSN From 1945, top management determined that the Brazilian engineers should be trained by American engineers on a supervised OJT basis.[114] However, at the plant level, the OJT evolved in a different way. Several of CSN's engineers refused to work under American supervision. They were probably conditioned by a nationalistic view against the Americans.[115] A few interviews within and outside CSN also suggested that some engineers were given office-based administrative tasks rather than operational tasks on site.[116] Supervised OJT was also formally recommended to the operators as a way of sharing knowledge with the American technicians.[117] However, the functioning of the training on the sites suggests an incoherence with formal expectations: 'The Brazilian operators soon realised that, with a few exceptions, the Americans were under-qualified for their duties. . . The operators did not trust them. There were constant misunderstandings in the instructions because of language barriers. . . This situation continued for years.'[118]

This problem, referred to in numerous interviews in CSN, suggested that poor communication between instructors and operators usually led them to totally misunderstand instructions. In addition, some operators realised that some of the American instructors were managing to have their contracts extended by virtually preventing the operators from learning. On some occasions, it was noticed that the operators were asked to stay away from the unit when a problem-solving activity (e.g. equipment repair) was to take place.[119] This evidence on the OJT for engineers and operators indicates total incoherence between top and plant levels as far as the functioning of the learning processes was concerned. This may have contributed to mitigating against the knowledge-conversion processes as whole.

Shared problem-solving

USIMINAS Interactions between individuals to solve problems began in the construction phase and evolved into the operations phase in the 1960s. Initially, they took place through intense sharing of experiences between the Brazilian and the Japanese engineers.[120] During this time a large group of Japanese were at USIMINAS, where they shared their tacit knowledge to the extent that it was almost 'squeezed' out of them, as the Brazilians were determined to learn how to solve intricate problems:

> It was quite common to see Brazilian engineers across the plant, with a board on their lap, 'interviewing' the Japanese. . . They took verbatim notes of what the Japanese said and tried things out in their units. If it did not work, they returned to clarify points. They kept repeating this until they had learnt how to solve the problems themselves.[121]

Shared problem-solving began even before start-up, when individuals used to discuss problems of raw materials supply, and evolved into the production phase in the form of regular *weekly meetings*. These meetings involved individuals within production units solving problems of bottlenecks in production, quality, costs, and 'capacity-stretching'. Workers began to share and articulate their tacit knowledge as they reviewed the crews' performance and discussed improvements based on individual experience, observations, readings, and conversations with the Japanese.[122] Informal interactions also took place in the start-up and initial absorption phase, as USIMINAS began to encourage friendly interactions between engineers and operators, thereby encouraging those individuals to share their different experiences.[123] In the light of Table 3.2, this suggests a good functioning of the knowledge-sharing process.

CSN As was announced by top management in the 1940s and early-1950s: 'the engineers should have a social function as managers and educators of the operators'.[124] However, at the plant level, those activities functioned differently. From the construction phase, a knowledge polarity began to develop in CSN. On the one side, there were the qualified Brazilian engineers who could speak English and share knowledge with the Americans. On the other side, there were the non-experienced, illiterate, and poorly-trained operators who did not have access to the engineers.[125]

From the start, CSN engineers were viewed as powerful, individual 'knowledge owners', and their interactions with operators became infrequent. Realising that their knowledge could be a source of competitive advantage, the engineers deliberately developed a practice of individual knowledge retention. As engineers had learnt their own way of solving production problems, they created their own production procedures. These were neither shared with their peers nor with the operators. This behaviour diffused to other professionals (e.g. managers, superintendents, supervisors) as a way of securing and/or achieving higher positions.[126]

In 1947, a financial reward system was created, known as the 'Giraffe'.[127] It was hoped that this would stimulate an increase in production volume and lift the workers' morale.[128] However, in their anxiety to increase production volume, the operators used to cause break-downs of the equipment. By 1949, they had begun literally to hide production 'procedures' to prevent the following crews from achieving the same production volume.[129] 'The system contributed to strengthening the focus on production volume and knowledge retention by individuals. It became a negative practice in CSN.'[130] Indeed the incentive system seems to have contributed to reinforcing a practice that had already been developed in CSN by that time: deliberate individual knowledge retention. This practice contributed to mitigating the functioning of knowledge-sharing processes among individuals. In other words, it contributed to exclusive knowledge-accumulation at the individual level.

6.3.2 Knowledge-codification Processes

The processes described below were mostly present in USIMINAS (1956–72). In CSN, however, as confirmed by several interviews, casual meetings, and access to archival records, knowledge codification processes were virtually absent during the 1938–53 period.

Overseas-trained engineers as unit leaders

USIMINAS From 1963, technical units gradually began to be headed by overseas-trained engineers (e.g. the 'Samurai'). This seemed a purposeful way of taking gradual and effective control of the plant from the Japanese. These trained personnel began to articulate their tacit knowledge through supervised OJT, informal accounts about their training programmes and daily operational instructions, and elaboration of training modules. Through such practices, elements of their tacit knowledge were made accessible to other individuals in different parts of the plant.[131]

CSN   Individuals who had been involved in overseas training were given management positions.[132] For example, most of the nine engineers who had been involved in project design activity in the McKee office became managers of critical operations areas. Although these individuals may have made some efforts to articulate their knowledge this was not sufficient: (a) they were limited in number; (b) the knowledge-acquisition processes were limited in variety and intermittent in intensity; and (c) the practice of individual knowledge retention was present in the plant. This suggests that the strategy of giving technical positions to trained engineers as a way of encouraging knowledge articulation was absent in CSN.

Systematic documentation

USIMINAS   From 1962, most of the recommendations given by the Japanese about routine operation and improvements on production began to be systematically documented. Each operation unit organised their own record books. These records also included the USIMINAS individuals' own experiences in process, product, and/or equipment improvements (e.g. BF revamping practices). The initiative was primarily concerned with the retention of knowledge within the company and its collective availability. Leadership seemed to have played a key role in encouraging its documentation.[133]

CSN   The absence of knowledge documentation in CSN was stressed by several interviewees in the company during the pilot and fieldwork. Indeed several pointed to the lack of documentation as one of the contributors to low operating capability in the initial phase.

Translating and adapting foreign documents

USIMINAS   From 1963, some suppliers' publications (e.g. manuals and blueprints accompanying the equipment and machinery) began to be translated and/or adapted into simpler internal manuals in a language accessible to the foremen and operators. Performed by USIMINAS engineers, this practice contributed to the diffusion of critical codified external technical knowledge and to the collective understanding of the technology principles.[134]

CSN   Interviews and documents in the company indicated that the practice of translating foreign documents by engineers did take place in CSN. However, the evidence indicates that there was no systematic commitment to turning them into manuals accessible to operators.[135] Therefore codification was absent.

Internal seminars

USIMINAS This mechanism had been established by 1962 on an informal basis. By the late-1960s, it had become systematic. Engineers, technicians, and managers returning from local and/or overseas training or conferences, were encouraged to lecture to other individuals. In addition, operational individuals were encouraged to talk about their problem-solving experience informally. These types of sessions not only contributed to the articulation of the elements of those individuals' tacit knowledge; they also contributed to the application of that codified knowledge in the operational units.[136]

CSN As revealed in interviews in CSN that practice was absent in the company.

6.3.3 Summary of the Cross-company Differences

In the light of Table 3.2 and the above empirical evidence, this subsection summarises the cross-company differences in features of the knowledge-socialisation and codification process during the start-up and initial absorption phase in USIMINAS (1956–72) and CSN (1938–53). These differences are summarised in Table 6.3.

As far as knowledge-socialisation processes are concerned, Table 6.3 indicates that the three processes were present in both companies. However, in USIMINAS in-house training of operators was preceded by basic training, while in CSN this was absent. In addition, in-house training (course-based) seemed to have reached a wider number of operators in USIMINAS than in CSN. This may be associated with the different concerns with operators' qualifications that each company demonstrated in the recruiting process. The evidence suggests that USIMINAS had diverse types of OJT (e.g. foremen/engineers, operators/Japanese, engineers/Japanese, operators/Brazilian engineers). In CSN, in contrast, OJT seemed limited to operators/Americans.

In addition, although the three processes were present in both companies they also differed in terms of intensity, functioning, and interaction. For example, the in-house training programme in USIMINAS was continuous and functioned in a disciplined way. In contrast, in CSN it was fairly improvised and on a one-off basis (e.g. the emergency course). Another key difference in their functioning is that most of the in-house training practices in USIMINAS took place before the engineers and operators engaged in critical technical activity (e.g. the training in the steel shop). This practice was absent in CSN. The way the course-based training, the OJT, and the shared problem-solving worked in USIMINAS seemed to have been influenced by the intensity and functioning of the knowledge-acquisition processes (e.g. overseas training, importing exper-

Table 6.3 Summary of the differences in key features of the knowledge-conversion processes during the start-up and initial absorption phase between USIMINAS (1956–72) and CSN (1938–53)

	VARIETY		INTENSITY		FUNCTIONING		INTERACTION	
KNOWLEDGE-SOCIALISATION PROCESSES	**USIMINAS**	**CSN**	**USIMINAS**	**CSN**	**USIMINAS**	**CSN**	**USIMINAS**	**CSN**
In-house training (course-based): basic	Present (diverse)	Absent	Continuous	–	Good	–	Moderate	–
In-house training (course-based)	Present (diverse)	Present (limited)	Continuous	Intermittent	Good	Poor	Moderate	Weak
On-the-job training (OJT) • For engineers • For operators	Present (diverse)	Present (limited)	Continuous	Intermittent	Good	Poor	Moderate	Weak
Shared problem-solving • Operations problem-solving • Meetings	Present (diverse)	Present (limited)	Continuous	Intermittent	Good	Poor	Moderate	Weak
KNOWLEDGE-CODIFICATION PROCESSES								
Overseas trained engineers as units leaders	Present (diverse)	Absent	Continuous	–	Good	–	Moderate	–
Systematic documentation	Present (moderate)	Absent	Continuous	–	Good	–	Moderate	–
Translating and adapting foreign documents	Present (moderate)	Absent	Continuous	–	Good	–	Moderate	–
Internal seminars	Present (moderate)	Absent	Continuous	–	Good	–	Moderate	–

Source: Own elaboration based on the research.

tise to lead training, etc.). In other words, the evidence suggests a strong inter-action between the processes. In contrast, the way those three sub-processes of knowledge socialisation worked in CSN seemed only weakly influenced by the knowledge-acquisition process (e.g. limited and intermittent overseas training, limited and poor functioning of importing expertise for training).

As far as knowledge-codification processes are concerned, four processes have been described in this chapter. As seen in Table 6.3, all four were present in USIMINAS but none in CSN. Although USIMINAS had, in general, a moderate diversity of practices related to those processes, deep concern with the importance of knowledge codification was present. The start of this process seemed to have been influenced by (1) the knowledge-acquisition processes (e.g. the intense contacts with steelworks overseas); and (2) the intensity of knowledge socialisation that started in this phase. In sum, in USIMINAS the key features of the learning processes were manipulated in a more deliberate and effective way than in CSN. In USIMINAS some mechanisms built up in the start-up and initial absorption phase reflected a concern with the knowledge-acquisition process not only to accumulate the capabilities to use the plant, but also particularly to build the capabilities to engage successfully in innovative activities in further phases – or the 'technological autonomy' approach. The evidence from CSN, however, suggests a very low level of concern with technological capability-accumulation in the long term.

NOTES

1. It should the noted that the 1956–76 period in USIMINAS was studied in depth in Dahlman and Fonseca (1978). However, this chapter moves a bit further in relation to that study by investigating more aspects of the paths and the learning processes.
2. Some of those members had advocated a designed capacity of 2 million tpy rather than 0.5 million tpy of steel.
3. Metallurgical engineer, former engineer at a large long-steel maker and a stainless steel maker and a lecturer on metallurgy at the São Paulo Polytechnic School (1943–46). He also participated in the project planning of a large flat steel maker in the state of São Paulo.
4. USIMINAS, Annual Reports 1958–66.
5. Interview with USIMINAS's former president. See also USIMINAS, Annual Reports 1959-66. The term 'technological independence' or 'autonomy' came up during different interviews meaning the in-house development of technological capability.
6. This derived from the joint initiative, in 1966, of the Brazilian Government and the World Bank to fund a study on the Brazilian steel industry through BAHINT and BNDE. Together with BNDE, the BAHINT specialists were given the authority to approve or reject expansions proposals.
7. Interview with USIMINAS's former president. See also BNDE/BAHINT, 'Plano Siderúrgico Nacional (1968–69), Brasília, 1969.

8. Interview with the head of the engineering superintendency. The engineering nucleus consisted of 12 engineers and was headed by an experienced Brazilian engineer, Mr Guatimosin, former head of USIMINAS's purchasing office in Tokyo.
9. Between 1966 and 1971 the average USIMINAS debt/sales ratio was 3.15. See USIMINAS, Annual Reports 1966-72.
10. Account by former operations director in USIMINAS/Fundação João Pinheiro (1988). This publication, written in Portuguese, entitled 'USIMINAS tells its history', is a set of papers in which eight former top managers provide an account of their experiences in coping with problems involved in managing their units.
11. A study made by USIMINAS's president estimated that the use of local coal, because of its low quality, would raise the pig iron production costs by about US$12 a tonne which, in turn, would raise the steel ingot production costs by about US$11. See A. Lanari Jr. 'Consumo de carvão nacional na siderúrgia', *Geologia e Metalurgia*, n. 27, 1965.
12. Interview with USIMINAS's former president. This was a challenging target since the most efficient BFs in Japan were achieving coke rates of 450kg, US furnaces were achieving 550kg to 570kg, and Indian BFs approximately 900kg. See Baer (1969).
13. Interview with a former manager of USIMINAS's Research Centre.
14. Company's documentary records: 'USIMINAS: Dados Operacionais dos Altos Fornos', IPQ/IQT (Technical Division of the BF Dept), September 1997.
15. Interview with an organisation analyst and with the corporate planning manager.
16. Interview with USIMINAS's former president.
17. See USIMINAS: organisational charts (1966–72). See also account by former manager of the Industrial Relations Dept in USIMINAS/Fundação João Pinheiro (1988). It should be noted that original organisational charts of USIMINAS for different years were gathered during the fieldwork. However, due to space limits they (or their simplified form) are not included in this book.
18. Until then, USIMINAS had been organised on a 'functional' basis.
19. Account by USIMINAS's former manager of the industrial relations dept in USIMINAS/Fundação João Pinheiro (1988*)*.
20. As referred to in Chapter 2, similar practices had been found in other Latin American plants, see Maxwell (1982).
21. See technical paper by the then manager of the reduction area, N. Nakamura, 'Desenvolvimento da técnica de produção de gusa', *Metalurgia ABM*, vol. 87, no. 21, 1965.
22. See 'USIMINAS lança Campanha Zero Defeito', *USIMINAS Revista*, vol. 2, no. 3, pp. 76–9, 1971. It should be remembered that the technique had been introduced in the US, in 1962, and in Japan, in 1968, indicating how up to date USIMINAS was in terms of production organisation activities.
23. Interview with a former manager of USIMINAS research centre.
24. See USIMINAS, Annual Report 1970.
25. See USIMINAS, Annual Reports 1969–72.
26. See USIMINAS, Annual Report 1972.
27. Interview with the plant general manager.
28. Interview with USIMINAS's former president. See also USIMINAS, Annual Reports 1962–70.
29. USIMINAS, Annual Reports 1956–66.
30. USIMINAS, Annual Reports 1956–66.
31. USIMINAS, Annual Report 1964.
32. Interview with a product research manager. See USIMINAS, Annual report 1966-69.
33. See USIMINAS, Annual Reports 1966–72. Interview with a product research manager.

34. Interview with the former manager of research centre. See USIMINAS, Annual Reports 1967–72.
35. See USIMINAS, Annual Reports 1969–70.
36. See Usiminas-Notícias, *USIMINAS Revista*, vol. 2, no. 3, p. 80, 1971.
37. Interview with a researcher (research centre). See also USIMINAS, Annual Reports, 1963–69.
38. Interview followed by plant tour with a technician of the technical division of the coke ovens and blast furnaces. See paper by the co-ordinator of the BF revamping 'Reforma do Alto Forno da USIMINAS', *USIMINAS Revista*, vol. 2, no. 4, pp. S/9–18, 1971.
39. See 'Reforma do AF teve tempo recorde (internacional) de 45 dias'. *USIMINAS Revista*, vol. 2, no. 3, p. 1, 1971.
40. Interview followed by plant tour with a technician of the technical division of the coke ovens and blast furnaces. See published paper by the then head of the pig iron metallurgy section and the head of the blast furnace dept: 'Eliminação do Cascão no Alto Forno da USIMINAS', *USIMINAS Revista*, vol. 2, no. 4, pp. S/2–8, 1971.
41. Interview followed by plant tour with a technician of the technical division of the coke ovens and blast furnaces.
42. Interview with a CSN's former president. Interview with USIMINAS's former president.
43. Brasil. Comissão Executiva do Plano Siderúrgico Nacional. Relatório Final, Rio de Janeiro, 1941. This document reflected the Getulio Vargas administration's view on the Brazilian steel project.
44. See Brazil/United States Steel Products Co., *Iron and Steel Industry,* Rio de Janeiro, October, 1939; Baer (1969); Gomes (1983).
45. The US government sought to avoid the threat of German participation in the Brazilian steel project and to secure, in case of war, the use of the Brazilian strategic points by US troops. See Baer (1969) and Gomes (1983).
46. Interview with USIMINAS's former president.
47. The office consisted of 90 of Arthur McKee's engineers, one independent American steel expert, and six Brazilian engineers. See Soares e Silva (1972). This is a historical account provided by a former president of CSN.
48. He was the president of a private bank, the vice-president of a docks company, and the vice-president of the Economic and Financial Council of the Federal Government.
49. This was Mr Macedo, a lieutenant of the army, who had been trained in metallurgical engineering in France (1925–30) with training in Italian steel companies (1932). He was a former member of the Brazilian Military Mission (1933–35).
50. See CSN; Annual Reports 1943–60. CSN: organisational charts (1945–1956). Although the raw materials department had been created, it engaged in commercial activities only. By 1949, the only organisational unit to support the operations lines was the metallography department processes and inspection unit, whose activities focused on the rolling mills only.
51. CSN's house newspaper, *O Lingote*, March 1952.
52. Interview with former USIMINAS's president. It should be noted that a large part of CSN's engineers came from the Army Engineering School, the 'locus' of the nationalist opposition to the foreign investors in Brazil during the 1940s, see Baer (1965)
53. Interview with a former foundry engineer. Interview with USIMINAS' former president.
54. Interview 1 with the research centre manager. Interview with a BF engineer.
55. Interview with CSN's former president. Also, CSN's documentary records; CSN (1996).
56. The year of 1946 was not considered.
57. The high utilisation rate of the steel shop during the 1950–53 period seems to have been associated with the reconstruction of the refractory linings. The document is not clear about who exactly performed that activity. The evidence suggests, however, that it was a type of one-off improvement. CSN (1996).

108   *Technological Learning and Competitive Performance*

58. See CSN, Annual Reports 1942–47. That facility was built in the southern state of Santa Catarina.
59. Sherwood (1966).
60. CSN, Annual Report 1943–60.
61. Several interviews at CSN.
62. Interview with a former Foundry engineer. See 'CSN: 50 anos transformando a face do país', *Metalurgia ABM*, vol. 47, no. 394, pp. 112–14, March/April, 1991.
63. A graduate in electrical engineering from the Federal Engineering School of Itajubá, in 1936, with strong experience accumulated in large projects in electric companies.
64. Interview with USIMINAS's former president. Interviews within the personnel development centre.
65. That was Mr Masao Yukawa. USIMINAS, Annual Reports 1963–67.
66. Interview with training manager at the Fundação General Edmundo Macedo Soares e Silva (FUGEMMS). CSN, Annual Reports 1946–60. CSN's archival records.
67. Interview with USIMINAS's former president.
68. Interview with USIMINAS's former president. Also, account by former head of the industrial relations dept. in USIMINAS (1988).
69. USIMINAS, Annual Reports 1967–70.
70. CSN, Annual Reports 1941–47.
71. Interview with a training manager of FUGEMSS. CSN's archival records.
72. USIMINAS, Annual Reports 1959–69.
73. Account by the former industrial relations superintendent in USIMINAS/Fundação João Pinheiro (1988).
74. Relatório da Comissão Executiva do Plano Siderúrgico Nacional.
75. Interview 2 with a foreman of the steel shop.
76. Interview followed by casual meeting with the former TIC manager. She organised and ran the TIC from 1966 to 1992.
77. Interview with USIMINAS's former president. Interview followed by casual meeting with the former manager of the technical information centre. When appointed as TIC's manager, she had a postgraduate degree in mechanical and electrical engineering. While running TIC she used to teach metallurgical engineering at the Federal University of Minas Gerais (UFMG) and to publish technical papers. At the time of the interview, retired from USIMINAS, she was running her own consultancy for specialised executive training.
78. Interview followed by casual meeting with the former manager of TIC.
79. Ibid.
80. Idid.
81. Interview with USIMINAS former president.
82. Ibid.
83. Interview with USIMINAS former president. See also USIMINAS, Annual Reports 1956–66.
84. Interview with a manager of the technical assistance unit.
85. Interview with USIMINAS' former president. Account by a former 'Samurai' and former Industrial relations superintendent in USIMINAS/Fundação João Pinheiro (1988).
86. Interview with USIMINAS's former president. Interview with a manager of the technical assistance unit. Also *USIMINAS Revista*, vol. 1, no. 1, 1970.
87. See Soares e Silva (1972)
88. CSN, Annual Reports 1946–60.
89. This type of training, developed in the US during World War II, sought to develop supervision skills.
90. Interview with a training analyst of the selection and training centre. USIMINAS Annual Reports 1962–70.

91. Ibid.
92. Interviews with USIMINAS engineers and managers.
93. USIMINAS, Annual Reports 1956–69.
94. Interview with a manager at FUGEMMS.
95. Interview with USIMINAS's former president. USIMINAS, Annual Reports 1956–69. The building of a working university had been planned. Although not operationalised, this gives an indication of how audacious USIMINAS's learning initiatives used to be.
96. CSN, Annual Report 1943.
97. Interview with a former general manager of the research centre.
98. Several interviews with managers and engineers in CSN.
99. Interview with USIMINAS's former president. Interviews with engineers in the plant.
100. Interview with a training manager at FUGEMMS. That issue was referred to in other interviews in CSN.
101. Account by a former development superintendent in USIMINAS/Fundação João Pinheiro (1988).
102. Interview 1 with a training analyst.
103. USIMINAS, Annual Reports 1956–66.
104. Paper by three USIMINAS engineers: Fusaro, V. L. et al. (1965), 'Resumo de um ano de produção na aciaria LD da USIMINAS', *Metalurgia ABM*, vol. 21, no. 9, pp. 459–69.
105. CSN, Annual Report 1944. Interview with a former foundry engineer.
106. CSN, Annual Report 1944. See also Soares e Silva (1972).
107. Interview with a former foundry engineer.
108. Interview 2 with a retired steelmaking shop engineer.
109. CSN, Annual Reports 1946–54.
110. Escola Técnica Pandiá Calógeras (CSN's archival records). Interview with an engineer of CSN's quality promotion centre.
111. Interview with USIMINAS's former president. Account by a former development superintendent USIMINAS/Fundação João Pinheiro (1988).
112. Interview with USIMINAS's former president.
113. Account by a former development superintendent in USIMINAS/Fundação João Pinheiro (1988).
114. CSN, Annual Reports 1945–49.
115. Interviews with managers in CSN. Interview with USIMINAS' former president.
116. Interviews with managers in CSN. Interviews with former engineers from CSN and USIMINAS.
117. CSN, Annual Reports 1946–49.
118. Interview with an operator whose father used to work as operator with the Americans. Such events were repeatedly recalled in the town of Volta Redonda by individuals who had worked in CSN or in some way had been related to the company, and in papers from that time. Low comprehension of the English language by the operators usually led them to do inappropriate tasks.
119. Interview 2 with the steel shop foreman.
120. Interview with an engineer from the technical division of power and utilities unit. Also, USIMINAS, Annual Reports 1956–63.
121. Interview with an engineer from the technical division of the power and utilities unit.
122. Account by a former development superintendent in USIMINAS/Fundação João Pinheiro (1988).
123. Interview with USIMINAS's former president.
124. Soares e Silva (1972).
125. Interview and casual meeting with a blast furnace engineer and a hot strip mill technician.

126. Ibid. Interview with a former foundry engineer.
127. This was a reference to the ascending curve of profit sharing which was reminiscent of the neck of a giraffe. CSN, Estatutos da CSN (artigo 48), 1949. Also, CSN's internal newspaper *O Lingote*, several numbers, 1953
128. Low morale was associated with high personnel turnover rates (from 1942 to 1946 there were 41,654 admissions and 38,751 dismissals) and with the fact that workers were being denied their rights to breaks, since CSN was an area of 'military interest'. Federal decree no. 11,087, December 1942; Soares e Silva, E. 'A formação técnica do brasileiro', *Carta Mensal*, CNC, Rio de Janeiro, August, 1979.
129. Interview and casual meeting with a blast furnace engineer and a hot strip mill technician.
130. Interview with the general manager of the CSN's research centre.
131. Account by a former industrial relations superintendent in USIMINAS/Fundação João Pinheiro (1988).
132. CSN, Annual Reports 1943–53.
133. Interview with a former general manager of USIMINAS research centre.
134. Account by the former industrial relations superintendent in USIMINAS/Fundação João Pinheiro (1988). Interview with a retired hot strip mill technician.
135. Interviews with managers and engineers.
136. Interviews with managers and engineers.

7. The Conventional Expansion Phase

This chapter covers the 1973–89 period for USIMINAS and the 1954–89 period for CSN, focusing on the paths of technological capability-accumulation and on the knowledge-acquisition and knowledge-conversion processes.

7.1 TECHNOLOGICAL CAPABILITY-ACCUMULATION PATHS

In the light of the framework indicated in Table 3.1, this section traces out the technological capability-accumulation paths followed by USIMINAS and CSN during their conventional expansion phase. In particular, the section is concerned with whether and how any innovative capability was accumulated.

7.1.1 USIMINAS: 1973–89

During this phase USIMINAS moved into the accumulation of technological capabilities beyond Level 4 across all five technological functions: investment (facility user's decision-making and control and project engineering), process and product organisation, product-centred and equipment. These technological capabilities were accumulated in a consistent way and at a fast rate through the Expansions I, II, and III, as indicated by the evidence in the following subsections.

Investment activities
As indicated below, from 1973 to 1989, USIMINAS moved from Level 4 routine (pre-intermediate) capability into the accumulation of Level 6 innovative (high-intermediate) capability for investment activities over its three conventional expansions.

Expansion I By late-1972, USIMINAS began to operate under improved macro-economic conditions permitting the execution of this expansion. Because USIMINAS did not have the capabilities to do this expansion wholly independently, technical assistance in project design and facilities specifications was obtained from NSC.[1] This was done through the 'engineering agreement'. Assistance in securing finance, and acquisition and supply of imported equipment was provided by Nippon-USIMINAS through the 'general agreement', that both had signed in 1970.[2] In that year, USIMINAS created the expansion group, consisting of 12 engineers, to engage in project engineering activities with NSC. By 1971, this unit was upgraded into the engineering department with 60 engineers. Avoiding full delegation to the Japanese team, the unit engaged actively in project design, negotiation with suppliers, and other engineering activities.[3]

Expansion II   By 1972 USIMINAS had been able to source, on its own, the funding necessary for its expansion.[4] USIMINAS was recommended to be technically assisted because of its inadequate capabilities to carry out that expansion independently. As a result, in July 1972, a 'contract of engineering consulting' was signed with NSC.[5] In parallel, a goal had been set by top management that USIMINAS should do at least 50 per cent of this expansion, although only 30 per cent was accomplished. This also sought to meet the goals of the National Steel Plan and the IS policy.[6] Although assisted by NSC, USIMINAS engaged actively in the project engineering activities by preparing the basic and the detailed plans, construction cost estimates and construction schedules. Presentation of the equipment specifications to the bidding parties, bid evaluations, decisions on suppliers and technical discussions with them were done together with NSC. The installation, testing and start-up were assisted by NSC, with USIMINAS in control of every activity.[7] This suggests that USIMINAS was heading towards the building up of its own project engineering capability.

In 1973, USIMINAS had its engineering department upgraded into the general engineering superintendency, with an increase from 60 to 260 individuals. The development superintendency was created to co-ordinate investment activities. Two years later, the engineering superintendency was again upgraded by the creation of the process engineering division (installations and detailed engineering) and the basic engineering division to engage in more complex investment activities.[8]

Expansion III   In the late-1970s, USIMINAS's president championed the idea that USIMINAS should do 100 per cent of this third expansion. However, it was advocated by other groups that the detailed and basic engineering services should still be technically assisted. In response, the dominant group, led by the president, managed to persuade the opposing groups to push the engineering superintendency to do the whole Expansion III, 'to be confident and learn more'.[9] This event indicates the critical role played by leadership in pushing USIMINAS into innovative capability building. This role consisted of setting of a challenging goal for the engineering teams. This expansion increased the installed capacity from 2.4 million to 3.5 million tpy.

To cope with the different problems, deriving from a much larger expansion, USIMINAS drew on its own project engineering capability accumulated over the two previous expansions. Several informal matrix organisational arrangements were created to tackle them.[10] USIMINAS managed to do the whole Expansion III (e.g. full control and execution of feasibility studies and basic engineering of individual facilities) without technical assistance.

In parallel, USIMINAS began to provide technical assistance, in project finance, to other steel companies. In 1975, the technical assistance unit was created to centralise and expand technical assistance activities. This suggests that USIMINAS was organising itself to engage in technical assistance activities. By 1981, USIMINAS was providing full technical assistance in investment activities to other steel companies. This began by the signing of the 'general agreement' with AÇOMINAS, in January 1976. USIMINAS provided assistance in technical and economic feasibility studies, detailed and procurement engineering, basic engineering, and guidance for operational start-up of a 2 million-tpy steel plant.[11] Over the 1980s, USIMINAS intensified technical assistance activities to other companies like COSIPA, CST, and ACESITA.

From 1972 to 1988, 46 steel companies were technically assisted by USIMINAS in Brazil and Latin America through 162 contracts. Through these contracts, technology sale/purchase ratios of 2 to 1 were achieved by USIMINAS, helping to finance its own investments.[12] The evidence suggests that in less than 30 years, USIMINAS had been able to reverse its condition as an absolute importer into a provider of high-level technical assistance in investment activities. By the late 1980s USIMINAS was able to do basic engineering for a whole plant independently and was in full control of investment activities (e.g. feasibility studies, search, evaluation and selection). The evidence also suggests that USIMINAS had been systematically providing technical assistance on these activities to several steel companies. In other words, in the light of Table 3.1, this suggests that by the late 1980s USIMINAS had accumulated Level 6 innovative capability for investments.

Process and production organisation activities
During the conventional expansion phase (1973–89), USIMINAS moved into the accumulation of Level 5 innovative (intermediate) capabilities for process and production organisation activities. This accumulation was characterised by the continuous creation and upgrading of organisational units and engagement in innovative activities on a increasingly independent basis.

In 1973, when the new facilities of Expansion I began to be installed, USIMINAS had its production support units substantially upgraded. The number of technicians in the metallurgical control and inspection dept increased from 215 in 1970 to 497 in 1973. These units intensified the provision of technical support to the operational units and engaged in metallurgical studies and continuous process improvements.[13] In parallel, the number of individuals in the research centre increased from 81 in 1972 to 322 in 1980.[14] In 1974, the information systems dept was upgraded into the information superintendency to concentrate on production planning and control (PPC) activities. Initially technically assisted by NSC and USS, this unit prepared the pluri-annual information plan and the integrated system for production planning and control (SIPCP). From the late 1970s, the activities of the SIPCP began to be wholly planned and executed by USIMINAS.[15]

In late-1972, USIMINAS created the process control group to engage in process automation activities. This led to the first automation master plan technically assisted by NSC. In 1974, the process control group was upgraded into the automation unit and the number of its technicians increased from 4 to 93 by 1987. From 1972 to 1988 USIMINAS engaged in a series of process automation activities across different operational units. Some of these activities were initially technically assisted and then based on USIMINAS's own capabilities.[16]

In the early-1980s, in response to the energy crisis of 1979, USIMINAS's metallurgical dept engaged in efforts to reduce fuel consumption. Efforts involved the manipulation of the granulometric distribution in the sintering process. In association with the foundry, changes to the equipment were also made (e.g. elevation of the layer height and control of the internal pressure in the furnace).[17] As a result of these efforts, the consumption of coke oven gas (COG) was reduced from 7.8 to 4.8 Nm^3/tonne within one year.[18] This evidence suggests that the accumulation of its own capabilities for process and production (and equipment) permitted USIMINAS to respond effectively to the energy crisis, since new efforts on process and production organisation improvements were undertaken.

Another example from the early-1980s, was the efforts to reduce the ferrous oxide (FeO) content in the sinter. This was done by improving the preparation of the coke fines (e.g. through strainers). The efforts contributed to reducing consumption of coke fines from 66kg/tonne in 1978 to 52kg/tonne in 1988 in the sinter machine 2. From 1985, the development of a mathematical model for the calculation of the raw materials mix by the automation unit with the research centre and the sinter plant, contributed to reducing the sinter dispersion: calcium oxide (CaO < 0.15) and silicon dioxide (SiO_2 < 0.10).[19]

As far as the evolution of production volume during Expansions I and II (1973-78) was concerned, the production of pig iron and molten steel increased more than 100 per cent. This was associated with the start-up of new facilities but also USIMINAS's capability to achieve high capacity utilisation rates in the production units. By 1977, two years after its start-up, steel shop 2 had a utilisation rate of 104.6 per cent and this reached 180.6 per cent in 1980.[20] This suggests that the accumulation of 'capacity-stretching' capabilities over the 1962–72 period had contributed to the achievement of high utilisation rates in new vintages of the plant during the 1973–89 period.

In the early-1980s, when Expansion III was being implemented, the Brazilian economy plunged into recession, following a substantial devaluation of the currency. Following the failure of the stabilisation plan set up in 1986, inflation and interest rates went up, demand fell and the currency was again devalued, leading to chaotic macroeconomic conditions.[21] Those conditions negatively influenced the evolution of USIMINAS's production volume. In 1981, pig iron and steel production fell 30 per cent in relation to 1980. Nevertheless, on average, increases over nominal capacity reached 75 per cent in the three worst years of recession, 1981–83. These achievements seem to reflect the long-term accumulation of the 'capacity-stretching' capability in association with the development of automated process control and PPC capability.

From the mid-1970s, USIMINAS engaged in efforts to reduce the silicon (Si) content in the pig iron. This was one of the measures taken to reduce the coke rate and meet the quality demands of the steel shop.[22] Through the joint efforts of the blast furnace dept, the metallurgical dept, and the research centre (technical support projects), the operational stability of the furnace and the quality of the metallic charge were improved. This was associated with improvements in the sinter granulometry and in the coke reactivity control. The Si content was reduced from 0.75 per cent in 1972 to less than 0.4 per cent in 1988.[23]

By 1982, a dynamic and updated automated control system had been installed in steel shop 2. This included automated continuous sampling of the metal bath variables in the oxygen furnace and a pre-set system of equipment and calculations of alloy additions in the ladle furnace. The development of the process control system derived from some joint work of the automation unit, the research centre, and the metallurgical control dept. From the late-1970s, these units became involved in extensive experimentation, trials and development of mathematical models. Although these units consulted with NSC to solve specific problems, much of the automated system was developed drawing on USIMINAS's own capabilities. This contributed to reducing the average heat time (tap to tap) from 40 minutes in 1976–82 to 32 minutes from 1983.[24]

From the mid-1980s, USIMINAS has engaged in the integration of the automated process control systems with the PPC system. Different organisational units were involved in the activity. The production management unit prepared the automation plans with the automation unit. The engineering management unit engaged in the feasibility studies, purchase specifications, and project follow-up. The research centre with the automation unit, engaged in building the mathematical models for the software, based on computer integrated manufacturing (CIM) and artificial intelligence principles. By the late-1980s, USIMINAS had achieved integration between the PPC systems (e.g. order acceptance, production programming, materials forwarding) and the process control systems (reduction, steel shop, rolling mills, and energy control).[25] The key benefits of this type of integration were: improvements in process yield; reduction in operational costs; product quality improvements; productivity increase; and reduction of energy consumption.[26] The benefits for performance in USIMINAS are presented in Chapter 10. In the light of Table 3.1, by the late 1980s, USIMINAS had accumulated Level 5 innovative (intermediate) capability for process and production activities.

Product-centred activities
During the conventional expansion phase (1973–89), USIMINAS engaged in the building of innovative capabilities to do in-house product design and development activities. By the end of this phase, it had accumulated Level 6 innovative capability for product-centred activities. While the design and development of its 'first-generation' steels, as it used to be said in the company's language, were 'technically assisted', the 'second generation' was wholly designed and developed in house.

'First-generation' steels (1973–79 approximately)   In 1973, following the launch of the research centre, USIMINAS engaged in systematic studies of the characteristics of new steels and their continuous 'reverse engineering'. In early-1974, the development of the USI-FIRE, a fire-resistant steel, provided the psychological boost for USIMINAS to set a product development plan to engage in more complex product activities.[27] From 1972 to 1978, nearly 25 new non-original steels for general use were developed, although technically assisted by NSC. USIMINAS then began to have its own product specifications, achieving a leading market position (e.g. for the car and construction industries).[28]

'Second-generation' steels (1979–89 approximately)   From the late-1970s, USIMINAS engaged in intense efforts to design and develop higher-valued and more complex steels, without technical assistance. These efforts involved the research centre (and its laboratories and pilot facilities) and the metallurgical dept, in the design activities; the steel shop and the rolling mills during the development stages (e.g. industrial trials); and the steel users (e.g. PETROBRAS) for tests and product application feedback.[29] The interviews and documents suggest that USIMINAS went through a painstaking process drawing on its own experience, capabilities, difficulties and failures accumulated over the development of the previous generation of steels. From the mid-1980s, complex steels, wholly designed and developed by USIMINAS, began to emerge from those efforts, in particular, the micro-alloyed ISI-IF, the USI-STAR, for car

making, and the extra-fine USI-BNR for packaging.[30] Although these steels
had also been developed elsewhere by technological frontier companies and
were therefore non-original, USIMINAS had been able to design, develop,
manufacture, and commercialise them drawing on its own capabilities. By
being able to carry out these activities in-house the evidence suggests, on the
basis of Table 3.1, that by the late-1980s, USIMINAS had accumulated Level 6
capability for products.

It took nearly 15 years for USIMINAS to develop the whole USI-R-COR
family (steels highly resistant to corrosion for car and domestic appliances),
and nearly 11 years to move from Level 5 to Level 6 capability for products. On
the one hand, it should be remembered that by the early-1960s the company
lacked even the basic capability for products; and engaging in the development
of new steels independently might have involved a wide range of problems that
had to be sorted out by the company. Therefore, these rates can be considered as
an effective achievement in USIMINAS. On the other hand, the evidence indicates that the way the company has moved up across those levels was not
straightforward, in other words, it was associated with deliberate and effective
in-house efforts.

Between 1979 and 1989, USIMINAS was awarded ten product certificates
(by local and international organisations) for the high quality of its new steels.
In addition, by the 1980s USIMINAS had introduced the 'quality system'
across the production lines. This system was built on the 'quality control' and
the 'assured quality' systems that had been introduced in the 1960s and 1970s,
respectively.[31] This indicates that, in association with the capability to design
and develop new steels, USIMINAS was able to manufacture and commercialise them under rigorous international and local quality standards. In other
words, the accumulation of the capability to develop new steels was running in
parallel with the capability to manufacture them efficiently (Level 2).

Some factors have stimulated USIMINAS's vigorous engagement in innovative product activities. One was the visionary leadership that, since start-up, had
been committed to making USIMINAS a leading and 'technologically independent' steel company in Brazil.[32] Another was USIMINAS's response to external
factors: the increased demand for lighter and thinner sheets (thickness < 0.20mm)
with high resistance and good stamping conditions, derived from the growth of
the domestic appliance and car industries in Brazil from the 1960s; and the increased demand for steels for offshore platforms and pipelines, derived from the
energy crises in the 1970s, which forced Brazil to explore its oil reserves in distant regions under adverse conditions (e.g. deep water).[33]

Equipment activities
This subsection indicates that during the 1973–89 period USIMINAS engaged
in the accumulation of innovative capabilities to undertake different and more

complex equipment activities. These were related to large equipment engineering. By the end of the conventional expansion phase (late-1980s), USIMINAS had accumulated Levels 5 to 6 equipment capabilities.

By the early-1970s, stimulated by the IS policy and its own leadership, USIMINAS had created Usiminas Mecânica SA (USIMEC).[34] This was to allow USIMINAS to design and manufacture its own equipment for its expansions, and to engage in the construction of metallic structures for steelworks, bridges, and general industrial structures.[35] USIMEC was organised by USIMINAS and its board and several technical positions were filled by USIMINAS's experienced engineers. USIMEC built on USIMINAS's existing equipment capabilities to accumulate the initial level of equipment engineering capability.

However, to engage in detailed and basic equipment engineering activities USIMEC had to draw on foreign technical assistance. In 1971, a 'memorandum of understanding' was signed with the Japanese company, Hitachi Zosen. This was followed by a 'memorandum of understanding' and a 'technical co-operation contract' signed with NSC in 1972 and 1974, respectively.[36] One year later, USIMEC's dept of design engineering was undertaking 'basic reverse engineering' of projects.[37] By 1975, USIMEC had manufactured the BF 3 (5,400 tonnes/day-capacity) for USIMINAS. Although technically assisted, the activity worked as a platform for the fast accumulation of higher levels of that capability. From 1975, new technical assistance agreements were signed (e.g. with the German Schalke; the American Ahlstrom; and the Austrian Andritz). They permitted USIMEC to engage in more complex detailed and basic equipment engineering. The relationship with that assistance was marked by close work, followed by continuous in-house efforts on reverse engineering of equipment engineering projects.[38]

By the late-1970s, USIMEC/USIMINAS had accumulated Level 5 innovative equipment capabilities. In other words, on the basis of Table 3.1, it was able to do basic and detailed equipment engineering of individual facilities (e.g. BFs), without technical assistance, though not whole steelworks. To engage in equipment engineering of whole steelworks, USIMEC's design engineering dept began to work jointly with technological frontier companies (e.g. via partnerships for specific projects) on the latest equipment developments to adapt them to different market demands.[39] By the mid-1980s, with nearly 4,000 employees, USIMEC had been awarded three special certificates which allowed the company to supply users like PETROBRAS.

In addition, USIMEC had become able to design and manufacture whole steelworks facilities and components to supply other industries (e.g. oil, cement, pulp and paper). Therefore, by the late-1980s USIMINAS/USIMEC were carrying out basic and detailed engineering and manufacturing of equipment for whole steelworks and components for other industries. In the light of

Table 3.1, the evidence suggests that the company had accumulated Level 6 innovative capabilities for equipment. By 1989, however, USIMINAS/USIMEC began to suffer the effects of the crisis in the Brazil's capital goods industry, as described in the next chapter.

7.1.2   CSN: 1954–89

During this phase, CSN sought to develop innovative capabilities across five technological functions. However, they were accumulated in an inconsistent way. Capability for investments was accumulated between Levels 4 and 5. Although innovative capability for equipment was accumulated at Level 5, innovative product-centred capability was accumulated only intermittently at Levels 3 to 4. Innovative process and production organisation capabilities were accumulated only at Level 3. In addition, CSN failed to complete the accumulation of Level 1 routine (basic) for process and production organisation and product-centred activities.

Investment activities
CSN moved into the accumulation of capability for investments between Levels 4 and 5 during Expansions I to III.

Expansions I and II   From 1954 to the early-1960s, CSN was involved in two expansions (the B and C Plans). In 1955, the number of engineers in the project unit increased from 37 to 75. CSN did some technical and economic feasibility studies and participated in equipment purchasing, while a large part of both detailed and basic engineering was delegated to the consulting firm Arthur McKee.[40] In 1961, stimulated by the IS policy, CSN upgraded its project unit into the technical planning superintendency to co-ordinate its expansions. In 1963, its subsidiary company, the Companhia Brasileira de Projetos Industriais (COBRAPI), was created to engage in project and equipment engineering activities. COBRAPI was organised by Arthur McKee who assisted the company until the early-1980s.[41]

Expansion III   In 1965, an expansion plan of 3.5 million tpy was prepared by COBRAPI/McKee but not approved by the BAHINT/BNDE group. This was followed by another plan of 2.5 million tpy that again was not approved by BAHINT/BNDE and the GCIS. The reason was that CSN was seeking to invest in the latest facilities which were not compatible with market conditions, and without demonstrating efforts to improve its existing capacity. It was then recommended by the GCIS that CSN's Expansion III should be done under a three-stage plan which became the 'D Plan' for 4.6 million tpy.[42]

Stage I   By 1966, COBRAPI had 650 employees. Its project engineering division was assisted by Arthur McKee. COBRAPI was able to secure finance from the US EXIMBANK and BNDE. It also was able to undertake about 70 per cent of the technical and economic feasibility studies and detailed engineering, but only about 20 per cent of the basic engineering activities. A large part of purchase management and basic engineering was done by Arthur Mckee.[43]

Stage II   This expansion was approved in the National Steel Plan on the basis of 'CSN's know-how in steel production and its adequate conditions for high productivity and efficient capacity utilisation'.[44] However, it should be remembered from Chapter 6 that until then CSN had been experiencing low capacity utilisation rates, although the company was seeking to accumulate innovative capability for project engineering. This inconsistent pattern of capability-accumulation evolved over this phase to become a key part of CSN's history. In Stage II, about 80 per cent of the technical and feasibility studies, purchase and detailed engineering were undertaken by CSN. About 60 per cent of basic engineering was undertaken by CSN/COBRAPI technically assisted by engineering firms like Arthur McKee. By 1975, the number of employees in COBRAPI had increased to nearly 1,000, and new facilities for calculation and design engineering had been installed.[45]

Stage III   In July 1976, a thorough revision of this expansion was made by the United States Steel Engineers and Consultants Company (UEC) to make the plan economically feasible. This involved searching for and selecting the latest product and process technology on the basis of large capacity, high-speed, continuous and automated operations and to obtain high quality products at a low cost.[46] Stage III completed the expansion to 4.6 million tpy. The involvement of CSN/COBRAPI in decision-making, and monitoring of the expansions activities was deeper in Stage III than Stage II. Indeed CSN/COBRAPI were engaged in rather more than 80 per cent of the basic engineering. However, the decision-making process and project engineering activities still received external assistance, for instance, from UEC and Arthur McKee.[47] The evidence does not suggest therefore that CSN was in full control of the expansion.

From the mid-1970s to the mid-1980s, CSN (engineering superintendency)/COBRAPI even provided intermittent technical assistance to other steel companies for their expansion activities (e.g. Usina Siderúrgica da Bahia, Companhia Siderúrgica do Nordeste, AÇOMINAS, CSBM among others). These activities involved installations, and some detailed and basic engineering. Although the provision of technical assistance was undertaken by CSN, this activity did not become systematic. In other words, CSN did not demonstrate any explicit pursuit of organising itself to provide technical assistance as a way of accumulating higher levels of investment capability.[48] The evidence suggests that during Stage III (mid-1980s) CSN had started to move from routine into innovative investment activities. In the light of Table 3.1, this suggests that CSN was between Levels 4 and 5 as far as the accumulation of capability for investments was concerned. Indeed the evidence suggests that CSN seemed to be aiming for the accumulation of capability at Level 5. However, by 1985, endorsed by CSN's top management, COBRAPI began to be transferred to the holding company SIDERBRAS. The idea was to make COBRAPI into a major provider of engineering services. However, this move had negative implications for CSN's investment capabilities, as described in the next chapter.

Process and production organisation activities

During the conventional expansion phase (1953–89), although CSN sought to accumulate innovative capability for process and production organisation, it did not move beyond Level 3, in other words, the capability to do minor and intermittent improvements. In addition, CSN failed to complete the accumulation of Level 1 routine capabilities for process and production organisation activities.

In 1954, import restrictions deriving from a new exchange rate policy made CSN shift from the use of imported coal (about 70 to 80 per cent), to control its coke rate, to local coal. As a result, the coke rate for BF 1 increased from 761kg/t in 1952 to 823kg/t in 1956. Despite the negative implications for its performance, CSN did not engage in significant efforts to bring the coke rate down and it waited until 1960 for a sinter plant.[49] This absence of systematic engagement in process improvement efforts reflects, on the one hand, CSN's limited capabilities to engage rapidly in innovative activities. On the other, it seems to reflect top management's limited initiative towards developing in-house capabilities to respond to further external factors.

During 1954, CSN's management board was changed three times. These changes were followed by the withdrawal of its president in October 1958, leading again to changes in several management positions.[50] These changes led to inconsistent corporate decisions and difficulties in co-ordinating the flow of production across the plant associated with conflicting instructions. Even then, the pressures to increase production volume continued.[51] From 1958, changes in CSN management positions became more frequent, as mentioned in different interviews in the company. As far as production growth is concerned, during Expansions I and II (1954–68), coke production grew by 53 per cent (1954–61), and fell by 2.6 per cent (1962–68). Pig iron production grew by 29 per cent (1954–61) and 38 per cent (1962–68) and molten steel production had a 96 per cent growth (1954–61) and a 15.6 per cent growth (1962–68). The evidence suggests uneven production growth. This was also reflected in the low and uneven capacity utilisation rates of the operational units. These rates might have been associated with the constant stoppages in the BFs.

Although the utilisation rates of the SM steel shop were relatively high, they resulted from intermittent rather than continuous 'capacity-stretching' efforts. Between 1955 and 1956, a group of engineers in the unit identified that the non-atomised air blown into the furnace and the high viscosity of the oil were limiting the furnace operation. Steam piping was introduced into the furnace to blow higher volumes of air, thus improving combustion. As a result, heat time was reduced by 30 minutes, increasing the number of heats per day and leading to higher capacity utilisation rates. From 1958, however, only minor and intermittent improvements took place in that unit. Further increases in production volume derived from the installation of furnaces 7 (1959) and 8 (1960) to stabilise operations.[52]

During Expansions I and II there was the formal presence of certain organisational units for the support for the production units (e.g. production planning and control, metallurgy dept), as seen in CSN's organisational charts from that period.[53] These units, in principle, should provide support for continuous improvements in the production lines. However, interviews with managers and engineers in CSN revealed that the engagement of those units in continuous improvements in the production lines was virtually absent. By the early-1960s, the research dept had been created, adding one more organisational unit to CSN. This unit was started in May 1962, with 10 engineers. Until the late-1970s most of its activities were related to solving routine operational problems for the production units ('trouble-shooting'). In other words, it did not engage in process (and product) improvement activity in a consistent and planned way.[54]

In 1972, as Stage I of Expansion III began to be implemented, CSN undertook its first large-scale corporate restructuring. It was led by the consulting firm Arthur D. Little.[55] The industrial engineering unit began to control production costs. The research dept was upgraded into the research superintendency, but it continued 'trouble-shooting'. The quality control unit engaged in studies of process parameters and product quality. Despite this restructuring, little improvement took place in process and production organisation activities. In other words, there was little engagement in systematic and continuous process improvement and some activities continued to be delegated to external consultants. For example, a standard cost control system was developed by a UEC consultant. However, it was not operationalised.[56] Until the late-1970s, a few managers were even suggesting the introduction of new production techniques (e.g. quality control circles – QCC). On the one hand, most of these suggestions were ruled out by top management since CSN was not 'mature enough' to introduce them. On the other, as mentioned in different interviews in CSN, new 'management packages' (e.g. 'management by objectives') were introduced at the discretion of individual managers. However, following the constant changes in middle-management positions over the 1970s and 1980s, the techniques were implemented on a one-off basis. On some occasions, they led to misunderstandings on the production lines because of inadequate explanation of their meaning and purpose.[57]

As far as automated systems activities are concerned, by the early-1970s CSN had contracted technical assistance from UEC to build its first automated system for PPC. From that time, although automated systems became available in CSN, they were acquired from other companies. It was not until the mid-1980s that CSN sought to engage the automation group, assisted by local suppliers, in systems development. By the late-1980s, CSN turned to NSC for assistance in the building of a master plan for informatics and automation.[58] This led to the structuring of the information technology superintendency. Therefore during the 1953–89 period CSN had very little engagement in the development of its own automated process control and production planning and control (PPC) systems.[59]

During Expansion III (1970s and 1980s), following the two major energy crises (1973 and 1979), the evidence suggests that CSN did not engage in systematic and significant efforts to reduce fuel consumption in the sinter plant or in the BFs. As a result, for example, the coke rate of BF 1 increased from 558 in 1972 to 643 in 1976.[60] Additionally, as a consequence of the expansion, the BOF shop started up in January 1977. This led to the shutdown of the Siemens-Martin shop in 1981. Interviews and production records in CSN revealed that during the first year of the steel shop, efforts to operate the unit efficiently and stabilise operations were either absent or intermittent. CSN seemed to pursue rapid increase in production volume only. As a result, for instance, in September 1977, following a mishandling of routine crane operations, the whole load of 'converter A' spilt over the shop damaging the installations and causing a 26-day stoppage.[61] This suggests a lack of basic routine capability. In addition, it was not until 1994 that the design capacity of the unit was reached. Among the reasons for this were the limited co-ordination between the production of the BFs, the steel shop, and the continuous casting mill. Because these three units operate with molten metal their operations influence each other.[62] The evidence suggests that, in the light of Table 3.1, by the late-1980s CSN had not completed the accumulation of Level 1 routine capability to operate the unit efficiently.

By the early-1980s, when the Brazilian economy went into recession, CSN had plunged into a deep financial crisis associated with uncontrolled operational costs. As far as production growth over the 1980s is concerned, some production units (e.g. BF 3) even had negative growth rates. Steel production grew by only 6.9 per cent (1980–81), fell by 2.2 per cent (1981–82) and rose by 24.2 per cent in the following year. In 1984, the steel shop had a negative production growth of 13.4 per cent in relation to 1983. This unstable production evolution was reflected in the utilisation rates of the operational units, when most production units operated below their design capacity, including the hot strip mill 2. As referred to by managers in the company, until the late-1980s, CSN's production growth was marked by 'ups and downs'.[63] The low utilisation rates in CSN might have been a reflection of the economic recession during the early-1980s. However, USIMINAS's experience suggests that if CSN had engaged in efforts in improvements on process and production organisation on a more continuous basis, the capacity utilisation rates would probably have been higher than those achieved.

Other factors within CSN may also have contributed to its low capacity utilisation rates. From 1986 to 1989, CSN went through financial and institutional crises that mitigated any efforts at capability building. From 1984, political interference was intensified in CSN (e.g. the appointment to key managerial positions on the basis of vested interests). These were combined with the presence of the radical wing of the Workers' Central Unit (CUT). From 1984 to 1989, there were at least two strikes each year.[64] In November 1988, following a 19-day strike, the plant was invaded by the army, three operators were shot dead, and the whole plant almost closed down.[65]

Product-centred activities

Although CSN engaged in innovative product activities in the conventional expansion phase (1953–89), these were basically characterised by only intermittent efforts on existing product improvements; and only intermittent efforts on design and development of new steels. Therefore innovative product-centred capability was accumulated, inconsistently, at Levels 3 to 4. The accumulation of Level 1 routine for process and production organisation and product-centred activities continued to be incomplete.

Until the early-1960s, complaints were made by CSN's users, particularly the car industry, on surface defects and poor steel composition of its plates.[66] In 1963, as a response to the start-up of a new local competitor (USIMINAS), CSN engaged in product improvement efforts. However, the evidence suggests that the way they were implemented did not lead to significant improvements. In 1964, the 'Quality Year' in CSN, over 100,000 analyses of hot and cold products were made, with nearly 5,000 special trials, and 25,000 chemical trials as part of the product improvement activities.[67] In 1965 and 1966, with the involvement of the industrial engineering dept and the methods and incentives dept, other types of steel were produced. This was done intermittently to meet the car industry's demands.[68] By 1967, this intensity of efforts began to weaken following different directions given by new management. The management gave priority to product quantity rather than quality, the 'volume-first practice'.[69] In 1968, the manufacture of higher-valued steels for the automobile industry began to be undertaken by technical assistance: UEC (to improve the quality and productivity of rolled products) and Armco Steel Corp. (manufacturing of galvanised sheets).[70] This suggests that CSN turned to external technical assistance instead of developing the capabilities to do those activities independently.[71] The evidence suggests that no substantial product improvement took place in CSN to face up to the new competitor.

As far as product development was concerned, in 1969, following a demand from the army, CSN had a team of researchers from the research dept to develop niobium steels. Systematic studies of new steel characteristics were undertaken from 1969 to 1970, followed by 'reverse engineering' in 1971. By 1973, the NIO-BRAS 200 (a steel for rails), and NIO-COR (a steel resistant to atmospheric corrosion) had been developed, based on the specifications of the COR-TEN steel, developed earlier in USS.[72] By 1977, NIOCOR-1 (skating hot rolled steel for structural use) was developed followed by NIOCOR-2, in 1978. Some of the industrial trials and the manufacture of these products were led by UEC.[73] This production innovation 'capability' (Level 3 to 4) resided in a few engineers. However, when they had all left the company by 1979 CSN was unable to develop niobium steels.[74]

In 1979, the new industrial director, a former middle manager, persuaded top management to re-start product development. As a result, the research superintendency engaged in this activity.[75] The unit began by studying the new steels' characteristics and by reverse engineering existing steels, technically assisted by UEC. From 1980 to 1985, CSN designed and developed a few steels for general use. However, some of those projects were further delayed and/or abandoned. This was associated with CSN's limited Level 1 routine process and product capability for the industrial trials and/or daily manufacturing. On some occasions, during the 1980s, new products were not even included in the manufacturing schedules.[76]

In 1984, product innovation efforts diminished when top management directed the research superintendency to focus exclusively on 'trouble-shooting'. Studies on new steels' characteristics continued but at individuals' initiatives.[77] It was not until 1988 that the research superintendency re-engaged in product development activities. From 1986 to 1989, new steels were developed including CSN's own specifications (e.g. CSN-COR and CSN-ZAR). They were more a result of the research superintendency's isolated efforts than any explicit corporate goal.[78] However, as pointed out by one engineer 'until 1990, it was frustrating to see poorly manufactured steels that had taken months to be designed in the labs'.[79] This suggests that although CSN had accumulated some Level 4 innovative product-centred capability, the accumulation of Level 1 routine capability to manufacture those products efficiently had not been completed.

Equipment activities

During the conventional expansion phase (1953–89) CSN moved into accumulation of innovative capability for equipment at Level 4. In the early-1960s, in response to IS policy, CSN engaged in innovative equipment activities through the creation of COBRAPI, whose equipment engineering division was organised and assisted by Arthur McKee.[80] In parallel, CSN created the Fábrica de Estruturas Metálicas (FEM) to make metallic structures.[81] By the end of Stage II, the participation of components produced in house had increased from 8 per cent to 17 per cent, as a result of COBRAPI's and FEM's activities.[82] Until the early-1970s, the revamping of the BFs was wholly done by Arthur McKee with little CSN engagement. From the mid-1970s, COPRABI actively involved itself in these activities although revamping was still led by suppliers or engineering firms. For example, revamping of BF 1 (1977–83) was led by Paul-Wurth, and BF 2 (1977–81) by Nippon Kokan KK.[83] The revamping of BF 3 (1984), although with a greater involvement from COBRAPI, was led by Nippon Kokan KK because CSN did not have the capability to do it alone.[84]

In the early-1970s, CSN engaged in detailed equipment engineering activities for Stage III. It was technically assisted by Arthur McKee and supported by the Grupo Villares, a Brazilian capital goods manufacturer. In 1978, COBRAPI undertook the complete detailed engineering of the metallic structures of the boiler of the new sinter plant in CSBM.[85] By 1978, COBRAPI began to engage in basic equipment engineering through licensing foreign equipment technology (e.g. from Dwight-Lloyd and Paul-Wurth). One year later, COBRAPI was able to manufacture large equipment, following given specifications, but not to design it. To do this, COBRAPI began by the 'reverse engineering' of certain basic equipment projects. In 1980, in association with Paul-Wurth, COBRAPI undertook the design, detailed and installation engineering of the CSBM BF 5.[86]

In 1982, following the shutdown of CSN's coke plant's battery 1, COPRAPI designed and manufactured a new coke battery which started production in 1990.[87] By the late-1980s, COBRAPI was engaging more actively in large equipment revamping in relation to the 1970s. Although it was able to carry out large equipment manufacturing, its engagement in basic equipment activities was still incipient. In the light of Table 3.1, this suggests that CSN/COBRAPI had accumulated Level 4 innovative capability for equipment. In addition, the company seemed to be heading towards the accumulation of Level 5. However, by the late-1980s, the transfer of COBRAPI to SIDERBRAS and the crisis in Brazil's capital goods industry had begun. These factors had negative implications for CSN's equipment capabilities, as described in Chapter 8.

7.1.3 Summary of the Cross-company Differences

The evidence outlined above indicates that both USIMINAS and CSN engaged in the development of innovative technological capabilities over the conventional expansion phase. However, the way, and rate at which, those capabilities were accumulated in the two companies were different. By the late-1980s, USIMINAS had accumulated Levels 5 to 6 across all five technological functions within less than 30 years of plant start-up. In contrast, CSN, after more than 40 years, had accumulated capability between Levels 4 to 5 across two technological functions and, intermittently, Levels 3 to 4 across three other technological functions. In other words, CSN followed a slow and inconsistent accumulation path as opposed to that of USIMINAS. In particular, the differences are as follows.

Investment capability

USIMINAS deliberately engaged in the accumulation of capabilities to carry out its expansion activities independently. As a result, the whole of Expansion III was planned and executed drawing on its own capabilities. USIMINAS was stimulated by its leadership and by the IS policy to pursue this accumulation. CSN also engaged in the accumulation of investment capabilities, as a result of the IS policy, but lacked a leadership initiative like that of USIMINAS. While USIMINAS accumulated Level 6 innovative capability for investments, in less than 30 years, CSN accumulated capability between Levels 4 and 5 only after about 40 years. In addition, USIMINAS had an explicit goal and organised itself (e.g. by creating the technical assistance unit) to sell technical assistance in project engineering. In contrast, CSN did not organise itself to do that and did not engage systematically in that activity.

Process and production organisation capability

Within less than 30 years, USIMINAS was able to accumulate Level 5 innovative capability. In contrast, CSN, after about 43 years, had not moved beyond Level 3 innovative capability for process and production organisation. In USIMINAS the accumulation path was marked by continuous process improvements and permanent upgrading of organisational units. In contrast, in CSN it was marked by scarce and intermittent process improvement efforts, In addition, although CSN had built organisational units that were similar to USIMINAS, those units barely engaged in continuous process and production organisation improvements in the operational units. By the late-1980s, USIMINAS had integrated its PPC and process control systems, drawing on its own capabilities. At that time, CSN still had not accumulated the basic capability to develop its own automated systems. In addition, USIMINAS responded to the energy crises by engaging in additional process improvements. In contrast, although CSN might have had some concern about the effects of those crises on its performance, the concrete responses produced in terms of innovative efforts on process improvements were limited.

Product-centred capability
USIMINAS engaged vigorously and continuously in the accumulation of capabilities to design and develop its own steels. On most occasions, it was pushed by top management's explicit corporate goals. In contrast, CSN engaged only intermittently in those types of activity. This may reflect the top management's inconsistency as far as the development of in-house innovative product capability was concerned. When USIMINAS had developed its 'second-generation' steels drawing on its own capabilities, CSN was still engaging in the early stages of product development activities, which were preceded by some earlier intermittent projects. Thus, USIMINAS was able to accumulate, within less than 30 years, Level 6 innovative capability for products, while CSN took nearly 45 years, to accumulate product capability at Level 4 only. In addition, USIMINAS could draw on its own basic routine process and product capabilities to have its new steels efficiently manufactured. In contrast, CSN's new steels were poorly manufactured reflecting the incomplete accumulation of Level 1 capability for process and product capability.

Equipment capability
Both USIMINAS and CSN were stimulated by the IS policy to engage in equipment engineering activities. However, although USIMINAS started later than CSN, it was able to accumulate, within less than 30 years, Level 6 innovative capability for equipment. In contrast, CSN accumulated, within about 40 years, the capability at Level 4.

7.2   KNOWLEDGE-ACQUISITION PROCESSES

In the light of Table 3.2, this section describes the knowledge-acquisition processes during the conventional expansion phase in USIMINAS (1973–89) and CSN (1954–89).

7.2.1 External Knowledge-acquisition Processes

Pulling in expertise from outside

USIMINAS   In the early-1970s, USIMINAS sought to bring new experts into the company like a 'project champion' from NSC. He was hired as 'chief-adviser' until 1976 and then as technical counsellor from Japan.[88] In 1972, the first PhD engineer was hired by USIMINAS to create the special unit for steel application with some technicians from the German firm Gute Hoffnungs Huette Sterkrade AG (GHH).[89] During the 1970s and 1980s technicians from NSC were also invited to give presentations to the engineers in USIMINAS.[90] From the late-1970s, local and international steel researchers were invited to give talks to individuals in the research centre and other support units[91] bringing new tacit knowledge on product development and process improvements into USIMINAS. Therefore USIMINAS continued diversifying the types of expertise brought into the company. The functioning of this process, as mentioned in different interviews, contributed to triggering innovative activities in the company (e.g. new ideas on product development brought by 'project champions' and a new project for steel application in the construction industry).

CSN   CSN did not seek to hire steel experts to head its technical positions. Instead, retired engineers (e.g. from UEC) were hired on a temporary basis.[92] Until the late-1960s, engineers were still recruited from the Army Engineering School and given key technical positions. From the early-1970s, a great part of CSN's engineers were recruited from the Federal Engineering School of Itajubá. Without previous experience in the steel industry or in the management of a large steel company, they brought little tacit knowledge into CSN.[93] During the 1970s, experts in technical and/or managerial areas were invited to give conferences within CSN. However, most of these conferences proved fruitless: 'The levels of talks were so advanced that the individuals could not follow them . . . they were too far away from CSN's needs'.[94] The conferences organised in the 1970s were costly for CSN but their benefits were limited. They could have produced better results if they had been simpler and matched more with the operators' needs.[95] Although CSN continued to hire new engineers, the variety of the expertise was still limited. This suggests that the functioning of this process contributed very little to triggering innovative activities in the company.

Recruiting operations individuals

USIMINAS   From the 1970s, the recruiting practice was integrated with the in-house training programmes. In other words, most of the operators were recruited from the 'Apprentices Training Centre' in USIMINAS. Some operators continued to be recruited from SENAI. 'To operate a large plant like this we needed individuals with a minimum knowledge to assimilate instructions.'[96] In addition, the practice of coupling the recruiting procedures with further initial training on the sites continued over the 1970s and 1980s.[97] These mixed recruiting practices (e.g. previous training, interviews and tests of aptitudes, further adaptation courses) probably contributed to improving the operators' abilities to engage in their jobs. This seems to have been influenced particularly by the functioning of the 'apprentices training centre', as described later in this chapter

CSN   From the 1960s, some operators were recruited from CSN's technical school and from SENAI. Although that training mechanism was present in CSN, its contribution to improving the operators' abilities seemed limited since the low level of concern of CSN with the qualification of operators continued until the late-1970s and there was no adequate site training following the recruitment.[98] Therefore there was only a slight improvement in this mechanism in the expansion phase in relation to the start-up and initial absorption phase.

Channelling of external codified knowledge

USIMINAS   The TIC was upgraded from 12 individuals in 1970 to 35 in 1988, recruited under rigorous criteria.[99] By the late-1970s, TIC had systematised nearly 1,600 technical papers published world-wide. By 1988, TIC had 12 of its own publications based on external published knowledge circulating within USIMINAS. They were written in a 'journalistic-and-provocative way'. In addition, they were widely spread across several units. 'Those papers had to reach individuals at the lowest points in the plant.'[100] In addition, interviews with engineers and managers in the plant (and outside the company) stressed the relevant contribution of TIC to different improvements and innovative projects in the 1970s and 1980s. The functioning of this process also contributed to: (1) accelerating and increasing the access of different individuals to external codified knowledge; (2) pointing out 'knowledge gaps' by describing technical achievements in other plants; (3) stimulating the performance benchmarking practice; and (4) triggering continuous intensity in knowledge-acquisition and knowledge-sharing processes, in other words, more interaction with other learning processes. By 1987, TIC had a collection of 24,450 books, 9,857 patents, 90,508 technical norms and specifications, 27,043 other publications, and 788 periodical subscriptions.[101]

From 1972, individuals in the general engineering superintendency began to document the latest developments in project and equipment engineering in advanced steel companies (e.g. detailed layouts, organisational charts).[102] From 1974, individuals in the research centre library were engaged in the acquisition of several types of technical literature.[103] This indicates a redundancy of that mechanism across the company.

CSN   By the 1980s the library in CSN had a number of technical publications. Units like the research centre also had a number of those publications. Although this type of codified knowledge was present in CSN there was no mechanism for its systematic dissemination across the company. This suggests that access to that knowledge was limited to a few individuals.

Overseas training

USIMINAS   Until the late-1970s, in particular, USIMINAS had groups of 50 engineers and technicians constantly sent overseas for one to twelve month's training.[104] This was a kind of 'rotating overseas training' which sought 'to enrich the knowledge'.[105] It led to a continuous process of acquisition of tacit and codified knowledge by individuals from different areas of activity. From the early-1970s, USIMINAS had operators included in overseas training before they became involved in new technological activities. In 1973, a group of 12 individuals (engineers, technicians and operators) was sent to Japan for practical training in the operation of the continuous cold rolling mill. This was four months before its start-up in USIMINAS.[106] In late-1975, a group of 30 engineers, technicians and operators was sent to Japan for a three months' training on the operations of the new plate mill. This was six months before its start-up. They were followed in 1977 by another group of 62 operators and technicians for training in the reduction, steelmaking, and rolling areas.[107] These groups stayed together during the whole training programme, leading to an intense knowledge sharing on the sites and informal knowledge sharing (e.g. in the hotel). This suggests a strong interaction between knowledge-acquisition and knowledge-conversion processes. Up to the mid-1980s, more than 400 USIMINAS individuals at the technical level, had been trained overseas.[108]

As a way of engaging in innovative equipment engineering activities, USIMINAS had nearly 400 USIMEC technicians and engineers sent overseas, during the 1970s and 1980s. They engaged in six to ten months' training and/or technical visits.[109] In the early 1980s, engineers from the research centre were sent overseas for six months' training on equipment research (e.g. automation). This was done under a three-year training agreement between USIMINAS and the Canadian Stelco.[110]

During the expansion phase, the overseas training process was also upgraded by sending engineers to postgraduate courses. In the 1980s, 18 engineers were sent overseas to follow MSc courses in France, Belgium, UK, and US. And three engineers were sent to Belgium for a PhD programme.[111] These individuals were sent with some corporate guidelines about the knowledge to be acquired. However, these training activities and USIMINAS's knowledge needs for innovative projects were not tightly connected. 'In some cases, engineers were sent overseas as a motivation factor rather than for training in one specific project for user X.'[112] This indicates a weakness in the functioning of this mechanism. Apart from this, as indicated in this section, new mechanisms were added to the overseas training process as compared to the 1960s. The practice of sending plant individuals overseas (mixing engineers and operators, before engaging in new activities, and continuously over the years) suggests, on the basis of Table 3.2, a good functioning and continuous intensity of the overseas training process over the 1970s and 1980s in USIMINAS.

CSN From the early-1970s, CSN began to send engineers to overseas training. This was an initiative from the head of the research dept, who had persuaded top management.[113] Until then, only isolated and intermittent overseas training used to take place (e.g. the group of eight engineers of the Siemens-Martin steel shop who were sent to the US in the early-1950s).[114] In 1971, CSN had sent three engineers overseas for an MSc course at the University of Sheffield (UK). From 1971 to 1976, an average of six engineers annually went to join postgraduate courses in the US, UK, France and Belgium. By the late-1970s, this number had dropped to about three per year. However, as revealed in different interviews in CSN, such training was not part of a long-term strategy for CSN's own capability development. In addition, 'engineers used to choose the course and the country of their own convenience'.[115]

From the late-1970s, about twenty engineers per year were sent overseas for five to twenty-day training. By 1988 the number had dropped to two.[116] Until the late-1980s, operators were excluded from overseas training. Another curious practice in CSN was that the 'best' technical individuals did not participate in overseas training.[117] This suggests that they had become indispensable to the units, reflecting the intense knowledge accumulation at the individual level. In other cases, specific engineers were given priority in the overseas training (e.g. courses, conferences). As a result, a small number of engineers had their knowledge substantially upgraded while others did not. The trained engineers then became the 'source of expertise' in the plant.[118] These practices indicate problems in the functioning of the process (e.g. dominance of unclear criteria; preference for individual rather than collective knowledge accumulation; limited access of operators). Although new practices were added in relation to the start-up and initial absorption phase, the variety was still limited (e.g. exclusion of operators and other technical individuals) and the intensity intermittent.

Participation in conferences and related events

USIMINAS   During the 1970s–1980s period, USIMINAS's engineers, technicians, and researchers continued to be encouraged to participate in local and international seminars, congresses, symposiums, round-tables and related events (e.g. ABM, ILAFA, International Iron and Steel Institute (IISI)). These events contributed to updating and expanding their mindsets and tacit knowledge through interaction with individuals from different backgrounds from within or outside the steel industry as well as reading new technical papers.[119] From the early-1970s, individuals were also encouraged to present their own papers at such events (e.g. the 'capacity-stretching' experience). These contributed to triggering interactions with other specialists leading to new insights.[120]

CSN   Up to the late-1970s, only a few engineers participated in such events. This may reflect the limited and intermittent overseas training, as described earlier. From the mid-1980s, engineers in the reduction area, on the basis of their own initiative, began to participate in the quarterly meetings on the reduction process in the Brazilian Metallurgy Association (ABM). During these events, production problems and solutions from different steel companies in Brazil were discussed. This permitted the acquisition of external tacit and codified knowledge. However, it did not reflect a corporate concern with the enhancement of the knowledge-acquisition process.[121]

Using technical assistance for knowledge acquisition

USIMINAS   New contracts were signed, particularly the TA-2 and TA-3, over the 1970s and 1980s, respectively. They involved the dispatch of NSC's personnel to USIMINAS and the training of USIMINAS individuals in NSC in Japan. Other key contracts with NSC were the Test and Research 1 and 2 (TR-1 and TR-2) to exchange knowledge on research on new steels development and process improvements.[122] USIMINAS's individuals continued with the active relationship with technical assistance teams. In other words, by working together, observing, imitating, systematic note-taking, informal meetings with the consultants and persistent questioning until every single point had been clarified therefore avoiding full delegation of activities.[123] Through a 'rotation' of foreign technical specialists, USIMINAS seemed to be deliberately adapting the technical assistance teams to meeting its different needs over time. Underlying those practices was the explicit commitment to building in-house capabilities and further transforming it into USIMINAS' own technical assistance.[124]

CSN   In 1969, a five-year technical assistance contract, renewed until 1979, was signed with UEC. As a result, from 1970 to 1975, 226 engineers went to the US and 257 US consultants came to CSN. However, virtually no engineer was sent to the US in 1976. Between 1977 and 1979 only five were sent.[125] This indicates an intermittent intensity of that process. The activities related to those technical contracts were more related to routine operational activity than innovative projects (e.g. process, equipment or product). Over the 1980s, technical assistance contracts with local institutions for innovative activities became more varied.[126] These sought to cover areas in which CSN had inadequate capability like product development and process improvement. The case of the hot strip mill (HSM) 2 suggests that the use of technical assistance as a learning process functioned poorly, as described below.

The HSM 2 was purchased in 1979 from Hitachi and included a training for CSN's engineers and technicians in NSC in Japan. The start-up was in 1981, but it was not until 1989 that CSN's individuals undertook this full training, although, by 1986, the whole package had been paid for. When Hitachi handed over the unit's operation to CSN, newly graduated engineers were suddenly appointed to operate it. They hardly knew the basic principles of that technology, contributing to a succession of low rates of capacity utilisation (e.g. 64.8 per cent, on average, from 1982 to 1987) and rates of equipment availability (e.g. 66.1 per cent during that same period).[127]

Interviews within CSN suggested that the presence of technical assistance was not limited over the 1970s and 1980s. However, the poor functioning of the relationship between CSN and external assistance suggests the absence of systematic efforts by CSN's individuals to learn from foreign consultants in order to build in-house capabilities.[128] This may reflect the absence of systematic efforts by CSN to sell its own technical assistance in the long term.

Interaction with suppliers and users

USIMINAS   The practice of paying regular visits to suppliers (e.g. the Brazilian mining company Companhia Vale do Rio Doce – CVRD), which had started in the 1960s, became more intense in the expansion phase. This was particularly done through the individuals from the reduction area.[129] During the 1970s and 1980s, USIMINAS's individuals engaged in intense interaction with the users. One type of interaction consisted of USIMINAS's individuals solving problem for different users. For instance, in the early-1970s, researchers and engineers of the rolling mills did several studies and experiments solving the problems of phosphatisation coating application in sheets used by Volkswagen. In the early-1980s, USIMINAS had individuals from the rolling mills, metallurgical dept, steel shop, and research centre adjusting the silicon content in the steel USI-CAR-50 being used by General Motors. At that time, even NSC did not have the answers. Later, NSC drew on USIMINAS's knowledge to solve the problem. These interactions forced individuals to acquire knowledge by discussing the steel application with the users and doing experiments.[130]

Another type of interaction was to gather feedback on product applications from key users. From 1980, when USIMINAS intensified efforts on the development of the API steels (for offshore platforms), PETROBRAS became its main user. PETROBRAS in turn had developed its own capabilities for steels testing, providing key feedback to USIMINAS's individuals: 'We used to make three or four phone calls to USIMINAS' people in a day. . . they were quicker and more effective than memos. . . We gave them feedback on the plates' mechanical properties and chemical composition with suggestions to improve the processing.'[131]

As a result, USIMINAS's individuals acquired tacit knowledge (e.g. sharing their problems, experiences with the users) and codified knowledge (e.g. access to the users' reports on steel application and exchanging technical papers). These interactions contributed to USIMINAS reducing the level of nitrogen in the plates for offshore platforms.[132] The evidence suggests that diverse forms of interaction with users were present. In addition, their functioning seemed good enough to lead to the acquisition of diverse elements of tacit and codified knowledge. It also seemed to trigger interaction with internal knowledge-acquisition processes (e.g. experimentation).

CSN During the 1980s, interactions between researchers and users took place intermittently during tests for product development. These seemed to reflect the intermittent innovative product activities during that time, as referred to earlier in this chapter. Interactions between operational individuals and users were more scarce.[133]

Providing scholarships for research and teaching

USIMINAS In parallel with the on-going support for research institutions that began in the 1960s (e.g. Fundação Gorceaux, Instituto Costa Sena), USIMINAS began to build tailored postgraduate courses for its engineers. In 1973, 17 individuals of the automation unit were given scholarships for a postgraduate course in industrial automation at the Federal University of Minas Gerais (UFMG). In 1978, an MSc course was created in UFMG for 12 laboratory analysts, technicians, and engineers of the research centre. And, in 1989, a PhD programme began to be designed with the São Carlos Federal University (UFSCAR) for three engineers.[134] Therefore the variety of this process was diversified. However, as with the overseas postgraduate training, this mechanism seemed to lack a tight practical connection with the company's knowledge needs, except for the automated systems training.[135]

CSN In 1978, an agreement was signed with the School of Metallurgy of the Fluminense Federal University (UFF) and CSN. The company began to provide practical support to the school (e.g. fun.ding to recruit foreign researchers, purchase of laboratory equipment and physical infrastructure). In turn, from the early-1980s, the school began to provide graduate metallurgical engineers for CSN. However, by the mid-1980s, support from CSN had weakened. This was associated with the financial crisis and top management's decision to limit support.[136] It was not until the early-1990s that support was restructured, as described in the next chapter.

Educational infrastructure in the community

USIMINAS By the late-1970s, nearly 15 years after the building of community schools (see Chapter 6), USIMINAS began to recruit a considerable number of individuals with basic and secondary education for its apprentices' training centre to be trained as operators. By the mid-1980s, the state of Minas Gerais took over the support of the first two schools, while USIMINAS continued to support the other two.[137] Thus the investment in these schools in the 1960s began to pay benefits for the company in the 1970s. The good functioning of this mechanism is suggested by its use by USIMINAS as part of the long-term training of the operators.

CSN   In 1951, the education assistance department was created in CSN to support the primary and secondary schools in the community of Volta Redonda. By 1957, five secondary schools and 45 primary schools were being supported by CSN. They also were expected to provide courses for the illiterate operators. However, the operators were not encouraged by CSN to follow them. Most of the schools had been built by the state or federal government. In the 1980s, scarce support was provided by CSN and the support was taken over by local government.[138] Despite the presence of these schools, they seemed not to be used by CSN as a way of improving the skills of its operators.

7.2.2   Internal Knowledge-acquisition Processes

USIMINAS   Individuals began to acquire different types of knowledge internally by undertaking new and more complex activities on a increasingly independent basis. In the early-1970s, USIMINAS even made explicit the corporate view on the way internal activities should be done: 'The company has continually to tighten the screening and assessment of its decisions and actions referring to production, costs, quality and other operations activities. . . by systematic investigation of facts, elaboration of operational models, experimentation, analysis of results, and their subsequent revision.'[139]

This view was probably reflected in several of the improvements on process and production organisation, equipment, and products described earlier. One of the key implications of engaging in improvements across the plant on a more continuous basis was the acquisition of greater understanding of the principles involved in the technology. This continuous pattern of 'learning-by-doing' innovative activities seemed to have been coupled with the acquisition of the necessary knowledge, via training, to engage confidently in new technological activities ('learning-before-doing'), as described in Chapter 6 and earlier in this chapter. This suggests that the external and internal knowledge-acquisition processes were contributing to individuals' engagement in technical activities beyond the existing levels of capability. In other words, a strong interaction between the learning processes began to develop.

USIMINAS also engaged in in-house knowledge-acquisition processes by continuous research being undertaken by individuals in the research centre laboratories in association with individuals in the operations units. The number of individuals in the research centre increased from 30 in 1972 to 359 in 1988. From the early-1970s, researchers were interacting with individuals in the operations and support units (e.g. the metallurgical dept) to implement technical support projects. During the expansion phase, there were, on average, 90 technical support projects annually. The centre had, on average, 100 development projects annually.[140] On average, 30 per cent of these efforts were related to innovative product activities over the 1980s, a period in which USIMINAS developed several products without technical assistance. About 30 per cent of the efforts were on raw materials, and 23 per cent on process improvements.

The diverse variety of research topics suggests, on the one hand, that individuals were given great freedom to do research (e.g. physical resources, time, choice of topics).[141] On the other, it suggests a wide focus for internal research activities. It should be added that, by the mid-1970s, USIMINAS's president led the idea of building another centre to do basic research 'to answer further more complex plant questions'. However, the idea was not operationalised. Among the reasons was the lack of corporate consensus.[142]

The research activities were based on an annual plan linked to a five-year plan. From the mid-1970s, these plans were elaborated by the research centre on the basis of top management's guidelines. They were further reviewed by the plant general manager, who set the project priorities.[143] Among the implications of the internal knowledge-acquisition processes, USIMINAS had 152 patents awarded from 1973 to 1989. Among them 18 were awarded overseas for their great originality.[144] In addition, this also reflects the accumulation of technological capability taking place in USIMINAS.

CSN During Expansions I to III (1954 to the late-1980s), plant individuals did not engage systematically in the acquisition of new tacit knowledge by doing more complex technological activities. Indeed 'until the mid-1980s, only intermittent minor adaptations were done in the plant by isolated individuals. . . the plant was operated on a "turn-key" basis'.[145] These turn-key activities may have contributed to giving rise to a pattern of 'learning-by-doing' simple routine with poor performance activities. They seem to reflect and have led to poor understanding, by the individuals on the site, of the principles involved in the technology.

CSN's research centre was created in the early-1960s at the initiative of a foundry engineer, who had managed to persuade top management of its value.[146] This bottom-up and individual initiative seems to reflect the low concern of CSN's top management with in-house capability building over the 1960s. The research centre was physically upgraded in the late 1970s. During the 1980s the centre had the facilities to conduct research on products, process, equipment, raw materials, and laboratories. Despite its installed capacity, only about 30 per cent of the research efforts were concentrated on development projects. About 70 per cent were concentrated on 'trouble-shooting' activities.[147] As a result, over the 1980s the centre was 'under-utilised', as mentioned in different interviews in CSN.

Until the late 1980s, the unit was not given adequate strategic guidance for medium and/or long-term targets for product and process innovation research.[148] The intermittent support contributed to slowing down or even delay-ing innovative activities. Among the indications of the limited variety and intermittent intensity of the internal knowledge-acquisition processes, were that only nine patents had been obtained locally by CSN by the late-1980s.[149]

7.2.3 Summary of the Cross-company Differences

In the light of Table 3.2 and the above empirical evidence, this subsection sum-marises the differences in the key features of the external and internal knowl-edge-acquisition processes during the conventional expansion phase for USIMINAS (1973–89) and CSN (1954–89). The cross-company differences in external knowledge-acquisition processes are summarised in Table 7.1.

As far as external knowledge-acquisition processes are concerned, Table 7.1 indicates that nine processes were present in USIMINAS and eight in CSN. Although similar processes were present in both companies, in USIMINAS most of them had a diverse variety. In contrast, most of the processes in CSN had limited to moderate variety. For instance, USIMINAS engaged in the up-grading of powerful mechanisms (e.g. systematic channelling of external codi-fied knowledge). In contrast, CSN continued with its conventional library. While CSN only hired newly graduated engineers or army engineers, USIMINAS sought to pull in NSC's 'project champions', steel experts, and PhD engineers. In other words, USIMINAS pulled in more diverse expertise than CSN. Additionally, USIMINAS engaged in more varied types of interac-tions with users than did CSN.

In addition to the differences in variety, Table 7.1 indicates that the processes also differed in terms of intensity, functioning, and interaction. For instance, while USIMINAS's overseas training was implemented on a continu-ous basis, and included operators, CSN's was implemented intermittently and excluded operators. In USIMINAS engineers were sent with reasonable direc-tions about the knowledge to acquire and usually before engaging in new tech-nical activities. In contrast, the evidence in CSN suggests the dominance of unclear criteria, scarce overseas training before new technical activities, and exclusion of operators. In other words, the evidence suggests a poor/moderate functioning of the processes. This suggests that the overseas training process in CSN had a weak influence on other knowledge-acquisition processes, for ex-ample, the internal knowledge-acquisition processes like systematic improve-ments across the plant.

As far as internal knowledge-acquisition processes are concerned, the differ-ences in their key features are summarised in Table 7.2. It indicates that four processes were present in USIMINAS and two in CSN. Both companies were engaged in routine plant operations. However, the evidence suggests that in USIMINAS, individuals were engaged in continuous improvements in equip-

Table 7.1 Summary of differences in key features of external knowledge-acquisition processes during the conventional expansion phase between USIMINAS (1973–89) and CSN (1954–89)

EXTERNAL KNOWLEDGE-ACQUISITION PROCESSES	VARIETY		INTENSITY		FUNCTIONING		INTERACTION	
	USIMINAS	CSN	USIMINAS	CSN	USIMINAS	CSN	USIMINAS	CSN
Pulling in expertise from outside • Contracting 'project champions' (frontier companies) • Hiring PhD and newly graduated engineers • Inviting experts for in-company talks	Present (diverse)	Present (limited)	Continuous	Continuous	Good	Poor/Moderate	Moderate	Weak
Recruiting operations individuals • From in-house or outside technical schools	Present (diverse)	Present (diverse)	Continuous	Continuous	Good	Poor	Moderate	Weak
Channelling of external codified knowledge • Systematic dissemination of interpreted papers • Engineering and foreign plant designs	Present (moderate)	Absent	Continuous	–	Moderate	–	Strong	–
Overseas training • 'Rotating' groups of engineers • Focused programmes (engineers, technicians, operators) • Technical visits and observation tours • Postgraduate courses (MSc; PhD)	Present (diverse)	Present (limited)	Continuous	Intermittent	Moderate	Poor/Moderate	Moderate	Weak
Participation in conferences and related events • Attendance • Presenting technical papers	Present (diverse)	Present (limited)	Continuous	Intermittent	Moderate	Poor	Moderate	Weak
Using technical assistance • Rotating external specialists	Present (diverse)	Present (moderate)	Continuous	Intermittent	Good	Poor	Moderate/Strong	Weak
Interaction with suppliers and users • Paying regular visits • Solving users' problems • Applying users' feedback • Tests for product development	Present (diverse)	Present (limited)	Continuous	Intermittent	Good	Moderate	Moderate/Strong	Weak
Providing scholarship for research and teaching • Conventional grants • Tailored MSc and PhD programmes	Present (moderate)	Present (limited)	Continuous	Intermittent	Moderate/Good	Poor/Moderate	Moderate	Weak
Educational infrastructure in the community	Present (moderate)	Present (limited)	Continuous	Intermittent	Good	Moderate	Moderate	Weak

Source: Own elaboration based on the research.

Table 7.2 Summary of the differences in key features of internal knowledge-acquisition processes during the conventional expansion phase between USIMINAS (1973–89) and CSN (1954–89)

INTERNAL KNOWLEDGE-ACQUISITION PROCESSES	VARIETY		INTENSITY		FUNCTIONING		INTERACTION	
	USIMINAS	CSN	USIMINAS	CSN	USIMINAS	CSN	USIMINAS	CSN
Routine plant operations	Present (diverse)	Present (diverse)	Continuous	Continuous	Good	Poor	Strong	Weak
Continuous improvements on process and production organisation and equipment	Present (diverse)	Absent	Continuous	–	Good	–	Moderate	–
Knowledge-acquisition processes before engaging in innovative activities	Present (moderate)	Absent	Continuous	–	Good	–	Moderate	–
Innovative projects within research centre • Planning process for research activities	Present (diverse)	Present (limited)	Continuous	Intermittent	Good	Moderate	Moderate	Weak

Source: Own elaboration based on the research.

ment, processes and production organisation. In addition, a number of operators and technicians and engineers had been trained overseas. In contrast, in CSN the engagement of individuals in continuous improvements across the plant was limited. And their routine operations activities did not interact with continuous training. In other words, USIMINAS's innovative pattern of 'learning-by-doing' with 'learning-before-doing' was absent in CSN. This suggests that in USIMINAS's operations individuals were in a better position to learn more about the principles involved in the technology than CSN's. This contributed to the strengthening of the existing routine operations capability in USIMINAS. In contrast, this may have contributed to the incomplete accumulation of that type of capability in CSN.

Both USIMINAS and CSN built research centres. USIMINAS's received explicit top management guidance and support (e.g. by drawing-up research plans). In contrast, CSN's had its activities frequently limited by inconsistent top management views. This might reflect the bottom-up way in which the mechanism was created. In USIMINAS, research activities were planned systematically. This process involved top management, the plant, and the research centre. This may have contributed to a continuous intensity of the process of acquiring knowledge through innovative activities. In contrast, this type of planning process was absent in CSN. In sum, in USIMINAS the key features of the external and internal knowledge-acquisition processes in the expansion phase were substantially improved in relation to the previous phase. However, in CSN only slight improvements took place in those features in the expansion phase in relation to the start-up and initial absorption phase.

7.3   KNOWLEDGE-CONVERSION PROCESSES

In the light of the framework in Table 3.2, this section describes processes for converting knowledge from the individual to the organisational level in USIMINAS (1973–89) and CSN (1954–89), in other words, both knowledge-socialisation and knowledge-codification processes

7.3.1   Knowledge-socialisation Processes

There were various processes for the sharing of tacit knowledge between individuals in USIMINAS and CSN. As in Chapter 6, this subsection distinguishes three processes of in-house training: (1) in-house (basic); (2) in-house (course-based); and (3) on the job. Although training (1) would not be a knowledge-socialisation process in itself, it may have practical implications for the other two training processes and the building of basic operating capability in the company.

In-house training (course-based): basic

USIMINAS   Since the late-1960s the basic courses in the apprentices' training centre have constituted the initial stage of the training for operators and technicians. Individuals joined the centre at the age of 17. One of the conditions for being recruited to the centre was the completion of the eighth level in the official schools. This basic training involved mathematics, Portuguese, industrial design, science, notion of steelmaking, and physical education. Most of the trainees came from the community schools which had been built by USIMINAS. Since the 1970s the duration of the basic training has been 12 months. This constituted a basis to engage in further technical training.[150]

CSN   In 1956, CSN's industrial school was re-named the Pandiá Calógeras Technical School (ETPC).[151] In the 1950s and 1960s following the limited recruitment process the operators engaged in basic courses called 'generic'. They were 'basic industrial', 'job apprenticeship', 'emergency course' (a quick preparation), and initiation (a notion of CSN).[152] From 1956 to 1957, 1,876 trainees were enrolled. Some of them abandoned the programme and fewer than 50 became operators in CSN.[153] The evidence suggests that this training did not have a standard content that every trainee should follow. Most of the trainees had inadequate literacy and numeracy skills, suggesting that assimilation of the course content was poor.

In addition, after the recruitment process for operators, the engagement in further specific training was not a compulsory activity. As a result, most operators were released to sites without adequate basic training.[154] By the early-1980s CSN even sought to shut ETPC, as described later in this chapter. This suggests that the basic training for operators became even poorer during the 1980s.[155]

In-house training (course-based)

USIMINAS   From the early-1970s, a specific part of the training for future operators consisted of a six-month technical course at the apprentices' training centre. The course was based on a standard content involving mechanical, electric, and electronic maintenance, use of lathes, welding and steelmaking operation. The course consisted of 60 per cent of training within the centre, followed by 40 per cent on a supervised basis. Experienced engineers and technicians were incorporated as instructors on the training programmes.[156] Each trainee had two supervisors: one in the centre, another on the site. These daily training activities must have led to a good socialisation of tacit knowledge between the experienced instructors and the trainees. These trainees could assimilate the principles of the technology before starting work as operators. By the 1970s this training had become one of the key conditions to become an operator in the plant. In 1975, 155 individuals were apprenticed followed by 498 in 1976 (e.g. millers, turners, mechanicians, welders and steelmaking operators). From 1964 to 1980, the centre trained 60 per cent of the plant operators. In 1987, 100 per cent of maintenance and 80 per cent of steelmaking operators were supplied by the centre. In the 1980s former graduates from the centre in USIMINAS made up 27.5 per cent of employees. During the 1970-85 period, about 30 per cent of the 'operators of the year' were former graduates.[157]

In 1981, the existing recruiting and training department was upgraded into the personnel development centre (CDP). This incorporated the apprentices' training centre, although its structure was kept independent. To build the CDP, USIMINAS used the incentives provided by Law 6,297 (1975).[158] The number of individuals in the CDP (e.g. training analysts, instructors) grew from 75 in 1975 to 130 in 1988. The training activities covered steel operations, automated process control and production organisation, maintenance, and management techniques. The evolution of in-house training efforts during the 1976-88 period is outlined in Table 7.3.

Table 7.3 USIMINAS: evolution of training efforts during the expansion phase: plant (1976–88)

	1976	1978	1980	1981	1982[c]	1984	1986	1988
Number of trainees [a]	41,066	31,985	44,771	40,398	37,237	26,961	22,178	22,093
Training hours/individuals[b]	865,763	1,126,934	999,178	901,001	715,299	640,969	639,849	640, 245

Notes: (a) and (b) include every single participation of the same individual in different training programmes. (c) From 1982, the training programmes were grouped into sub-projects involving a specific training activity (e.g. reduction process) leading to an increase in the average duration of each training subproject. This justifies the fall in the number of trainees (per-capita) from 1982.

Sources: Records of the USIMINAS's personnel development centre and syllabuses of the training programmes for 1989.

From 1981 to 1987, these training activities involved, on average, 55 per cent of individuals from the production units, 12 per cent from equipment and installations areas and 33 per cent from the production support units. Other types of training programmes concentrated on specific activities like automated process control. From 1977 to 1980, before engaging in in-house systems development, USIMINAS had 236 individuals trained involving 1,929 hours. From 1981 to 1988, 623 were trained involving 1,087 hours. From 1977 to 1988, the number of process control systems developed in-house increased from 9 to 32,[159] reflecting the intensity of these training programmes. This also suggests the contribution of in-house training for innovative capability building.

By 1984, an 'after-training' exercise was created to assess the effectiveness of the existing in-house training. One of the criteria was to assess whether or not individuals had become able to contribute to innovative activities. By 1986,

26.1 per cent of the lead operators and 44.6 per cent of other operators had been re-trained.[160] The evidence in this subsection suggests that bringing individuals from different backgrounds together in a continuous way contributed to socialising (and expanding) their tacit knowledge through contacts with instructors and other trainees. They must have contributed to spreading tacit knowledge, documented in the training materials, leading to a collective understanding of USIMINAS practices.[161] This also suggests a strong interaction between knowledge-socialisation and codification processes.

CSN   In the 1950s and 1960s the ETPC provided a six-month technical course for individuals to engage in the operations units. The course involved mechanics, electricity and metallurgy. That training must have led to a knowledge-socialisation between instructors and trainees. However, the training was not a key condition for becoming an operator in CSN, and not all the trainees joined CSN.[162] In the late-1950s a three-year course for technicians had been created in ETPC. This involved practical lessons in which interactions between trainees and instructors took place.[163] However, although a training based on knowledge-sharing was present, it sought to form technicians not operators. This suggests that training of operators was again not given adequate importance by CSN.

In 1962, 30 courses, including TWI, were given to 994 individuals. And the 'programme for supervisors' was created by the training and recruiting unit. In 1965, five courses were given to 30 supervisors on a one-off basis.[164] In 1975, CSN created the Fundação General Edmundo de Macedo Soares (FUGEMMS)[165] as an independent unit for training activities, incorporating ETPC. Formally, the training sought to improve the performance of operational individuals. CSN used the fiscal incentives provided by Law 6,297 (1975) to build this unit. By 1976 about 1,150,000 hours of training had been developed.[166] Despite the volume of hours, the training efforts only reached the levels of engineers and administrative personnel. Again, there seemed to be low concern with the training of shop-floor workers. 'Although a brand new training facility had been built, it barely met the needs of the operators. . . CSN seemed more interested in the fiscal benefits of that Law rather than in the training of the operator.'[167] In addition, the evidence does not indicate that this number of training hours was aimed at building capabilities for innovative activities (e.g. development of in-house systems). They seemed more related to routine operational and administrative activities.

By the late-1970s, training specialists were contracted to train groups of operators and technicians. These people would then replicate the courses for their peers on the site. However, the operators' low skills to assimilate the contents combined with problematic knowledge-sharing on site contributed to mitigating the functioning and intensity of this training.[168] In 1979, CSN sought to close down the ETPC in order to reduce costs. Following a reaction from the Volta Redonda community and some politicians, the decision was changed. However, in the early-1980s, CSN did shut FUGEMMS's activities as a cost-reduction measure. It was not until the late-1980s that they were re-started.[169] These actions made in-house training intermittent, contradicting the efforts for training over the 1970s. The evidence also suggests that, although present, the variety of this process was limited and its functioning was poor. In addition, this in-house training in CSN did not reach the operators adequately. As a result, it did not seem to have contributed to accelerating the rate of accumulation of either routine (Levels 1 and 2) or innovative capability (Level 3 and beyond).

On-the-job training

USIMINAS   From the early-1970s, OJT for operators began with classes on-site based on the training materials (modules). These were prepared by the training unit (CDP from 1981) and the plant engineers. During OJT, the operator was supervised by engineers and technicians and then released to do the routine crew work under the lead operator's supervision. The knowledge-socialisation related to that training started with the preparation of the teaching materials (e.g. interaction between the engineers and technicians who designed the modules). Then it continued during the operationalisation of OJT on the site (e.g. through oral, written and body language, observation and imitation of experienced individuals). This ended up in the daily knowledge sharing within crews (with operators and lead operators).[170]

From 1973, the OJT for engineers began to be combined with *job rotation* practices. Before being settled in a production unit, the engineer was trained on an OJT basis in different operations and support units (e.g. the metallurgical dept). They interacted with experienced engineers through observation, imitation, participation in meetings and involvement in problem-solving activities, informal face-to-face or telephone conversations, etc.[171] The modules of OJT for engineers and supervisors were designed by engineers trained overseas. They covered six critical issues in detail: production control (e.g. analysis of production evolution, production planning); equipment control (e.g. maintenance, components replacement); technical control (e.g. updated production organisation techniques); personnel management (e.g. training, productivity assessment); cost control; and materials control.[172] Therefore OJT contributed to socialising knowledge on diverse operations issues.

CSN   Following their recruitment, most operators received some basic safety instructions and were released to the sites. They were formally 'expected' to work closely with, and learn from, the more experienced operators.[173] However, their low educational level and the absence of an adequate basic training limited their assimilation of instructions. In addition, continuing with the practice of hiding operational 'procedures' from their peers, as mentioned in Chapter 6, they were also not encouraged to share their experiences on how to operate the facilities more efficiently, but to increase production volume at any cost. This seemed to have been influenced by the 'production-volume' practice and the 'Giraffe' reward system.[174] This indicates inconsistency between top and plant levels as far as the functioning of OJT was concerned. By the late-1960s, CSN even sought to reverse the 'production volume' practice by intensifying OJT activities to improve product quality to compete against USIMINAS. However, in several units OJT never got off the ground.[175] The focus on production volume, a feature of CSN's behaviour, seemed to keep constraining the OJT process, although the company, on that occasion, sought to avoid it. Poor functioning OJT also took place in support units, for instance, the central laboratory (1950s to late-1980s): 'That [OJT] was an amateur training. We did not have any guidance or basic rules for the observations and training on the job. We were told to "find" someone experienced, stay by his side, and go learning'.[176]

By the late-1970s, the knowledge gap between operators and engineers had become even wider. On the one side, there were the technically qualified engineers who barely interacted with the operators. They even used to have stars on their hard hats, as in the armed forces, to differentiate them from the operators.[177] On the other, there were the semi-literate and poorly-trained operators who, until the late 1980s, were not supposed to talk to the managers, superintendents, or engineers. The start-up of the steel shop, in 1981, is illustrative of this sort of weak interaction and the poor functioning of OJT, as described below.

In October 1981, a group of technicians from Nippon Kokan and USS arrived in CSN to start the steel shop. Earlier that year, a group of 15 of CSN's engineers had undertaken a 45-day training in the US on the basic operation of that unit. When they returned, they trained a group of 45 engineers and technicians, but not the operators. These were be trained on an OJT basis. During the installations, start-up and test of performance the Japanese technicians sought to talk, in English, with the operators. But the operators' assimilation was limited by the language barrier and poor previous training. The Americans had refused to talk to the operators and interacted only with CSN's managers and engineers. These, in turn, did not train the operators in the principles involved in the technology, leading to knowledge accumulation only at the engineers' level.[178]

Shared problem-solving: meetings

USIMINAS From the early-1970s, the practice of *daily meetings* was added to the *weekly meetings* schedule. They evolved into a systematic practice within each production unit taking place either early in the morning or early in the afternoon. They involved as large a number of individuals as possible (e.g. managers, engineers, technicians and lead operators) to review the performance indicators throughout the crews of the previous day, discuss the causes of breakdowns, internal and users' complaints. Each meeting generated further actions to sort out problems and to meet corporate goals, giving rise to an 'informal communication system'.[179] This contributed to spreading viewpoints and common understanding of the innovative process, product, and equipment activities (mental models) and the skills and technical insights (technical elements of tacit knowledge) across the plant.

CSN By the mid-1970s, it was clear that engineers were retaining their tacit knowledge as competitive advantage to achieve and sustain senior positions. When a engineer managed to adapt a production process or find a new way to solve a problem, this knowledge was barely socialised with other individuals in the units (or codified).[180] By the mid-1970s, key technical positions were dominated by a group of engineers, known as the 'Itajubá group'. This group deliberately sought to retain technical knowledge within its boundaries.[181] CSN's top management seemed to cover up,[182] contributing to mitigating any knowledge socialisation. As a result, operators and foremen did not know how to cope with the basic problems in the existing and new facilities derived from the expansions: 'If anything went wrong, say, in the BF or rolling mills, the whole crew panicked. . . They had to phone experienced engineers at about 3.00 a.m. to solve a problem. They came up, solved the problem, but the foremen and operators were never told how it was solved.'[183]

This problem was recalled in different interviews in CSN during the pilot and main fieldwork. Stories revealed that on some occasions engineers even knew they would be called out during the night. On the one hand, these facts reflect the deliberate accumulation of knowledge at the individual level. On the other hand, they indicate the poor functioning and intermittent intensity of the knowledge-socialisation process as a whole in CSN.

Interactive problem-solving practices

USIMINAS In the early-1970s the standard cost system was introduced in USIMINAS. The way it functioned during the 1970s and 1980s became a powerful process by which tacit knowledge was socialised through problem-solving interactions. The method involved four dimensions: observation (e.g. evolution of operations performance); analysis (e.g. breaking the observations into elementary operations); information (e.g. searching specialised literature and benchmarking); and comparison (e.g. USIMINAS versus other steel companies' indicators, done by TIC's individuals).[184]

Once bottlenecks were identified by the production individuals, they asked for co-operation from individuals in the industrial engineering dept to break the operation into elementary pieces, so that de-bottlenecking alternatives could be studied. The individuals from the metallurgy dept and quality control were involved to help in these tasks. Some types of problems involved individuals from the maintenance dept. Individuals from the engineering dept were also involved since their knowledge of the intricacies of the process and product equipment helped to tackle bottlenecking and improvement problems. If the plant's problems became more complex, the research centre's individuals were involved in their analysis (e.g. factors influencing increase in coke rate, steel grading or product defects).[185] This functioning, which evolved into the 1980s, may have contributed to a spread of the company's common practices, views, and ways of solving problems. It started at the individual level moving into group and plant levels.

CSN   In the 1970s a standard cost system was designed in CSN by a consultant engineer. In principle, this could have led to interactions for problem-solving. However, the system was not operationalised.[186]

Knowledge-sharing links

USIMINAS   From 1972, following the creation of the research centre, one researcher began to be allocated to each operational unit on an informal basis. Top management sought to avoid the rise of conflictive relationships between researchers and the operational individuals. Although, initially, the heads of the production units had not welcome the researchers' presence on their sites, the researchers gradually became involved in the daily and weekly meetings in each production unit. Problems were then taken to the research centre and discussed with other researchers, which contributed to speeding up the problem-solving in the production units.[187]

From the mid-1970s, individuals in the production units had an increasing demand for technical support projects involving researchers. In turn, researchers could rely on their co-operation for product development activities (e.g. industrial trials).[188] Through those links an intense knowledge-sharing by individuals with different technical backgrounds took place. From the late-1970s, this practice was followed by the engineering superintendency by allocating one engineer in each production unit. This led to daily interaction between individuals of different expertise involving maintenance, expansion problems and needs (civil works, equipment, etc.) and equipment performance.[189] However, it was not until the 1990s that these interactions were fully strengthened, as described in the next chapter.

CSN   In the early-1950s, when the quality control unit and the central laboratory were created, their individuals were expected to develop informal links with those in the operational units. However, until the mid-1970s, reflecting the disputes taking place at top management level, bitter conflicts developed between the individuals from these units.[190] Managers and engineers from different production units and from the laboratory blamed each other for product quality problems. In the 1970s, heads of certain production units even used to deny the access of the laboratory technicians to 'their' units to collect samples.[191] Between 1976 and 1986, when CSN engaged in product development activities in the research centre, top management downplayed the role of the operations units and their individuals in that activity. Indeed CSN failed to create any mechanism to stimulate co-operation and integration between them and the researchers.[192] As a result: 'Our product development projects were boycotted by engineers and managers in the operational units. . . Industrial trials were delayed or simply ignored. . . On some occasions, they even denied the researchers' access to the units.'[193]

By the 1980s CSN had gone through different knowledge-acquisition processes. However, it failed to engage effectively in knowledge-socialisation processes. This might reflect the absence of adequate mechanisms: 'By that time [1970s and 1980s] we had highly qualified engineers and technicians in the plant, research centre and labs, and the latest equipment, but we did not have an "organisation" to bring them together.'[194]

That 'missing organisation' reflected CSN's failure to bring different elements of tacit knowledge together into a workable system. This situation became worse, in the early-1980s, when the Itajubá group's new policy reduced foreign technical assistance, made severe cuts in overseas training, and shut down the training facilities.[195] From 1981 to 1983, low morale spread across CSN and several experienced engineers and project-leader researchers left the company. They took substantial knowledge with them since it had been accumulated exclusively at the individual level. As a result, in 1982, when expansion Stage III was being implemented, newly graduated and inexperienced engineers were recruited to head production units. This led to confused procedures in the operational units.[196] From 1982 to 1985, individuals in the operations units were severely pressed to reduce costs, to increase production, and to improve process and product performance indicators, but without adequate training. This set of events became known in CSN as the 'massacre of managers'. It led to managers covering up problems, further mitigating knowledge sharing.[197] It was not until 1989 that this situation began to be reversed, as described in the next chapter.

Team-building

USIMINAS This practice was greatly encouraged through the 'principles of organisation' of USIMINAS, formalised in 1972. Individuals from different areas of expertise and different functional areas were encouraged to work together on innovative activities through project teams, matrix practices, commissions, permanent committees, and temporary job rotations.[198] Over the 1970s and 1980s, several teams were created within those types of structures (e.g. the personnel committee, 1975; patents commission, 1976; product innovation committee, 1978).[199] Up to the late-1980s, teams still needed an organisational structure within which to work. This indicates that although present, and triggering knowledge socialisation, teams were not yet spontaneous therefore their functioning was moderate.[200]

CSN   Interviews with engineers and managers in CSN revealed that the building of teams was not encouraged in CSN until the early-1990s. This might reflect the limited knowledge socialisation taking place in CSN during the 1950s–1980s period.

Dissemination of lead operators

USIMINAS   Lead operators were created by the late-1960s. During the 1970s and 1980s, each operational unit had at least three lead operators. A basic condition to becoming a lead operator was accumulated operational experience and the ability to interact with other operators. In addition, they usually received specific training on leadership skills. They also went through an after-training assessment.[201] Their interactions with crew members developed in diverse ways (e.g. oral instructions, OJT, job assessment, bridging communication between managers, engineers and other operators). This contributed to the socialisation of their tacit knowledge with other operators in the operations units.

CSN   It was not until the 1990s that the position of lead operator was developed in CSN.

7.3.2 Knowledge-codification Processes

The processes described below were mostly present in USIMINAS (1973–89). In CSN, however, as confirmed by several interviews, casual meetings, and access to archival records, knowledge-codification processes were absent during the 1954–89 period.

Systematic standardisation of production practices

USIMINAS   Standardisation of production practices was organised within each operational unit through the standard cost system. It began by homogenising production through two practices: (1) *process-specific practices* (detailed description of the characteristics of the equipment, the production process, and the way individuals should interact with them to generate a product); and (2) *product-specific practices* which involved the detailed description of the product characteristics (e.g. composition, quality, chemical analysis, packaging and transportation), and detailed operations sequence, which should be known before making the product.[202] They sought to ensure that the same practice would be followed over time and across different crews. They became the standards whereby USIMINAS's own way of doing things like process, production organisation, product and equipment activities were codified. The creation of those practices contributed to making engineers, technicians and operators write down different elements of their accumulate tacit knowledge and experience.

As part of the standardisation process, USIMINAS began to document the recommendations given by the Japanese missions from start-up (1962) – which totalled nearly 3,000. Then it moved into the documentation of improvements in process, product and equipment (e.g. new procedures for revamping) led by its own individuals, in other words, its own tacit knowledge. This became a critical source of codified knowledge for USIMINAS's individuals in different production units.[203] In 1975, a group of eight to ten engineers was created and trained in the auditing dept to follow the execution and updating of these practices.[204] Indeed several interviews in USIMINAS revealed the strong concern of the company with the codification of production practices and its continuous updating.

By the late-1970s, the practice had spilt over into the engineering superintendency with the creation of the appropriability of project engineering design system (ADEP). This system documented and made available all the approved engineering designs and their analysis, and improvements in detailed and basic engineering. By the early-1980s, ADEP had become the largest bureau of engineering designs in Brazil and 'critical to control, support and accelerate improvements in the plant'.[205] USIMINAS deliberately engaged in the standardisation practice because 'individuals should be clear about their activities; there should be no mystery'.[206] Additionally, it should contribute to knowledge retention within USIMINAS and to facilitating further knowledge-acquisition by new operators: 'Knowledge should be accumulated collectively, not stored exclusively in the minds, drawers, and shelves of particular managers. . . [Therefore] the complete loss of knowledge, following the withdrawal of individuals, could be minimised.'[207]

This systematic documentation contributed to codifying important elements of individuals' tacit knowledge that otherwise would have remained tacit. It also contributed to creating USIMINAS's own 'codified proprietary knowledge'. However, until the late 1980s, that documented knowledge was available only to the individuals within each unit. This indicates a partial weakness in the functioning of this process.

CSN   Several interviews and casual conversations in CSN stressed that standardisation of production practices were absent in CSN until the early-1990s.

In-house elaboration of training modules

USIMINAS   As referred to in previous subsections, the training modules for course-based and OJT for operators and engineers were designed by USIMINAS's own engineers and technicians from the early-1970s. Most of them had been trained overseas,[208] suggesting an interaction between the learning processes. This also suggests that the processes of designing these modules contributed to the codification of their tacit knowledge. Being applied in different training programmes, access to this codified knowledge by different individuals, contributed to triggering the conversion of individual into organisational learning.

CSN   In CSN site engineers were not encouraged to design training modules over the 1950s–1980s period. This was usually done by the instructors at ETPC. The absence of this practice seems to have been influenced by the limited knowledge-socialisation process during that period.

Reporting from external training

USIMINAS   From the early-1970s, as individuals returned from their local and/or overseas training, they were asked to prepare a detailed report of their activities. This was made available to individuals in the units to serve as a source of information to consult about process, product and equipment improvements activities.[209] For instance, in 1976, a group of five individuals in the hot strip mill on their return from a two-month training trip to Japan, prepared a report and teaching materials to train other individuals in the unit about how to cope with welding problems. They made changes that brought the cracking indices near to zero.[210] However, this seemed to be one of the few exceptions. In general, the reporting process was not formally linked to its immediate implementation as an innovative project. These types of reports were made available to other individuals in the company.[211] It was not until the 1990s that the functioning of the process was improved.

CSN   When engineers returned from external training they were asked to report on their activities. The report was handed over to the heads of the units. However, the reports were literally stored in on their shelves and in drawers, becoming inaccessible.[212] This fact, reveals the low concern in the company with the knowledge-codification process.

Internal seminars

USIMINAS   Internal seminars evolved from an informal practice (1960s) into a formal and continuous practice during the 1970s and 1980s. From the early-1970s, individuals were systematically encouraged to give talks, conferences, and write papers. By the late 1980s, there were different events taking place across the plant on an annual and/or monthly basis. For example, the symposium of the steel shop, symposium of the rolling area, the internal symposium of the coke ovens and blast furnaces, research centre's first technical symposium.[213] As mentioned in different interviews in USIMINAS, underlying those events was an explicit concern with 'institutionalising the knowledge' that individuals had acquired through experience and training. These mechanisms triggered codification of tacit knowledge that had been acquired (e.g. through overseas training) and therefore a strong interaction between the processes.

CSN   During the 1970s and 1980s, events with the purpose of making tacit knowledge available to different individuals were absent in CSN.[214]

Learning phrases and symbols

USIMINAS   From the early-1970s, USIMINAS's leadership intensified the spread of corporate views by breaking them into simpler phrases and creating metaphors. They contributed to articulating tacit knowledge throughout the corporation like the view on in-house technological capability-building: 'the corporation has to be able to adapt imported technologies and even create new ones... and the technological capability is proprietary to the company and does not reside in individuals only, but in the integral collectiveness of the corporation.'[215]

Over the 1970s, engineers were encouraged to go on overseas training accompanied by the phrase 'we have to enrich the knowledge'. Under this phrase, engineers were constantly encouraged to go overseas for different types of training processes.[216] The phrase 'there is always a better way of doing things' contributed to pushing individuals to making continuous improvement efforts (e.g. 'capacity stretching').[217] By the late-1970s, the phrase 'we don't buy black boxes' had been disseminated across the corporation. This contributed to articulating the concern with the development of in-house capabilities to adapt, improve, and even sell the imported technology. It was also symbolised by the 'USIMINAS cycle of technological development': selection, acquisition, absorption/adaptation, development, and commercialisation of the acquired technology.

CSN   This type of mechanism was absent in CSN over the 1950s–1980s period. Interviews in CSN indicated that, during the 1970s, some managers sought to disseminate certain 'fashionable' managerial terms. However, they only puzzled the operators because of inadequate communication. In other cases, this dissemination was ignored because of the intermittent way by which these managerial practices were implemented.

7.3.3 Summary of the Cross-company Differences

In the light of the framework in Table 3.2 and the above empirical evidence, this subsection summarises the differences in the key features of the knowledge-socialisation and knowledge-codification processes during the conventional expansion phase for USIMINAS and CSN. The cross-company differences are summarised in Table 7.4.

As far as knowledge-socialisation processes are concerned, Table 7.4 indicates that eight processes were present in USIMINAS and only three in CSN. Although three processes were present in both companies, their variety in CSN was less diverse than in USIMINAS. For example, USIMINAS developed OJT processes involving intense knowledge socialisation between individuals (e.g. site training and job rotation) and incorporating experienced engineers in the training programmes. This forced an interaction with other knowledge-acquisition and knowledge-conversion processes. In contrast, in CSN, OJT was limited or even absent in some units, despite top management expectations. In addition, the OJT process in CSN functioned poorly and on a intermittent basis.

In USIMINAS, in-house training programmes (e.g. in the apprentices' training centre and CDP) had a continuous intensity and good functioning. The evidence also suggests a strong interaction with knowledge-acquisition processes (e.g. in-house training led by overseas trained engineers). In contrast, in-house training in CSN (e.g. ETPC) was implemented intermittently and with poor functioning. In addition, this process had a weak interaction with the knowledge-acquisition processes. Strikingly, both USIMINAS and CSN used the incentives of Law 6,297 (1975) to build training facilities. USIMINAS used them to improve the functioning and variety of its in-house training. In contrast, in CSN, little improvement took place over the 1970s, and in the 1980s, in-house training became even more intermittent.

As far as knowledge-codification processes are concerned, Table 7.4 indicates that five processes were present in USIMINAS that were absent in CSN. USIMINAS developed powerful mechanisms (e.g. production standardisation) that continued over time with good functioning. In CSN these processes were absent. On their return from overseas training USIMINAS individuals were asked to prepare a report that was made available to other individuals. In CSN, in contrast, this process was absent. In sum, during the expansion phase USIMINAS improved the key features of the knowledge-conversion processes in relation to the previous phase. As a result, the company was able to achieve an effective conversion from individual learning into organisational learning. In contrast, CSN did not improve these features of the knowledge-conversion processes. As a result, the company continued with the accumulation of knowledge at the individual level.

Table 7.4   Summary of the differences in key features of knowledge-conversion processes during the conventional expansion phase between USIMINAS (1973–89) and CSN (1954–89)

KNOWLEDGE-SOCIALISATION PROCESSES	VARIETY		INTENSITY		FUNCTIONING		INTERACTION	
	USIMINAS	CSN	USIMINAS	CSN	USIMINAS	CSN	USIMINAS	CSN
In-house training (course-based): basic	Present (diverse)	Absent	Continuous	–	Good	–	Moderate	–
In-house training (course-based)	Present (diverse)	Present (moderate)	Continuous	Intermittent	Good	Poor	Moderate	Weak
On the job training (OJT) • For engineers • For operators • Job rotation coupled with OJT	Present (diverse)	Present (moderate/ limited)	Continuous	Intermittent	Good	Poor	Moderate	Weak
Shared problem-solving • Operations problem-solving • Meetings (daily, weekly)	Present (diverse)	Present (limited)	Continuous	Intermittent	Good	Poor	Moderate	Weak
Interactive problem-solving practices	Present (diverse)	Absent	Continuous	–	Good	–	Moderate/ Strong	–
Knowledge-sharing links	Present (moderate)	Absent	Continuous	–	Moderate/ Good	–	Moderate	–
Team building	Present (diverse)	Absent	Continuous	–	Moderate/ Good	–	Moderate	–
Dissemination of lead operators	Present (diverse)	Absent	Continuous	–	Good	–	Strong	–
KNOWLEDGE-CODIFICATION PROCESSES								
Standardisation practices • Production practices (across plant) • Project engineering designs	Present (diverse)	Absent	Continuous	–	Good	–	Moderate	–
In-house elaboration of training modules	Present (moderate)	Absent	Continuous	–	Good	–	Moderate	–
Reporting external knowledge-acquisition processes	Present (moderate)	Absent	Continuous	–	Good	–	Moderate	–
Internal seminars	Present (diverse)	Absent	Continuous	–	Good	–	Strong	–
Learning phrases and symbols	Present (moderate)	Absent	Continuous	–	Good	–	Strong	–

Source: Own elaboration based on the research.

NOTES

1. From 1972, NSC consisted of a set of Japanese steel companies including Yawata, who assisted USIMINAS in the start-up and initial absorption phase. Hereafter it will be referred to as NSC.
2. USIMINAS, Annual Reports 1969–70.
3. Interview with USIMINAS's former president. Interview with the head of USIMINAS's engineering superintendency.
4. That US$105 million loan came from the International Bank for Reconstruction and Development (IBRD).
5. See paper by the then head of the engineering department, A. Assi, 'Relacionamento entre USIMINAS e NSC'. *I Reunião do Comitê Misto SIDERBRAS-NSC* [undated].
6. Interview with USIMINAS's former president. See *Plano Siderúrgico Nacional*, BNDE/BAHINT, 1969.
7. Account by a former development superintendent in USIMINAS/Fundação João Pinheiro (1988).
8. Interview with the head of the engineering superintendency. See also *Metalurgia ABM*, Special Edition, ABM, 1987.
9. Interview with USIMINAS's former president (who led that challenge). Interview with the former head of the engineering superintendency.
10. USIMINAS, Princípios de Organização, 1972–88.
11. Interview with the manager of USIMINAS's co-ordination of technical assistance. Interview with the head of the general superintendency of engineering.
12. See 'USIMINAS 25 anos', *Metalurgia ABM*, Special Edition, 1987.
13. Account by a former industrial relations superintendent in USIMINAS/Fundação João Pinheiro (1988).
14. See 'USIMINAS 25 anos', *Metalurgia ABM*, Special Edition, 1987.
15. Collective interview with the manager and two engineers of the automation unit. Also, 'Evolução das Atividades de Automação'. USIMINAS, Unidade de automação. [internal publication of the automation unit]
16. Ibid.
17. The granulometric distribution is associated with the coke preparing equipment therefore changes in the equipment were needed.
18. See 'USIMINAS 25 Anos', *Metalurgia ABM*, Special Edition, 1987.
19. Interview with process research manager at USIMINAS's research centre. See also 'USIMINAS 25 Anos'. *Metalurgia ABM*, Special Edition, 1987.
20. See 'Aciaria da USIMINAS: história', mimeo [undated, internal publication of the steel shop].
21. See Baer (1994)
22. The Si content is directly associated with the fuel consumption in the blast furnaces. The reduction of Si is associated with the operational stability of the blast furnace which, in turn, is associated with the quality of the metallic charge.
23. Interview with the manager of the technical division of coke ovens and blast furnaces.
24. Interview with the manager of the technical division of the steel shop.
25. Collective interview with the manager and two engineers of the automation unit. Also, 'Evolução da Automação na USIMINAS', mimeo [undated internal publication]; 'USIMINAS 25 Anos', *Metalurgia ABM*, Special Edition, October, 1987; Account provided by an engineer of the automation unit in *Metalurgia e Materiais ABM*, vol. 49, no. 414, February, 1993.

26. See 'Automação e controle de processo por computador' in 'USIMINAS 25 Anos', *Metalurgia ABM*, Special Edition, 1987
27. Interview with the technical assistance unit manager, who led the development of USI-FIRE.
28. Interview with a former general manager of USIMINAS's research centre.
29. Interview with a product development manager, USIMINAS's research centre.
30. Interview with a former general manager of USIMINAS's Research Centre.
31. Interviews with managers and engineers. Also, 'A qualidade na USIMINAS' [Copy of overheads of internal presentation, undated].
32. Interview with a former general manager of USIMINAS's research centre. Interview with a marketing manager.
33. Interview with USIMINAS's development director.
34. Later UMSA, as explained in Chapter 8.
35. Interview followed by plant tour with UMSA's capital goods business manager.
36. 'Breve história da USIMEC' [undated internal document].
37. Ibid.
38. Interview followed by plant tour with UMSA's capital goods business manager. Also 'Breve história da USIMEC' [undated internal document].
39. Interview with UMSA's manager of quality guarantee.
40. Interview with a co-ordinator of special projects, CSN's general superintendency of technology. Interview with the general superintendent of engineering.
41. Ibid.
42. Also Relatório BAHINT/BNDE. Plano Siderúrgico Nacional, BNDE, 1966.
43. Interview with a co-ordinator of special projects, CSN's general superintendency of technology. Also *Evolução da engenharia siderúrgica brasileira: a experiência da COBRAPI*, II Reunião do Comitê Mixto, San Nicolas, Argentina, May 1982.
44. See 'Plano Siderúgico Nacional', BNDE, 1974 [abridged].
45. Interview with a co-ordinator of special projects, CSN's general superintendency of technology. Also *COBRAPI. Evolução da engenharia siderúrgica brasileira: a experience da COBRAPI*. II Reunião do Comitê Mixto, San Nicolas, Argentina, May 1982.
46. See 'CSN: 50 anos transformando a face do país', *Metalurgia ABM*, 47(394), 116–17, 1991.
47. Interview with a co-ordinator of special projects.
48. Interview with a co-ordinator of special projects, CSN's general superintendency of technology. Other interviews in CSN.
49. Interview followed by a tour around the reduction area with a blast furnace engineer, a former researcher from the research centre. Also CSN, Annual Reports 1953–57.
50. Three different boards ran the company: from January to April, from May to September and from October to December. See CSN, Annual Reports 1953–57.
51. Interview with the general manager of CSN's research centre.
52. Interview # 1 with a former foreman from the steel shop. Also CSN (1996).
53. CSN, Organisational charts (1950s–1960s).
54. Interview with a process researcher from the research centre allocated in the blast furnace.
55. See CSN, Annual Report 1973.
56. Interview with two technicians of the organisation and methods dept.
57. Interview with two technicians of the organisation and methods dept. Other interviews with engineers and managers in CSN.
58. 'CSN: 50 anos transformando a face do país', *Metalurgia ABM*, vol. 47, no. 394, 1991.
59. Interview with a former head of the Foundry. Interview with two CSN's organisation analysts. Also CSN, Organisational charts (1980s).
60. CSN (1996).
61. Interview with a retired steel shop foreman. Also CSN (1996).

62. Interview with a manager at the research centre. Interview with a technician of the steel shop.
63. Interview with a blast furnace engineer. Also 'DO: a qualidade é obrigação de todos', *Nove de Abril*, XIII (150), February, 1990 [internal newspaper of CSN].
64. Interview with CSN's former president.
65. Interview followed by tour in the reduction area with a blast furnace engineer, former researcher at the research centre. Interview with the same engineer.
66. Those complaints were expressed in a annual meeting of the Brazilian Materials and Metallurgy Society by two engineers of the quality control unit of Willys Overland Brazil, speaking on behalf of the local automobile industry. See J. B. Hucke and E. D. Neto, D. (1960). 'Defeitos de aços nacionais para a indústria automobilística', *XV Congresso Anual da Associação Brasileira de Metais*.
67. CSN, Annual Reports 1963–65.
68. See CSN, Annual Reports 1964–66.
69. Interview # 2 with a product development superintendent.
70. Interview # 1 with a manager of CSN's research centre. Also CSN, Annual Report 1968.
71. Interview # 2 with a product development superintendent.
72. Interview with CSN's product development superintendent and with a engineer from the cold strip mill.
73. Ibid.
74. Interview # 2 with a product development superintendent.
75. The industrial director was Mr Leonards who, for several years, had been trying to persuade CSN's top management to support in-house product development activities.
76. Interview with the product development superiendent and with a hot strip engineer.
77. Interview with a product development superintendent.
78. Interview with a manager at CSN's research centre.
79. Interview with a product researcher at CSN's research centre. The issue of lack of capability to manufacture newly developed steel efficiently came up in different interviews.
80. Similar to COPRABI, FEM was 98 per cent owned by CSN.
81. Interview with the co-ordinator of special projects, CSN's general superintendency of technology. Also 'CSN: 50 anos transformando a face do país', *Metalurgia ABM*, Vol. 47, No. 394, 1991.
82. 'CSN: 50 anos transformando a face do país', *Metalurgia ABM*, vol. 47, no. 394, 1991.
83. Interview followed by tour in the reduction area with a blast furnace engineer. See also 'CSN atinge a marca dos 100 millhões', *M&M Metalugia e Materias*, vol. 53, no. 467, July, 1997.
84. 'CSN atinge a marca dos 100 millhões, *M&M Metalugia e Materias*, vol. 53, no. 467, July, 1997.
85. See *COBRAPI. Evolução da engenharia siderúrgica brasileira: a experiencia da COBRAPI*, II Reunião do Comitê Mixto, San Nicolas, Argentina, May 1982.
86. Ibid.
87. See also 'CSN atinge a marca dos 100 millhões, *M&M Metalugia e Materias*, vol. 53, no. 467, July, 1997.
88. See USIMINAS Notícias, *USIMINAS Revista*, vol. 7, no. 14, 1976.
89. See USIMINAS, Annual Report 1969. This engineer was Mr F Tschemmernegg. He was awarded his PhD in 1968 from the Technical School of Graz, Austria. He had developed research on the use of steel on aerodynamics stability of suspension bridges.
90. Interview with a training analyst in the personnel development centre. See USIMINAS Notícias, *USIMINAS Revista*, vol. 4, no. 8, 1973.
91. Interview with a former general manager of USIMINAS's research centre.
92. Interview with a product development researcher.

93. Interview with training manager at FUGEMMS. Interview with two of CSN's organisation analysts.
94. Casual meeting with engineers from the reduction area.
95. Ibid.
96. Account by the former head of the industrial relations dept. See USIMINAS/Fundação João Pinheiro (1988).
97. Interview # 1 with a training analyst of the personnel development centre.
98. Interview with a training manager from FUGEMMS. Interviews with other engineers and managers.
99. Those involved knowledge of foreign languages, aptitudes to search and disseminate knowledge.
100. Interview and casual meeting with the former manager of the technical information centre
101. Interview and casual meeting with the former manager of the technical information centre. Also 'O Centro de Informações Técnicas: inovação no conceito de informação,' *Metalurgia ABM*, Special Edition, October, 1988.
102. Interview with installation engineer in the engineering superintendency.
103. Visit to the USIMINAS research centre library.
104. Interview with USIMINAS's former president.
105. Ibid.
106. Interview with an operations analyst of the cold strip mill.
107. Interview with a training analyst at the personnel development centre.
108. Interview with the manager of the personnel development centre. Account by a former industrial relations superintendent in USIMINAS/Fundação João Pinheiro (1988).
109. 'Breve história da USIMEC' [undated internal document].
110. Interview with a equipment engineer at the research centre.
111. Interview with the manager of the R&D equipment development and laboratories of the USIMINAS research centre.
112. Interview with the research centre's general manager.
113. This was Mr Leonards, the first postgraduate engineer in CSN.
114. Interview # 1 with a former foreman of the Siemens-Martin steel shop.
115. Interview with a blast furnace engineer, former process researcher at CSN's research centre.
116. Interview with a training manager at FUGEMMS. Also CSN, Annual Reports (1970–92)
117. Interview with CSN's former president. Interview with USIMINAS's former president. Interview with a blast furnace engineer, former researcher at CSN's research centre.
118. Interview followed by casual meeting with the former head of the foundry. Casual meeting with a BF engineer.
119. Account by the then USIMINAS project engineering superintendent, in 'Geração de Tecnologia – o exemplo da USIMINAS', *USIMINAS Revista*, vol. 1, no. 2, pp. 14–20, 1970 (X Congresso Latino Americano do Ferro e Aço, Caracas, 1970).
120. Interview with the manager of the technical information centre.
121. Interview followed by casual meeting with an engineer from the blast furnace, a former researcher from the research centre, who engaged in these meetings from the early-1980s. This point also came up in other interviews.
122. Account by a former industrial relations superintendent in USIMINAS/Fundação João Pinheiro (1988).
123. Interview with the manager of technical assistance co-ordination. Interview with a planning analyst of energy and utilities (USIMINAS energy centre).
124. This practice was mentioned in different interviews and also archival records.
125. Interview with a manager of CSN's research centre.
126. Interview with a product development superintendent of CSN's research centre.

160   *Technological Learning and Competitive Performance*

127. Interview with a project development manager of the engineering superintendency. He was one of the engineers appointed to lead the operation of the HSM 2. Plant tour with an engineer of the hot strip mill 2. Also 'Laminador the Tiras a Quente # 2 da CSN: 15 anos de operação priamndo pela qualidade'. CSN/LTQ2 [undated internal document]
128. Also, interview with CSN's former president.
129. 'USIMINAS 25 anos', *Metalurgia ABM*, Special Edition, October, 1988.
130. Interview with USIMINAS's former president. Interview with the head of the hot strip mill technical division.
131. Interview followed by laboratory tour with an engineer of the division of steel corrosion research of the PETROBRAS's research centre (CENPES).
132. Interview with a USIMINAS product researcher. Interview with the head of the technology commercialisation division of the PETROBRAS's research centre (CENPES).
133. Interview with the product development superintendent.
134. Collective interview with a manager and two engineers of the automation unit. Also 'Evolução das Atividades de Automação', USIMINAS [internal publication of the Automation Unit]. Interview with General Manager of USIMINAS's research centre.
135. Interview with the research centre's general manager.
136. Interview with the adviser to the superintendent director of the CSN's steel area. Interview with a researcher at the Fluminense Federal University's School of Metallurgy at Volta Redonda.
137. Interview with a training analyst in the personnel development centre.
138. Interview with a training manager at FUGEMMS. Also CSN, Annual Reports 1953–72.
139. See 'Estudos de Administração da USIMINAS: Planejamento e controle empresarial', *USIMINAS Revista*, vol. 1, no. 1, pp. 39–43, 1970.
140. Interview with a former general manager of the research centre.
141. Interview with a product research manager at the research centre.
142. Interview with USIMINAS's former president. The interview also suggested that, if this idea had been operationalised, the next step would have been to build an overseas research facility to be in closer contact with the latest knowledge on steel research. This gives an idea of the long-term view and how audacious that leadership's ideas used to be.
143. Interview with the former general manager of the research centre.
144. Interview with USIMINAS's manager of the technical assistance co-ordination. See also *Metalurgia ABM*, Special Edition, October, 1987, pp. 104–29.
145. Interview followed by tour in the reduction area with a blast furnace engineer, a former researcher at the research centre.
146. This engineer was the Ukranian, Mr Wladymir Krywickyj, who had been working for 10 years at CSN in the areas of foundry, metallography, and thermal treatments. Interview with a process research engineer who, at the time of the interview, was allocated in the reduction area.
147. Interview with a manager of the research centre
148. Interview with product development superintendent at CSN's research centre.
149. Interview with the general manager of the research centre
150. Interview # 1 and # 2 with a training analyst of the personnel development centre. See also Gerência de Seleção e Desenvolvimento de Pessoal, Nov. 1997.
151. Decree 38,776, 27 February 1956.
152. CSN, Annual Reports 1955–56.
153. CSN, Annual Reports 1956–61.
154. Interview with a training and development manager and with a training analyst at FUGEMMS.
155. Ibid.

156. Interview # 1 and # 2 with a training analyst of the personnel development centre.
157. Interview # 1 and # 2 with a training analyst of the personnel development centre. Also, syllabuses of the professional training programmes (1987–1989).
158. Interview with the manager of the personnel development centre. Tour around the personnel development centre with a training analyst. On the basis of Law 6,297 companies were given government financial incentives to invest in training facilities.
159. See *Metalurgia ABM* Special Edition, October, 1987. Also 'Evolução do efetivo *vs* sistemas implantados' [1 page, undated] USIMINAS, Unidade de Automação [automation unit].
160. See USIMINAS's personnel development centre: 'Apresentação do Programa de Capacitação Profissional' (1987).
161. Syllabuses of the professional training programmes (1989).
162. CSN, Annual Reports 1955–68.
163. CSN, Annual Reports 1957–63.
164. Ibid. Search into CSN's archival records.
165. That was a foundation created by CSN in honour of General Macedo, its first technical director.
166. CSN, Annual Reports 1975–78.
167. Interview with a training and development manager and with a training analyst at FUGEMMS.
168. Ibid.
169. Ibid. Interview with CSN's former president.
170. Ibid.
171. Interview with the manager of the metallurgical dept. Interview with # 2 with a training analyst in the CDP. Account by a former industrial relations superintendent in USIMINAS/Fundação João Pinheiro (1988).
172. Published interview with a former industrial relations superintendent in USIMINAS/Fundação João Pinheiro (1988).
173. CSN, Annual Reports (1959, 1963, 1967, 1970, 1972).
174. Interview with # 2 with a Steel Shop foreman. Interview with a manager at CSN's research centre. It should be remembered from Chapter 6 that that was a financial reward system created by CSN in 1948.
175. Interview with a manager at CSN's research centre
176. Interview followed by laboratory tour with the supervisor of the central laboratory.
177. Interview with two CSN's organisation analysts.
178. Interview #1 with a foreman of the steel shop.
179. Interview with the head of the technical division of the steel shop.
180. Interview with the project development manager in the engineering superintendency. Interview with a cold strip engineer. Interview with a blast furnace engineer.
181. Interview with a project development manager in the engineering superintendency. This group referred to the engineers recruited from the Federal Engineering School of Itajubá.
182. Ibid.
183. Interview with a blast furnace engineer.
184. Interview with USIMINAS's former president. Also presentation given by USIMINAS's president in A. Lanari Jr., 'A empresa siderúrgica no processo científico e tecnológico', *Simpósio sobre C&T em universidade, instituto de pesquisa e empresa*', FUNDEP, Belo Horizonte, 27 April 1977. Also description by the superintendent of control: R. F. Ramos, 'Custo padrão na USIMINAS', *USIMINAS Revista*, vol. 4, no. 8, pp. 45-9, 1973. The principles of the standard cost system had originally been developed in the US. However, when the Japanese saw it implemented in USIMINAS they took the practice to the Japanese steelworks.
185. Ibid. Interview with plant general manager.

186. Interview with two technicians of the organisation and methods dept.
187. Interview with the development director. Interview with the head of the personnel development centre.
188. Interview with the development director. Interview with the head of the personnel development centre. Casual meeting with a engineer of the products research laboratory (research centre), allocated to the steel shop.
189. Interview with an installations engineer in the engineering superintendency.
190. See CSN, Annual Reports 1950–63. Also CSN's internal newspaper *O Lingote* [several numbers]. Interview with a technician from the hot strip mill 2.
191. Interview followed by laboratory tour with the supervisor of central laboratory.
192. Interview # 1 with an engineer from the hot strip mill. Interview # 1 with a product development superintendent.
193. Interview # 2 with a product development superintendent.
194. Interview with a technician from the hot strip mill.
195. Interview with a project development manager in the engineering superintendency.
196. Interview # 1 with a blast furnace engineer. This information also came out in the joint interview with a product development superintendent and a hot strip mill technician.
197. Interview with a manager in the engineering superintendency. This individual was one of the young engineers appointed, at that time, to replace old heads of division. As a state-owned company, managers were not fired promptly, but they could suffer administrative prosecution.
198. See USIMINAS. 'Princípios de Organização', 1972 [Internal document].
199. Interview with the manager of the technical assistance co-ordination.
200. Interview with the manager of USIMINAS's personnel development centre and former researcher in USIMINAS's research centre.
201. Interview # 1 with an analyst of USIMINAS's personnel development centre and syllabuses of the professional training programmes 1985–89.
202. See 'Custo padrão na USIMINAS', *USIMINAS Revista*, vol. 4, no. 8, pp. 45–9, 1973.
203. Interview with the manager of the technical information centre. Interview with the manager of the technical division of the coke ovens and blast furnaces. The interview was followed by in-situ view at those initial archives where knowledge began to be documented.
204. Interview with the head of the metallurgical department.
205. Interview with an installations engineer in the engineer superintendency, who participated in the creation of ADEP.
206. Interview with USIMINAS's former president.
207. Interview with the former head of USIMINAS's research centre.
208. Interview with the manager of the metallurgical dept. Interview # 2 with a training analyst (CDP).
209. Interview with the head of the personnel development centre.
210. Interview with a retired technician from the hot strip mill.
211. Interview with the head of the personnel development centre.
212. Interviews with a blast furnace engineer, with a hot strip mill engineer and with a engineer from the engineering superintendency.
213. See *USIMINAS Revista* (several numbers)
214. Interview with the training manager. Also *O Lingote* (various numbers).
215. Interview with USIMINAS's former president. Published conference given by USIMINAS's president in Lanari Jr 'Seminário Universidade-Empresa', *USIMINAS Revista* vol 1, no. 2, pp. 9-13, 1970.
216. Interview with USIMINAS's former president.
217. Talk given by USIMINAS's president during internal event in *USIMINAS Revista* vol. 6, no. 2, 1975.

8.   The Liberalisation and Privatisation Phase

This chapter describes the paths of technological capability-accumulation followed by the two companies during the liberalisation and privatisation phase (1990s). In parallel, it describes the learning processes underlying those paths. From the early-1990s, the two case-study companies began to operate within a new set of economic conditions. For this reason, the chapter starts with a brief overview of some features of these conditions.

By March 1990, when the Collor government took office, inflation in Brazil had reached a monthly rate of 81 per cent. Radical anti-inflation actions were taken, through the 'Collor Plan'. Although contributing to a temporary decline in inflation, it led to a fall of 4.4 per cent in the GDP in that year. Gradual and steady liberalisation of the exchange rate was adopted, combined with a set of actions to de-regulate and open the economy to foreign competition (see Baer, 1994). In addition, the National Privatisation Programme was created to sell off large state-owned companies. Privatisation was viewed as part of a long-term programme based on the liberalisation process leading to novel conditions within which companies began to operate (see BNDES, 1994; Saravia, 1996; Suzigan and Vilela, 1997).

An industrial policy was created to prepare the economy for world competition. In April 1990, the Industrial and Foreign Trade Policy (PICE), consisting of several programmes, sought to stimulate the development of industrial technological capability. These programmes also involved fiscal and credit incentives. The Brazilian Programme of Quality and Productivity (PBQP) consisted of: (1) subprogrammes to diffuse new management and production organisation techniques (e.g. TQC/M, JIT) in manufacturing industries; and (2) the upgrading of institutions for manufacturing quality control (ibid.).

In April 1990, Law 8,031 was passed leading to the creation of the Privatisation Committee. BNDES was given the task of selecting the companies to privatise and implementing the privatisation process. Steel was chosen as the first industry to be deregulated and privatised. As a result, the regulator of prices and investments, CONSIDER, and the state-owned holding and regulator SIDERBRAS, were extinguished. This led to the end of IS policy in steel. As part of the strategy to guarantee a successful start to the privatisation pro-

163

cess, USIMINAS was chosen as the first company to be sold. This was based on its competitive performance and technological capability (ibid.).

8.1   TECHNOLOGICAL CAPABILITY-ACCUMULATION PATHS

8.1.1   USIMINAS: 1990s

USIMINAS's path of technological capability-accumulation evolved in a different way as compared to the conventional expansion phase (1973–89). Instead of moving into the accumulation of higher levels of innovative technological capabilities, USIMINAS focused on the deepening, extending, reorganising and also routinising of activities within the existing Levels 5 and 6 innovative capability across all five technological functions. This was done as USIMINAS responded to the new set of conditions in the early-1990s.

Investment activities
Between 1990 and 1997, USIMINAS sought to sustain Level 6 innovative capability for investments. Drawing on its own capabilities, it engaged in the updating of its plant to make it compatible with world trends. Certain investment activities were reorganised as a response to the new conditions. By the early 1990s some activities had been effectively routinised, as indicated by the evidence below.

In 1991, in response to the foreign competition and environmental pressures, USIMINAS created the technology updating plan. This sought to reduce costs via improved process control, minimise the energy balance, reduce pollution, and enter new markets[1] (e.g. pulverised coal injection – PCI – for the BFs). At the time of the fieldwork for this research, USIMINAS was independently undertaking feasibility studies, to acquire the Corex process whereby pig iron is produced without coke ovens.[2]

By 1995, the plant optimisation plan (POP) had been prepared with the joint efforts of the engineering unit, the research centre, and the technical assistance unit. This plan sought to achieve continuous plant upgrading to respond to market needs, taking advantage of new opportunities. The implementation of POP led to a greater integration between the engineering unit and the operations units compared with the conventional expansion phase.[3] For example, the steel shop was involved actively in the installation of the vacuum degassing system, from pig iron studies and technology selection to installations and start-up.[4]

In addition, by the early 1990s, as mentioned in different interviews, an integrated way of carrying out investment activities had become part of USIMINAS's daily routine. This 'routinisation' seems to have helped the company cope with the complexities and uncertainties of investment decision-making and implementation activities. An example of this is the process for the purchase of the new cold strip mill, as described in Box 8.1.

Box 8.1   Routinised investments activities: the new cold strip mill

In the early-1990s, USIMINAS decided to enter the galvanised sheet segment market. In 1992, the electrolytic galvanising line was purchased followed by the pickling line in 1994. The investment process started with the joint studies by the corporate planning unit and the marketing dept on the potential for cold-rolled products in the domestic automobile and household appliances industries. They had been interacting with TIC, the engineering unit, and the research centre to obtain details about the technology. The next activities were the responsibility of the engineering unit that did the basic engineering, interacting with the plant and the research centre, and the detailed engineering involving the plant. In parallel, the technical assistance unit negotiated with the suppliers the details of payment, lead-time, and conditions whereby USIMINAS could sell any improvements made in the technology. Following the 'go-ahead' by the technical assistance unit, the engineering unit was involved with the suppliers in the equipment manufacturing until its installation in the plant. Within three years, USIMINAS became the leader in that market segment. This effective routinisation contributed to integrating different investment capabilities residing in different organisational units with positive implications for USIMINAS's market performance.

Source: Interviews with an installation planning engineer in the engineering superintendency and with the manager of the technical assistance co-ordination.

In the early-1990s, USIMINAS kept its engineering unit centralised. As a response to external conditions, it was re-organised to become more effective. From 1993, some detailed engineering activities began to be provided by specialised engineering firms on a third-party contractual basis, under the full control and co-ordination of USIMINAS. This freed USIMINAS to focus on more critical and complex activities, particularly basic engineering (e.g. from opportunities identification to basic specifications) and overall project management (e.g. contracting suppliers, accompanying equipment projects and manufacturing etc.).[5]

In the procurement area, the company achieved swiftness and simplification in the purchasing process. This seems to have been influenced by less regulation and directives, as a result of privatisation and liberalisation. The time needed for contracting companies was reduced from 60 to 30 days and for suppliers, from 25 to 5 days. Substantial gains were secured through purchasing via consignment contracts, the introduction of JIT practices, and minimal inventory levels and materials costs.[6] Day-to-day activities in the engineering unit were improved by the elimination of some of the administrative tasks. Reductions in projects costs were achieved by routinised control over technical designs. And reductions in the maintenance time were achieved, for example, by instant access to technical information. This also was achieved through the greater involvement of operations units in the engineering services, detailed and basic engineering, and daily problem-solving (e.g. via e-mail).[7]

USIMINAS sought to intensify its technical assistance activities by drawing on its existing capabilities. Additionally, from 1993, technological alliances were formed with different companies in Japan and Europe to increase the variety and scale of technical assistance in Brazil and Latin America.[8] To do this, the technical assistance unit was reorganised. In 1992, one year after privatisation, the number of technical assistance sales increased to 12 customers/month as opposed to three/month in the 1980s. From 1992 to 1995, the average annual sale of technology was valued at US$10.9 million against US$2.6 million of purchase. Between 1990 and 1995, the average annual sales of technical assistance reached US$8.4 million compared with US$3.89 million during the 1980s and US$3.59 million during the 1970s. By 1997, for each US$1 of technology purchased, USIMINAS sold US$5.[9]

Through the 'USIMINAS Outlook 2000' programme, an investment goal was set whereby 80 per cent of the activities would be focused on steel and 20 per cent on steel-related businesses. As a result, USIMINAS engaged in: creation of subsidiaries in Brazil and overseas (e.g. Usimpex – Imports and Exports); acquisition, via consortia, of 80 per cent of the Argentinian SOMISA (later SIDERAR) to penetrate the common market of the southern cone (MERCOSUR);[10] acquisition of COSIPA, in August 1993, adding 3.9 million tpy of steel to USIMINAS's production capacity;[11] and acquisition of steel products distributors (e.g. Rio Negro SA and Fasal in 1992). These activities were co-ordinated by the new businesses unit. They represented a considerable increase in the scale and variety of activities in relation to the 1980s. Although they were all within the steel business, these activities may reflect the greater freedom for investment resulting from privatisation.

Process and production organisation activities

During the 1990s USIMINAS focused on the deepening and routinisation of the activities within Level 5 innovative capability for process and production organisation thereby sustaining this capability. As a response to the new market demands associated with the opening-up of the economy, USIMINAS also improved the efficiency of its plant operations to meet the latest international standards. As a result, USIMINAS also strengthened and routinised activities within Level 2 routine capability for process and production organisation.

In April 1990, the new president took the lead of the dominant group and USIMINAS regained experienced leadership.[12] With a strong and up-to-date background, this leadership not only prepared USIMINAS for privatisation; it also engaged in its reorganisation to compete in a commercial industry in Brazil and the open economy.[13] In late-1991, following the new shareholders' directions, USIMINAS went through a 're-engineering' process to reduce costs and increase profits to pay off their investments. This sought to wipe out some of the existing units and organisational flows. In practice, however, little change occurred. The areas responded by ignoring the new instructions and by carrying out their activities as before. This might have been a reflection of the consistency of USIMINAS's organisational system, and its 'esprit de corps', an invisible part of that system.[14] By late-1992, USIMINAS had managed to replace the shareholders that were not in line with the company's concern to sustain in-house technological capability.[15]

In 1995, USIMINAS engaged in a corporate reorganisation whereby the line-staff principle was maintained and new practices were introduced to achieve: (1) simplification of the organisational processes; (2) effective managerial information and personnel adequacy and rationalisation; (3) elimination of double functions and managerial layers; (4) greater delegation of responsibility and authority; (5) greater integration of the support units (e.g. automation unit, research centre) to the operational units; and (6) greater integration between the operations units.[16] Managers were given greater autonomy and were delegated more responsibilities and therefore wider leeway to act.[17] A technical division unit was created within each operational unit to study and review the process in detail, intensify the integration between the automated control systems and the automated PPC, and engage actively in continuous process and product improvements/development activities. The technical divisions began to play a critical role in achieving a symbiotic integration of the operations units with the support units.[18]

One managerial layer was eliminated leading to a more straightforward communication between top management and operational levels than in the 1980s. In the metallurgical dept the number of individuals was reduced from 550 in the late-1980s to 170 in 1997. Its role was changed to focus on: (1) introducing new methods to meet new local and international production norms; (2) inspecting production processes; (3) pointing out every failure in the plant and related units and reporting them to the presidency; (4) keeping users informed and updated about USIMINAS's steelmaking process to enhance steel application; and (5) co-ordinating and integrating, in association with the technical division, the activities of the production units.[19] Therefore the department became even more active than before. In response to the new market demands, USIMINAS engaged in the upgrading of its quality system into the continuous quality improvement process (PAQ). This led to the achievement of ISO 9000 certification in 1992. These efforts contributed to routinising USIMINAS's quality system thereby enhancing processes and products quality. These were wholly done by USIMINAS through the 'implementation plan for PAQ', involving 100 quality engineers and 60 auditors led by the metallurgical dept. This upgrading also permitted USIMINAS to achieve the QS 9000 certification in 1996.[20] The main benefits for process and production organisation were the greater reliability of the results of laboratory analyses derived from the development of new statistical studies for process, equipment inspection, measurement and tests.[21]

In 1993, the fourth technical assistance contract (TA-4) was signed with NSC, called the 'Daruma Project'. It involved plant assessment and led to 426 recommendations.[22] The corporate restructuring, the quality system upgrading, and the TA-4 all had implications for the automation unit.[23] It became more integrated with other support units than in the 1980s and in particular with the operational units. Teams from the automation unit were allocated to the reduction area to speed up the introduction and routinisation of new systems.[24]

The integration contributed to the routinisation of existing automated practices and the development of new systems to overcome discrepancies in the plant. Many of them had been highlighted by TA-4. By 1997, 61 new systems had been developed and 100 were expected by 2000, based on artificial intelligence logic.[25] The automated system for the BF 3, developed in-house, permitted a quick and safe diagnosis of irregular situations, reducing stoppage time, increasing the availability rate of the BF. It also improved the interaction between the operator and the process, leading to improved control associated with increased BF productivity.[26] Key process parameters could be manipulated from the control room by two or three operators and one crew supervisor without any rigid hierarchical boundaries. In the external area, about twelve operators became responsible for the clean-up, supply, and control of pig iron delivery for the torpedo car. In case of emergency, these operators could take over in the control room.[27] Drawing on this process automation capability, from 1993, USIMINAS intensified its provision of technical assistance to steel companies in Brazil and Latin America.[28]

As a result of the 'Collor Plan' (1990), pig iron and steel production decreased by 25 per cent over 1989. Moderate production growth was achieved from 1993, associated with the new stabilisation plan, the 'Real Plan'. Most production units operated above their design capacity. On the one hand, this might reflect the economic recovery, as a result of the implementation of the

'Real Plan', particularly from 1993. On the other hand, it could have been a re-flection of the long-term sustaining of the 'capacity-stretching' capability. For instance, by 1994 the steel shop was operating about 20 per cent above its nomi-nal capacity. Indeed at the time of the fieldwork, this 'stretching' capability had become so pervasive in the company that it was seen as being part of USIMINAS's daily routine.[29] In addition, high capacity utilisation rates might reflect the routinised integration between automated process control and auto-mated PPC systems. In other words, they might reflect the sustaining of Level 5 innovative capability for process and production organisation. However, dur-ing the 1990s USIMINAS did not move into process research coupled with en-gineering. Among the reasons were: (1) lack of corporate consensus about this activity and (2) the greater efforts being put into innovative product activities.[30]

Product-centred activities

During the 1990s, USIMINAS focused on the deepening and routinisation of activities within Level 6 innovative capability for products. It engaged in con-tinuous and incremental improvements to the chemical and mechanical charac-teristics of the steels developed in the 1980s to add value and improve their application by users. Unlike the expansion phase, only a few new products were developed. In parallel, USIMINAS sought to deepen its Level 2 routine capa-bility for products to meet the latest local and international quality standards.

In response to the new market conditions in the early-1990s, USIMINAS re-viewed its perspective on users by putting them at the centre of product activi-ties. In addition, the company concentrated its efforts on the automotive, domestic appliances, and construction industries. These actions were associ-ated with: (1) the new demands for quality resulting from the opening up of the Brazilian economy; (2) the growth of vehicle production in Brazil, from 914,000 in 1990 to an expected 2,140 million for 2000 and the contribution of the automobile industry to GDP: from 8 per cent in 1990 to 13.5 per cent in 1995; (3) the high potential for steel application in the construction industry in Brazil; (4) the world industrial trend towards intensifying product rather than process development activities; (5) the international trend towards strengthen-ing the links between the steel and the automobile industries.[31]

As a result of the PAQ implementation, from October 1992, a new system of ten-day delivery and a five-day-maximum solution for any customers' com-plaints was introduced.[32] USIMINAS's deliveries changed from 'any day within a month' to fixed dates, with greater speed, wider decision power in sales and price negotiation and therefore improved relationships with users.[33] As a result, customers' complaints rates decreased from 0.04 per cent in 1990 to 0.02–0.01 per cent in 1997.[34] In addition, the upgrading of its quality system led to (1) the inclusion of USIMINAS in the list of world suppliers for Ford, GM and Chrysler; (2) the implementation of anticipatory actions to meet the users' expectations in an objective way, reflected in the low customer complaints' rates;[35] and (3) the achievement of 26 new product quality certificates.

The new features of USIMINAS's innovative product activities reflected the way it responded to the new market conditions of the 1990s: (1) increased involvement with users towards increased feedback for continuous improvements in existing products; (2) 'application engineering' (i.e. adding new properties to products to improve their application); (3) 'simultaneous engineering' (i.e. the involvement of the steel designer in the users' product design projects); (4) greater involvement of the operations units in product development activities.[36] These efforts are reflected in the characteristics of its 'third-generation' steels. They consisted of new and more complex variations of the existing steels and the development of a few new ones.

For instance, the API (oil) steels became resistant to hydrogen-induced cracking (HIC). A balance between increased resistance to corrosion and a thinner coating was achieved for the USI-R-COR (household appliances) family. The existing USI-STAR (automobile) had its variations upgraded from 450/470mpa to 500mpa with improved chemical composition (e.g. carbon: from 0.06 per cent to 0.15 per cent; manganese: from 1 per cent to 1.4 per cent; aluminium: from 0.050 per cent to 0.037 per cent; silicon: from 0.40 per cent to 0.38 per cent) and mechanical properties (e.g. USI-STAR-350 = 232 LEmpa *vs.* USI-STAR-500 = 377 LEmpa). In other words, the steel became more resistant and more ductile. These improvements met the new automobile industry's needs for greater resistance and reduced weight. In October 1994, the high-resistant 'interstitial free' (IF) steel had its rejection and re-working indices reduced from 5 per cent to less than 1 per cent by eliminating the use of stamping oil. This was achieved through improved application methods involving the operations units, the users, and the research centre.[37] This is an example of benefit of the greater involvement of the operations units in innovative product activities during the 1990s compared with the 1980s.

Efforts were undertaken in association with users (e.g. Rockwell-Fumagalli) to develop high-resistant hot strip steels for car wheels. From 1993 to 1996, three new steels emerged from those efforts: the USI-RW-450 (490 LR-mpa) and the USI-RW-550 (596 LR-mpa). At the time of the fieldwork for this research, the third steel, the USI-RW-600-DP (600 LR-mpa) was being co-developed with the user. Rockwell-Fumagalli's demand for USIMINAS's high resistant steels increased from 7 per cent in 1990 to 38 per cent in 1996.[38] In parallel, USIMINAS became involved with other users in the application improvement of galvanised sheets and on users' development projects (e.g. GM's T3000 project). This suggests USIMINAS's response to the world trend of 'early vendor involvement' practice.[39] For benchmarking purposes, USIMINAS joined the Ultra Light Steel Auto Body (ULSAB) project in 1996 demonstrating its focus on the automotive industry.[40] ULSAB is a world-wide project, involving more than 30 steel companies, that seeks to build blanks for a lighter weight car body to compete against aluminium.

USIMINAS continued its efforts on the development of new variations of cryogenic steels with nickel, for pressure vessels. It also engaged in the development of isotropic steels.[41] New product developments were represented by: (1) 'dual phase' steels, for the oil industry, involving intense joint work with the steel shop and studies with the hot strip mill, permitting the export to the US via car wheels; (2) 'tailored blanks',[42] to meet the automobile industry's need for weight reduction; (3) 'sandwich' sheets which had, by 1997, not moved beyond bibliographic studies and were stopped. The reason for this was that USIMINAS's partner, responsible for the resin that is placed between the sheets, had pulled out of the project.[43]

In 1992, commercial cells were created, a flexible and informal organisational unit designed to pay special attention to individual clients' needs, including after-sales assistance. The cells involved different areas like the metallurgical (steel making, rolling units), and sales and marketing. They grew from three in 1992 to 12 in 1997.[44] In 1995, a sophisticated logistics system, the Usifast, was created to supply steels for the automobile and domestic appliances industries on a JIT basis. This responded to the opening-up of the economy and the demand from users to reduce area and volume of stocks. To do this, a fine co-ordination between several production units and product delivery systems had to be achieved. This suggests that there was an integration between process and production organisation and product-centred capabilities. Usifast began by supplying Fiat and, by 1997, it was supplying steels for different users (e.g. General Motors, Rockwell, Dako, Caterpillar) meeting their different logistical needs.[45] In 1995, the unit for steel application was upgraded into the new steel application unit, and the Usicivil service was created to promote steel use in the construction industry.[46]

Equipment activities

Here the focus is on key features of equipment activities in USIMINAS in the 1990s: the sustaining of Level 6 innovative capability for equipment engineering and the diversification of activities within Level 2 routine capability for equipment. This was achieved as USIMINAS responded to the crisis in Brazil's capital goods industry and to USIMEC's privatisation.

By 1989, USIMEC was taken over by BNDES as a preparation for privatisation. The number of employees was reduced from 4,000 in 1987 to fewer than 500 in late-1990. By downsizing the company, BNDES seems to have paid little attention to the implications for existing technological capabilities.[47] USIMEC was put up for auction in June 1991, but it was not sold.[48] USIMEC was then taken over by USIMINAS who sought to revitalise the company, being explicitly concerned with the sustaining of its technological capabilities. This action by USIMINAS implied: (1) a competitive market decision to fill a gap left in the industry following the crisis from the late-1980s; and (2) a way of diversifying its product activities to meet the new challenges from the automotive industry. In 1991, USIMEC was re-named Usiminas Mecânica SA (UMSA) and its board was taken over by USIMINAS's directors.[49] In 1992, UMSA went through a corporate restructuring led by Booz-Allen and USIMINAS who designed a strategic plan, the 'UMSA 2000'. By 1996, UMSA's new corporate structure had been consolidated with 2,200 employees in the plant. By 1993, UMSA's production quality system had been thoroughly reviewed leading to the award of ISO 9002 and 9001 certifications, which were reassured in 1996.[50]

Unlike the 1970s and 1980s, when the company was engaged on its own in equipment projects, in the 1990s, UMSA sought to build partnerships to carry out these activities. This seemed to be a way of complementing and sustaining its Level 6 innovative equipment capability. In March 1992, an agreement was signed with Cometarsa (from the Argentine group Techint) for joint equipment manufacturing to expand USIMINAS's activities in the MERCOSUR market.[51] In parallel, an agreement was signed with Hitachi, for joint manufacturing of mechanical equipment for rolling mills. Another agreement was signed with Chugai-Ro Corp. in order to gain new capital goods business in Latin America in re-heating furnaces and continuous annealing manufacture. With British Ahlstrom Equipments Ltd, UMSA sought to re-engage in the design and manufacture of calcining kilns for the cement industry.[52] In 1996, an agreement was signed with Butler Manufacturing Company, for the joint design and manufacture of metallic structures for the construction industry. In February 1997, the joint-venture Vamec Hidro Energética Ltda was created with Voest-Alpine MCE GmbH, to design and manufacture hydro-mechanical equipment.[53] In 1996, UMSA manufactured the whole USIMINAS's Continuous Casting 4 in association with Demag and Hitachi Zosen. By 1997, UMSA had regained its status as a leading designer, manufacturer and supplier of capital goods, on order, to the steelmaking and others industries (e.g. cement, pulp and paper, petroleum, hydro-mechanics).

USIMINAS also engaged in the diversification of equipment activities within existing Level 2 capability via UMSA. In 1994, the Usistamp project was created, with Fiat, to produce pre-stamped parts for car manufacture. A whole pressing line was removed from Fiat and set up in UMSA. Integrated with Usifast, Usistamp began to supply Fiat's factory through a sophisticated JIT delivery system. From 1994, the Usiblank system began to supply parts made from hot sheets or coils using oxycutting, plasma, or pressed mechanical cutting, for wheel discs, pressure vessels, machine tools, etc. This operation was integrated with Usifast. In 1995, Usicort was created to supply the automotive and other industries in the state of Minas Gerais, with 'tailored blanks' and platen.[54]

These projects illustrate USIMINAS's response to the world trend of 'de-verticalisation' in car manufacturers. To increase productivity (hours/vehicle) and reduce assembly space, they have been transferring the initial stages of car manufacture to steel companies.[55] As a result, 'instead of supplying one kilo of steel, we are supplying one kilo of bumpers'.[56] Therefore USIMINAS began to deepen and stretch its Level 2 routine capabilities for equipment, process and production organisation, and product-centred activities by taking on the manufacture of automobile parts and distributing them on a JIT basis. USIMINAS began by simply operating the equipment and following the specifications provided by Fiat. However, interviews and site observations suggested that from 1997 the company might have been moving into more complex activities with carmakers involving joint application, design and development of car parts (e.g. the USIMINAS/GM T3000 project).

8.1.2   CSN: 1990s

During the 1990s, CSN's path of technological capability-accumulation was characterised by the weakening of Level 4 innovative capabilities for investments and equipment. In parallel, CSN moved from Level 4 to Level 5 innovative capability for products, and from Level 3 to Level 4 innovative capability for process and production organisation. CSN eventually completed the accumulation Levels 1 and 2 routine capability for process and production organisation and products. In other words, CSN's path in the 1990s was a mix of backward and forward moves in technological capability-accumulation.

Investment activities

Levels 4 to 5 capability for investments in CSN were weakened during the 1990s. The transfer of COBRAPI to SIDERBRAS, which was extinguished in 1990, contributed to the deterioration of COBRAPI's Project Engineering Division. As a result, 'all that engineering structure, built over the decades, was rapidly dismantled'.[57] By 1991, COBRAPI had been taken over by its employees whose number had decreased from 600 in the mid-1980s to fewer than 100 in the early-1990s. By 1993 COBRAPI had been closed down.[58]

In 1991, about 30 engineers were chosen from COBRAPI to be brought back to CSN's engineering superintendency. This unit centralised project and maintenance engineering activities.[59] However, CSN had stopped doing a large part of the basic and detailed engineering in house, the 'packages', as they were called in CSN. In the 1990s, 'turn-key' practices became dominant. In other words, a large part of the engineering projects began to be provided by suppliers and specialised engineering firms rather than by CSN.[60] As suggested in

Chapter 7, by the mid-1980s, CSN seemed to be heading towards greater engagement in investment decision-making and project engineering activities, in other words, the accumulation of capability at Level 5. Interviews in CSN suggested that the problem was not the 'outsourcing' in itself, but the fact that CSN did not seek to deepen the existing capability for strategic investment activities (e.g. basic engineering and overall project management).

Another factor was that, following the corporate restructuring in 1996, the unit began to be decentralised into the 'business units' in the mill. As suggested in the interviews in CSN, from 1996 the engineering superintendency was given less and less support from top management. In addition, this decentralisation consisted of pulling apart areas of the superintendency based on criteria not totally endorsed by the unit. As pointed out by one chief engineer, 'in the steel industry the engineering activities and areas are all inter-related and have to stick together. CSN is going the opposite way.'[61] Therefore the *way* that decentralisation was done contributed to weakening the engineering unit.

Among the results of the decentralisation was the diminishing ability of the engineering area in CSN to decide on investment activities from an integrated perspective. For example, when CSN had to decide about the introduction of a dynamic control system in the steel shop, the engineering area failed to answer questions about market potentialities, the supply capacity of the reduction area, and the implications for the rolling mills in an integrated way. As pointed out by one manager, 'the engineering perspective became limited to their specific business-units.'[62] This may be reflected in the problematic purchase of a sub-lance for dynamic control in the steel shop which, by late-1997, was still not working effectively.[63]

Despite the weakening of its in-house investment capability, during the 1990s CSN experienced a substantial increase in the scale and variety of its activities. Following privatisation in 1993 CSN engaged in a technology updating and development plan to expand plant capacity to more than 5 million tpy of steel by 1999. The plan also included the acquisition of the latest technologies (e.g. PCI for BFs 2 and 3, electrostatic precipitator for the sinter plant, roughing line for HSM 2, etc.).[64] From 1995, CSN engaged in the acquisition of large companies (e.g. the electricity distributor, Light SA, the mining company CVRD, and a cement company), the control of the Sepetiba docks and the building of power generation utilities.[65] These decisions, made by the new shareholders, were not welcomed by the existing management that had taken over in 1990, which had been seeking to stick to steel and increase in-house technological capability. As a result of disagreements between the existing executive management and the new shareholders, in 1995, new management was appointed, one of the key actions of which was to carry out corporate restructuring by 1996.[66] This influenced the investment capability-accumulation path.

Process and production organisation activities
CSN eventually completed the accumulation of Levels 1 and 2 routine capability for process and production organisation. In parallel, CSN moved into the accumulation of Level 4 innovative capability for that technological function as follows.

By the late-1980s CSN was in deep financial crisis and had out-of-control production costs. It sought to contract a management consulting firm to remedy the situation. However, it could not afford to pay the consultancy fees. In addition, the federal government refused to provide any additional funding for the company. Among the responses to these constraints, the technology absorption group (TAG) was created in CSN in late-1989 by the new industrial director.[67] This informal organisational unit engaged in the dissemination of a concern about process and products quality, cost-reduction, and overall performance improvement.[68]

However, the appointment of CSN's new president, in March 1990, and the privatisation decision triggered a new succession of strikes, contributing to the worsening of operational problems. For instance, in March 1990, a 31-day strike stopped production in several units. CSN's new president was even given the authority to close down the company, but he engaged in CSN's recovery. By supporting TAG and keeping the industrial director, the president emerged as a leader who played a key role in re-organising CSN's basic production activities.[69] Following a trip by a group of engineers and top managers to some Japanese and US steelworks, the decision was made to introduce a total quality control and management (TQC/M) programme in CSN.[70] The TQM implementation went through three phases: introduction (1991–92); dissemination (1993–96); and weakening and recovering (from 1997).

Introduction   TQM was introduced on a top-down basis to guarantee its dissemination down to the operations units. This was led by the operations directorate and the newly-created quality promotion centre. A package of tools was prepared for introduction on a plant-wide basis: production standardisation, the problem-solving analysis and method (PSAM), PDCA, the Ishikawa diagram, Deming's 14 points, among others. One of the aims of their introduction was to achieve ISO 9000 certification.[71]

Dissemination   The basic strategy was to transfer the responsibility for process and product quality to the operations units. TQM implementation began in the hot strip mill, whose leadership had been involved in the preparation and pilot work. As a result, the number of accidents in that unit decreased from 89 in 1989 to 4 in 1992. Material rejection rates decreased from 3.9 per cent in 1991 to less than 2 per cent in 1992.[72] In contrast to the 1940s–1980s period, operators became actively involved in improvement activities. For example, in the cold strip mill the operators of the rolling bridge tackled the problem of the monthly 500 tonnes loss of coils associated with fractures causing stoppages in the annealing line. This problem was solved by applying PSAM/PDCA principles and standardising the handling practices.[73] By 1992, new tools were added to the TQM programme (e.g. the '5 S programme' and the 'routine management system'). As a result, in the cold strip mill equipment availability rates increased from 55.5 per cent in 1991 to 84.5 per cent in 1995. In 1993, TQM implementation was enhanced by the introduction of quality control circles (QCCs). The number of volunteer groups increased from 130 in 1993 to 250 in 1995.[74]

In 1992, CSN achieved its first quality certification, and in 1993 received the ISO 9002. TQM also sought to reduce the 'ups and downs' in production growth, dominant in the 1970s and 1980s.[75] As far as the 1990s was concerned, in 1990, pig iron production fell by 16.4 per cent and molten steel production fell 19.1 per cent as compared to 1989. On the one hand, this reflected the recession in Brazil's economy. On the other, it reflected the frequent strikes that occurred in 1990. It was not until 1992 that production growth became more balanced across production units. This might have been associated with the implementation of TQM. This was also reflected in the substantial improvement in the capacity utilisation rates, particularly from 1992. For instance, by 1994 the steel shop eventually absorbed its design capacity, if molten steel production is considered.

Another example was the HSM 2. In 1992, the HSM's design capacity was eventually absorbed. From 1992 to 1997, its high utilisation rates even rose to 131.2 per cent. Availability rate increased by 22.5 per cent from 1990 to 1997 and energy consumption decreased by 17.2 per cent. Particularly important, was the re-organisation of the way of carrying out daily tasks, as a result of TQM implementation. In 1992, software for equipment maintenance was adapted into a tool for daily inspection and control of process operation, quality, and critical points in the strip path. Defects and problems began to be analysed on a weekly basis to identify their causes and find solutions. Cards recording daily problems were handed between crews, speeding up problem-solving. Production process practices were standardised with the operators' involvement. The command of the control panel, previously in the hands of the crew supervisor, was given to the operators. As a result, the supervisor was free to co-ordinate and control the group's activities and the process. The operators were given wider freedom for decision-making, as a result of the new automated system which improved interaction between them and the equipment.[76]

Weakening and recovering   In late 1995, CSN's president, who had created TAG, and had championed TQM, left the company. Until then, he had been one of the most influential leaders in CSN's history.[77] Among the reasons for his withdrawal was constant disagreements with the new shareholders. One of them was over the idea of 'downsizing' CSN. By 1996, a huge corporate re-structuring had taken place, based on 'the replacement of a conventional 50-year-old management model. . . by an innovative structure'.[78] As mentioned in different interviews, the restructuring seemed concerned with improvements to financial performance in the first place. CSN was re-organised on a 'busi-ness-units' basis consisting of four sectors (energy, infra-structure, steel, and corporate centre). The steel sector was organised into four business units. Each unit became responsible for its own financial results that had to be delivered to the corporate centre. The director of the steel sector, who came from an industry other than steel, was given little decision-making power having to report fre-quently to the corporate council. This was seen in the plant as a constraint to daily decision-making on technological activities.[79] From 1996, the TQM programme received decreasing support from top management. This was seen in the plant as a way of top management dissociating itself from any successes of the previous management. In other words, this became a threat to TQM's continuity.[80]

As a result, plant managers were instructed to introduce whatever manage-ment tool they wanted as long as the production volume and financial goals were met. The quality promotion centre received much less attention from top management than in the early-1990s.[81] At the time of the fieldwork in 1997 and during a visit to the company in 1998, the operations units were implementing improvements in process and production organisation. However, the interviews and observations suggested that these improvements were being done in isola-tion, with varying intensity and differing types of tools across units. This was different from the 1990-95 period, when they were undertaken in a relatively coherent way. Initiatives to sustain and even routinise TQM practices were taken in the HSM 2 with the introduction of new tools (e.g. poka-yoke[82]), but these derived from its isolated entrepreneurial leadership. In the mid-1990s, the dissatisfaction of plant managers at the weakening of the TQM programme and at its possible negative implications for the interaction between operational units, was reflected in their comments during interviews: 'CSN is like a collec-tion of different "states" and each one pursues its own interests.'[83] The number of CCQ groups increased to 600 in 1996.[84] However, observations and meet-ings on the shop floor suggested that this reflected the operators' way of 'nest-ing' themselves in the CCQs as protection from being sacked. This was entirely different from the volunteer involvement in the early-1990s.[85]

As far as automated systems are concerned, although the information tech-nology superintendency was created in the late 1980s it engaged in only inter-mittent efforts in systems development in the 1990s. Only a few computer programs were developed in association with the process development superin-tendency at the research centre. In the early 1990s, this superintendency had been providing technical support to the operations units particularly the reduction area. However, by late-1996, the unit had been abolished.[86] These factors may also have prevented CSN from moving into more sophisticated process control activities independently. Nevertheless, from late-1998, top management, recognising the importance of the TQM to CSN's competitive performance – and probably as a result of listening to plant managers – sought to renew its commitment to the continuity of this programme throughout the company.

Product-centred activities
In the 1990s, CSN eventually completed the accumulation of Levels 1 and 2 routine capability and moved into the accumulation of Level 5 innovative capability for products. By 1990, as part of the activities of TAG and later, as part of TQM implementation, CSN engaged in the identification of problems associated with poor product quality (poor composition, scratches in the plates). The standardisation of production practices permitted the manufacture of different specifications with reasonable homogenisation as opposed to the 1940s-1980s period.[87]

The 'product inspectors', regarded in the plant in the previous decades as 'police', were eliminated. From 1991 product quality responsibilities were transferred down to the production lines. By 1993 CCQs had been introduced. These contributed to intensifying the operators' involvement in product quality control and improvements. For example, in 1994, a CCQ group in the hot coils mill modified the lay-out and the lighting of the production line. This contributed to improving quality control leading to an impressive decrease in customers' complaints rates: from 0.34 per cent in 1993 to 0.07 per cent in 1997.[88] By 1992 CSN had obtained the first product quality certificate, leading to the QS 9000 certification in 1997.

During the 1990s, particularly from 1996, HSM had been in the lead in the plant in the introduction of new product quality control and improvement practices (e.g. TQM, CCQ, poka-yoke, among others). Most of these techniques have been championed by its leadership. These efforts may have contributed to increasing the integrated yield rate from 89.4 per cent in 1992 to 91.9 per cent in 1994. The domestic market share of CSN's hot rolled products expanded from 26 per cent in 1991 to 41.2 per cent in 1995. However, from 1996, the implementation of those managerial practices seemed far less integrated throughout the plant and disconnected from any top management perspective.[89]

As far as product development activities were concerned, in 1993 the product development superintendency was created within the research centre. This permitted CSN to re-engage in product design and development. The unit began by reverse engineering given specifications and by undertaking laboratory tests. They were assisted by consultants and efforts were more continuous than the 1970s and 1980s.[90] From 1993 to 1997, CSN developed about 15 products in-house. For instance, some complex specifications were developed (e.g. API-X65) and also CSN's own specifications (e.g. the electrical CSN CORE). Although CSN had been developing API steels since 1984, it was not until the early-1990s that they began to be manufactured. This was associated with the failed integration between the research centre and the operational units over the 1980s, as described earlier in Chapter 7.

To cope with this problem during the 1990s, the 'transition phase', an informal organisational arrangement, was created. Following the hand-over of product manufacture to the operational units, it was monitored systematically by the research centre for 6 to 12 months until the product quality indicators had been stabilised.[91] However, since 1994, the manufacture of the CSN-CORE (a steel for electrical use), whose value is 30 per cent higher than any other cold rolled steel, has not been welcomed by the operational units. This suggests that integration between the operational units and the research centre for innovative product activities is still limited. From 1996, the 'production-volume practice', dominant in CSN during the 1950s–1980s period, seem to have re-emerged in the plant. This seems to have contributed to mitigating against the efforts to manufacture higher-value steels.[92]

As far as CSN's product strategy is concerned, although it dominated the domestic segment for galvanised plates, from 1946 to 1993, scarce efforts were made at product development in this area. A very small number of products have been developed since the 1980s. This may reflect CSN's ineffective competitive product strategy in relation to its main competitor in Brazil: 'When we re-engaged in the development of galvanised products in 1994 USIMINAS was far ahead and competing aggressively in that segment, although they had entered only in the early-1990s.'[93]

The 'CSN 2000 Plan', designed in 1990, was the first formal initiative whereby a clear product strategy was set, particularly the penetration of the domestic automotive industry. However, this strategy evolved from being focused and aggressive (1991–95) into being inconsistent and unclear (from 1996). CSN's share in this industry increased from 11 per cent in 1990 to 35 per cent in 1994. Reflecting its product improvement efforts, by 1995 CSN had also joined the ULSAB project for benchmarking purposes.[94]

However, from 1996, CSN's efforts to increase its market share in the automobile industry began to diminish. By 1997, its share in this industry had contracted from 35 per cent to 30 per cent.[95] From the plant's viewpoint, top management was diverting efforts into industries other than the automobile (e.g. construction) as a way of avoiding facing up to the competition from USIMINAS.[96] From 1996, the decision on the lines and types of products to be developed was made by the research centre, disconnected with any corporate guidance.[97]

Equipment activities

During the 1990s CSN's Level 4 innovative capability for equipment weakened. This was associated with three main factors: (1) following COBRAPI's transfer to SIDERBRAS, the equipment engineering division deteriorated and by 1995 it had been closed down; (2) the crisis in Brazil's capital goods industry from the late-1980s; and (3) CSN's decision, in the early-1990s to abandon large equipment design and manufacture, concentrating on the manufacture of standard, small, and low-cost equipment in the foundry and FEM.[98] This last was probably influenced by factors (1) and (2). However, by 1995, CSN transferred FEM to the company Inepar. By mid-1998, the foundry was transferred to a third company. In other words, CSN began to 'outsource' the capability to make cylinders, rolls, ingot moulds, and several other parts that it had been making itself since the 1950s. The 'outsourcing' of the foundry and FEM does not seem to indicate a problem for the innovative equipment capability in CSN.

However, the above evidence and interviews with managers and engineers in CSN indicate a substantial reduction in the scale of equipment activities in the company during the 1990s. CSN did not manage to retain and deepen a strategic part of the Level 4 capability for equipment that had been accumulated over the 1980s. In other words, the dramatic reduction in the scale of equipment activities contributed to putting CSN in a weak position as far as involvement in activities like equipment design, manufacturing, and large equipment revamping were concerned. Not sustaining a strategic part of this capability may have limited CSN's further involvement in other technological activities like investments (e.g. decision on equipment specification for expansions), improvement in process and production organisation and equipment (e.g. preventive maintenance and equipment revamping), and also improvements in operational performance.

8.1.3   Summary of the Cross-company Differences

This subsection summarises the cross-company differences in technological capability-accumulation paths on the basis of the evidence outlined in Sections 8.2.1 and 8.2.2. Although USIMINAS did not move into the accumulation of new levels of innovative capabilities, it was able to sustain, deepen, and routinise its Levels 5 to 6 capabilities across all five technological functions within approximately 35 years. In contrast, CSN's path of technological capability accumulation was characterised by a weakening of Level 4 capability for investment and equipment; a move from Level 4 to 5 innovative capability for products; and a move from Level 3 to 4 innovative capability for process and production organisation. In parallel, over 50 years, CSN eventually completed the accumulation Levels 1 and 2 routine capability for process and production organisation and products.

Investment capability

USIMINAS re-organised its engineering unit, keeping it centralised to make it more effective and better integrated with the operations units. In addition, USIMINAS sought to 'outsource' some of its engineering activities. This permitted the company to focus on strategic engineering activities (e.g. basic engineering and overall project management). USIMINAS even sought to make alliances with Japanese and European companies to deepen the capability for these activities. These actions contributed to (1) the maintenance of the capability to undertake strategic investment activities; and (2) an increase in revenues derived from the increased technical assistance in strategic investment activities.

Like USIMINAS, CSN sought to reorganise its engineering superintendency. However, unlike USIMINAS, CSN did not keep this organisational unit centralised and united. Instead, the way the re-organisation was implemented contributed to weakening and even dismantling the activities within the unit. In a similar way to USIMINAS, CSN sought to 'outsource' certain engineering activities. However, the deterioration of COBRAPI's project engineering division and the weakening of the engineering superintendency led to a substantial reduction in the scale of activities related to its existing Levels 4 to 5 capability for investments.

In other words, CSN moved into a weaker position to undertake complex investment activities. Interviews with managers and engineers in CSN suggested that although USIMINAS had also reduced its in-house engineering activities, it was in a stronger position than CSN to undertake and control strategic investment activities. In sum, during the 1990s, within 35 years, USIMINAS sustained, deepened and routinised its Level 6 innovative capability for investments. In contrast, CSN having taken about 40 years to accumulate capability between Levels 4 to 5 for investments, had this capability weakened.

Process and production organisation capability

In response to the new conditions in Brazil in the early-1990s, USIMINAS engaged in a company-wide re-organisation involving, in particular: the creation of a technical division within each operational unit; greater integration between the operational units themselves, and between them and the production support units (research centre, automation unit); and a review of the role of the metallurgical dept to guarantee effective integration. In addition, USIMINAS had its quality system upgraded through the PAQ implementation leading to new certifications (e.g. ISO 9000 and QS 9000). The company also engaged in continuous improvement and routinisation of its integrated automated process control and PPC systems. These activities suggest a deepening of Level 5 innovative capability and Level 2 routine capability for process and production organisation.

In response to its operational and financial crisis, the threat of close-down, and the new set of conditions in Brazil in the early-1990s, CSN engaged vigorously, from 1990 to 1995, in a review of its routine production practices. The review involved the introduction of new tools to improve process and production organisation (e.g. TQM). Those efforts, however, were negatively affected from 1996 by: (1) the company being organised into 'business units' resulting in the implementation of new tools in a non-integrated way; (2) reduction in top management's support for the TQM programme. In contrast, USIMINAS was able to strengthen the integration in the plant achieving a co-ordination across the units to deliver products on a JIT basis.

In contrast to USIMINAS, CSN did not have any organisational unit engaged in such integration and support for the operations units (e.g. the USIMINAS metallurgical dept). In addition, the process development superintendency was closed down in CSN preventing this unit from providing technical support for process improvement to the operational units. During the 1990s, USIMINAS was able to develop new automated systems in-house for PPC and process control. They contributed to increasing the integration between those two systems through the automation unit. It should be remembered that this type of integration has positive implications for operational performance.

By the 1990s CSN had not developed any integrated automated systems for PPC and process control. Although in the 1990s the information technology superintendency was present in CSN, this unit did not engage in continuous and significant systems development in the way the automation unit in USIMINAS had been doing since the 1970s. In addition, the evidence suggests that both USIMINAS and CSN reduced their process activities in the research centre, as described later in this chapter. However, while CSN made a dramatic reduction, USIMINAS kept strategic activities of process research, permitting continuous support to the production units. In sum, within 35 years USIMINAS had sustained and routinised Level 5 innovative capability for process and production organisation. Slower than USIMINAS, it had taken CSN nearly 50 years to accumulate Level 4 capability for process and production organisation.

Product-centred capability
In response to the opening up of Brazil's economy, new demands from users and the world trend towards greater interaction with the automotive industry, USIMINAS engaged in continuous efforts, particularly on upgrading its existing quality system to meet international product quality standards; adding value to its products by upgrading their chemical and mechanical properties; interacting with users in product design, development, and application activities; and increasing market share through a consistent product strategy.

Like USIMINAS, CSN sought to respond to these new conditions. However, it engaged in activities associated with capability for products at lower levels than USIMINAS. For example, CSN had built a basic product quality control system while USIMINAS already had its system routinised. CSN had to re-engage in product development activities at Level 4. In contrast, by the 1990s USIMINAS was deepening innovative product activities at Level 6. While USIMINAS had improved integration between the research centre and the operations units for product development, CSN still had difficulties in improving this type of integration. Additionally, USIMINAS had a consistent pursuit of increased market share in the domestic automotive industry. In contrast, CSN entered the 1990s with a similar pursuit but, from 1996, its product strategy became inconsistent. In sum, by the mid-1990s, within 35 years, USIMINAS had deepened and routinised its Level 6 innovative capability for products. In contrast, CSN had taken 50 years to accumulate this type of capability at Level 5.

Equipment capability

Although both companies were affected by the crisis in Brazil's capital goods industry from the late-1980s, they responded differently. For instance, USIMINAS engaged in the revitalisation of USIMEC (later UMSA). USIMINAS/UMSA engaged in a series of technological co-operation agreements with frontier companies as a way of sustaining the existing capability at Level 6. Indeed, in USIMINAS during the 1990s there was a reduction in the scale of innovative equipment activities compared to the 1980s. However, the capability that USIMINAS was able to sustain permitted the company: to undertake critical innovative equipment activities (e.g. large equipment engineering design and manufacturing for different industries); to assist its own plant on equipment activities (e.g. equipment improvement, large equipment revamping, 'total preventive maintenance', equipment components manufacturing, decision-making on equipment investment); and provide technical assistance for existing and new markets on equipment activities.

In contrast, the response produced by CSN led to a dramatic reduction in the scale of innovative equipment activities in the company. Indeed the scale of reduction in those activities in CSN was much more radical and larger than in USIMINAS. As a result, CSN moved into a weaker position in relation to USIMINAS as far as innovative equipment capability was concerned. In addition, USIMINAS further diversified its equipment capability by engaging in the manufacture of the early stages of car manufacturing and JIT distribution of car parts (e.g. through Usistamp, Usiblanks, and Usifast projects). In contrast, these activities were absent in CSN. In sum, by the mid-1990s, USIMINAS had sustained and deepened, within 35 years, Level 6 innovative capability. USIMINAS also obtained increased revenues from the continuous provision of technical assistance in equipment activities (e.g. revamping engineering for BFs and steel shop converters). In contrast, CSN's Level 4 innovative capability, which took about 40 years to accumulate, had been weakened.

8.2   KNOWLEDGE-ACQUISITION PROCESSES

8.2.1   External Knowledge-acquisition Processes

This subsection describes the processes for the acquisition of knowledge from outside the company. Each subsection focuses on how these processes worked in USIMINAS and CSN.

Pulling in expertise from outside

USIMINAS   As USIMINAS had already formed its internal expertise, outside experts were no longer hired for senior positions. Nevertheless, experienced steel practitioners, steel application researchers, experts in corporate management, and in other industries (e.g. automobile) continued to be invited to give talks in USIMINAS's internal seminars.[99] From 1993, MSc engineers began to be recruited, particularly to work in the technical divisions of the operations units. The idea was to speed up solutions by improving the practice of identifying, analysing, and interpreting problems.[100] Therefore USIMINAS began to hire individuals not only for problem-solving, but for problem 'framing', suggesting higher complexity of its technological activities. USIMINAS also sought to hire back a few engineers who had recently retired from the company. Most of them had been in key positions for more than twenty years.[101] Retired engineers from suppliers (e.g. Bardella) were hired into UMSA to provide new knowledge on equipment manufacturing.[102]

CSN   From 1991, CSN hired temporary independent consultants who had retired from other companies in Brazil and abroad. They were hired to assist in product development activities (e.g. from UEC and USIMINAS) and in the TQM introduction and implementation (e.g. from Toshiba and the Christiano Ottoni Foundation).[103] Newly graduated engineers were also recruited, particularly from UFF. In addition, CSN hired back some foremen who had retired from the company. However, direct-site observation suggests that some of these shop-floor individuals continued to be 'key knowledge sources' to solve basic production problems.[104] This might reflect the intensity of individual knowledge accumulation over the previous decades. Nevertheless, the evidence suggests that the variety of expertise pulled in by CSN during the 1990s became more diverse and the functioning of the processes improved in relation to the 1970s and 1980s.

Recruiting operations individuals

USIMINAS   The recruitment of operators from the in-house professional training centre (former apprentices' training centre) was intensified in the 1990s.

For 35 years, the centre had provided 50 per cent of USIMINAS's operators, and 75 per cent in 1996. In 1997, of USIMINAS's employees, 22.5 per cent had come from the centre. Among the 'best operators' (1992–96), 35 per cent had been former trainees. These proportions are quite substantial considering that 67 per cent of USIMINAS's employees were operators. The practice of recruiting individuals on the basis of a recommendation from an employee, preferably a relative, continued in the 1990s. As a result, about 70 per cent of USIMINAS's employees were relatives. This practice seemed to guarantee the reproduction of USIMINAS's knowledge over time, leading to a natural interaction between individuals through their families.[105]

CSN　In the early-1990s, reflecting the implementation of TQM, CSN sought to improve the qualifications of its operators. CSN sought to increase the number of operators recruited from the ETPC to replace operators who left the company during the 1991–93 period. By 1993, of the 716 operators recruited by CSN, 674 came from the ETPC. However, the number of operators recruited from ETPC fell from 415 in 1995 to 120 in 1996.[106] On the one hand, the updating of the contents of the courses in ETPC in the early-1990s contributed to improving the qualification of new operators recruited by CSN. On the other, the fall in the number of operators recruited from ETPC from 1996, might reflect the diminished support provided by CSN for the school after that year, as commented, in more detail later in this chapter.

Channelling of external codified knowledge

USIMINAS　By the 1990s, TIC had become an indispensable mechanism for USIMINAS's innovative activities (e.g. investment, product development, performance benchmarking, etc.). In the 1990s, TIC had accumulated wide experience in searching varied types of codified knowledge world-wide (qualitative and quantitative), interpreting them in the light of USIMINAS's goals, and distributing them among as large a number of individuals and units as possible. By 1997, TIC was producing a total of 11 publications on a daily, weekly, monthly, bi-monthly, and three-monthly basis. Through these publications, external codified knowledge was disseminated to be absorbed by about 2,200 individuals.[107] This contributed to (1) homogenising the 'language' used across the company; and (2) stimulating individuals to search and share knowledge thereby 'refreshing' their mindsets.[108] By 1994, TIC had become the largest specialised unit on the steel industry and of technology information in Latin America. Its activities had become so routinised that, from 1993, 'technical information' began to be sold by USIMINAS as a new product. Up to 1997 TIC had 230,000 references, a massive acquisition of external codified knowledge. From 1991 to 1997 individuals in the TIC undertook, on average, 6,000 consultations per year to acquire new codified knowledge.[109] When the first manager of TIC retired in 1992 its management was taken over by another metallurgical engineer.[110] This suggests that USIMINAS continued to rely on an individual with a strong technical background to lead this important mechanism.

CSN   In the early-1990s CSN sought to reorganise the handling of technical publications in the company. The technical information management was structured in a way that seemed to emulate USIMINAS's TIC. Although CSN invested in an automated database, the unit did not engage in a systematic flow of external codified knowledge into the operations units. In other words, many of its characteristics were those of a conventional library.[111]

Overseas training

USIMINAS   In the early-1990s, as part of the PAQ implementation, nearly 100 engineers and 50 quality auditing engineers were sent to the US to undertake training at the American Society for Quality Systems (e.g. observation tours in plants, laboratories, and product distribution systems). Managers, engineers, and operators continued to be sent overseas for short-term courses and technical visits to steel companies and equipment suppliers in Europe and Asia for training before operations start-up. From 1990 to 1997, about 300 individuals were sent overseas, 50 per year, on average.[112] In 1993, the technology absorption programme led to the restructuring of the functioning of overseas training. New criteria were created for sending individuals overseas. As a result, before going overseas a detailed plan had to be prepared consisting of: details about the training to be undertaken; its practical contribution to process, equipment improvement and, in particular, to product value and application; and the contribution to USIMINAS's existing knowledge. The plan was submitted to top management for approval. The criteria for approval became more based on a company-wide perspective rather than limited to the needs of specific units, as in the expansion phase.[113]

CSN   In the early-1990s, as of part of TAG and later of TQM implementation, overseas training was restored in CSN. The number of technical visits increased from six in 1990 to 106 in 1994. The overall number of individuals sent overseas increased from 108 in 1992 to 208 in 1994. From 1990 to 1994, CSN sent different groups of engineers overseas every six months, particularly to Japan, South Korea, Germany, Holland and the US. From 1996, following the take-over by the new management, the diversity of these practices and their intensity were reduced. The number of technical visits fell from 95 in 1995 to 17 in 1997. Other types of overseas training were encouraged, but on an intermittent basis as a cost-reduction measure.[114]

Participation in conferences and related events

USIMINAS Individuals continued to be encouraged to attend conferences promoted by ABM, IBS, ILAFA, in Brazil and Latin America, and the US, Europe, Japan and Australia (e.g. active participation in the ceramic application meeting in Australia, 1997), as a way of updating and expanding their technical knowledge. In addition, individuals continued to be encouraged to present their own papers at these events.[115]

CSN As a result of TQM activities in the early-1990s, CSN's employees were stimulated to increase their participation in the ABM, IBS and ILAFA's events in Brazil. They began to give presentations about their experience of making improvements in product and process and production organisation. In 1993, 14 technical papers were presented. In 1994, technicians from the hot strip mill even organised external events with ABM. This was a substantial improvement compared with the 1970s and 1980s.[116] In relation to the previous phase, the activities within this process became diverse and continuous.

Using technical assistance and agreements for knowledge acquisition

USIMINAS Following the signing of the TA-4 (the 'Daruma Project'), USIMINAS's individuals engaged in their already routinised way of working together with foreign technicians to absorb any new knowledge. In contrast to the previous phase, NSC delivered 'recommendations' only. Drawing on its own capabilities, USIMINAS was able to analyse them and decide whether and how to implement them. To achieve this, a recommendations implementation plan was prepared in each operational unit, this time led by the technical divisions. The TA-4 generated 426 recommendations. By March 1997, 367 had been implemented, 55 were being implemented, and four were being prepared for implementation. The process of implementing the new procedures, triggered a series of discussions and interactions between managers, engineers and operators[117] thereby further increasing knowledge socialisation. This suggests a strong interaction between the knowledge-acquisition and knowledge-conversion processes.

In the 1990s the building of agreements with technological frontier companies was intensified as a way of sustaining USIMINAS's Level 5 to 6 innovative capabilities; and acquiring new knowledge to expand and diversify its technical assistance activities in Brazil and Latin America. Agreements were made with European companies (e.g. British Steel, Thyssen). From 1992, new expertise in equipment engineering for UMSA was acquired through agreements with world-class equipment designers and manufacturers (e.g. Hitachi), leading to daily interaction with UMSA's individuals. From 1991, four new major technological agreements were built with local research institutions:

UFMG, the Federal University of Rio de Janeiro (UFRJ), São Carlos Federal University (UFSCAR), and the Technological Research Institute (IPT).[118]

CSN   From 1992, following the technological updating plan, a technology contracting programme was created. The number of technical assistance agreements increased from three in 1990 to 35 in 1993, of which 24 were international. As a result, in 1992, 14 experts from France and seven from the US came to CSN to provide technical assistance and training for individuals in different production units. This gave rise to the acquisition of new perspectives, mental modes and skills by CSN's staff. From 1990, CSN also intensified technological agreements with local universities and research institutions: from three in 1990 to 15 in 1993.[119] However, despite the presence of external technical assistance in CSN, several interviews in the company suggested that there continued to be an absence of any systematic pursuit of using technical assistance and agreements as a means of building CSN's own capabilities. In other words, CSN did not develop any clear commitment to transforming the assistance received into technical assistance to be sold to other companies. This may reflect the fact that during its lifetime CSN had had little engagement in the supply of technical assistance.[120]

Interactions with users

USIMINAS   In 1992, commercial cells were created to intensify interaction with users. Through them a schedule was created for visits, meetings, and discussions with users to gather their feedback about USIMINAS's product performance and application, delivery, and suggestions for further improvements. USIMINAS's individuals acquired substantial tacit and codified knowledge through shared experiences, insights, and new ideas, technical reports, and graphs of product performance, etc., triggering innovative actions throughout the company. In 1994, the electronic information system was created by USIMINAS, whereby a direct communication channel was set up with users triggering several interactions (e.g. daily phone calls, e-mails, and informal meetings for problem-solving in product development and application).[121] Another type of interaction was joint projects with users for product and processes improvement and/or development.

CSN   In 1992, as a result of TQM implementation, the annual event 'Week with the Customer' was created to promote talks and seminars with the users for the acquisition of product performance feedback. From 1993, the research centre increased its interactions with steel users (e.g. Volkswagen) through testing for product development. Although interaction between operational units and users increased in relation to the expansion phase (1970s–1980s), it was all at the discretion of each unit with no integrated effort. On some occasions, users asked for a product specification which they had helped to develop in CSN, but which was still unknown by the operational units.[122]

Providing scholarships for research and teaching

USIMINAS In the early-1990s new agreements were added to the existing ones. They were made with UFSCAR, UFRJ and the University of Campinas – UNICAMP (e.g. in industrial automation and information technology, metallurgical and mechanical and material engineering).[123] The project with UFSCAR for a tailored PhD programme, the design of which began by the late-1980s, did not materialise in the 1990s, However, by 1991, one engineer was undertaking a PhD programme in UFRJ (COPPE). Two others had started PhDs in UFRJ/COPPE and UNICAMP. From 1992, other agreements to support research were signed with the UFMG's Foundation for Research Development (FUNDEP/UFMG), the University of São Paulo's Foundation for Chemistry and Physics Support (FAFQ-USP) and UFRJ/COPPE.[124] In 1995, to upgrade the educational level of operators, an agreement was signed with the Brazilian Foundation for Educational Support (FUBRAE). This sought to homogenise the educational level in the company at the minimum of secondary level. By 1996, 190 individuals had completed this programme.[125]

CSN In the early-1990s, CSN sought to improve its links with the School of Metallurgy of UFF. The evidence indicates, however, that the programme evolved in an inconsistent way. In November 1993, CSN took the initiative of signing an agreement with the UFF's School of Metallurgy to implement a MSc course and other research activities. The idea was to recruit MSc engineers and develop joint research. The course started in 1994 with 13 engineers. Research activities involved the contracting of 12 PhDs, by UFF, to concentrate on teaching and research (from reduction to surface treatment). It also involved visits from foreign researchers. From 1996, following the take-over by the new management, CSN's support for that agreement began to weaken. By late 1997, from the 39 MSc graduates, only 17 had been recruited by CSN. Among them, four were sacked in 1998. As an indication of CSN's diminished commitment to the project, UFF began to move in a different direction, designing its own PhD programme with federal government support, and engaging in alternative projects with other companies.[126] This suggests that despite the presence and the good start of this learning mechanism, its functioning deteriorated over time.

Educational infrastructure in the community

USIMINAS In 1994, USIMINAS took over the control of the Colégio São Francisco Xavier, a school whose capacity was expanded to 3,000 pupils per year to provide primary and secondary education to the Ipatinga community, mostly employees' sons. These were seen as potential USIMINAS employees. In 1995, the USIMINAS cultural centre (Usicultura) was created encompassing several cultural activities, 'to open a new knowledge field in the community'.[127]

CSN　In the early-1990s, CSN restored the support for the Colégio Edmundo Macedo Soares. This has a primary and secondary school for 3,000 pupils. Apart from this, no other support was provided.[128]

8.2.2　Internal Knowledge-acquisition Processes

The research centre

USIMINAS　In the early-1990s, the activities in the research centre were reorganised. Until the late 1980s, it had been organised on a functional basis.[129] By the early-1990s, it was organised on the basis of 'research teams', involving a mix between project and matrix organisational arrangements.[130] The process units were reduced from three to one and efforts in product research were intensified. In the 1980s, the way of organising research activities reflected the mill's production flow. Researchers worked within well-defined functional boundaries. In the 1990s, individuals began to be brought together, on the basis of their expertise, to undertake specific projects with fixed time from start to finish. This reflected the more integrated way that activities were being carried out in USIMINAS in the 1990s compared with the 1970s and 1980s.[131] Additionally, individuals were assigned as a leader in one project and, in parallel, as a team member in another project. This is was done on the basis of his/her knowledge depth. This practice suggests a mix between internal knowledge acquisition (individuals doing their own research), knowledge socialisation and also codification (as they interacted with and transferred their ideas into projects and research reports and papers).[132] In other words, this suggests a strong interaction between the learning processes.

The number of people in the research centre was reduced from 389 in 1988 to 176 in 1997. This reflects the overall reduction in the number of employees in USIMINAS from 13,413 in 1990 to 9,210 in 1996.[133] However, the number of MSc engineers in the Research Centre increased from 19 in 1987 to 26 in 1997. The number of PhDs increased from three in 1990 to five in 1997.[134] Although USIMINAS reduced the total numbers of employees, evidence from the research centre suggests that the company sought to keep and/or increase the number of better qualified employees.

Despite the reduction in the number of employees, the average number of projects per year was maintained (above 100). The time to complete a product research project was reduced from 32 months in 1988 to 17 months in 1997. As a practice that had become routinised in USIMINAS by the 1990s, another five-year research plan was approved in 1994. Its preparation involved different units including operations. The plan focused on 16 research lines, among them iron and steelmaking (with lower residual contents), consumption of refractories, solidification process, products upgrading and development, and environmental protection.[135]

By 1997, 48 per cent of research efforts were being focused on products (expected 50 per cent by 2000), 33 per cent on process improvement, 14 per cent on raw materials and inputs, and 5 per cent on environmental protection.[136] In the early 1990s, about US$10 million was invested in the centre in the latest laboratory facilities. USIMINAS drew on fiscal incentives from the Programme of Industrial Technological Development (Law 8,611, 1993). From 1993, USIMINAS was investing 0.7 per cent of sales in research.[137] As a result of its re-organisation, the functioning (and behavioural) features of the research centre in the 1990s had changed from the 1970s and 1980s, as outlined in Table 8.1.

Table 8.1 Evolution of some features of USIMINAS's research centre

Functioning features	1970s–1980s	1990s
• Freedom to do research	wide	narrower
• Perspective on research	technical	technical and commercial
• Research project horizons	long-term	short-term plus practical benefits
• Research focus and topics	wide, varied	users' trends and needs
• Coverage of research activities (product, process, energy, raw materials)	diverse	diverse, but deeper on product
• Interaction between units in the centre	island-type	team-type
• Interaction with operations and other support units	good	good/excellent
• Interaction with users	intense	more intense
• Support from top management	strong	stronger
• Number of individuals	389 (1988)	176 (1997)
• Scale of projects (annual)	above 100	above 100

Source: Several interviews within USIMINAS.

In the 1980s, researchers were given complete freedom to investigate, for example, a new type of alcohol as fuel for the automobile industry. In the 1990s, however, they were given specific projects with a clear expected outcome (e.g. the isotropic steel project for the user 'x' to increase 'x' per cent in market share). In addition, researchers were encouraged to have a commercial and competitive approach to research and to treat the operations individuals as their equals therefore achieving a behavioural change.[138]

CSN In 1990, top management decided that the research centre should be shut down. However, the decision was reversed and support for it restored following lobbying by the researchers. In 1992, on its own initiative, the research centre reorganised its activities. Up to the late-1980s, it was a combination of functional with matrix organisation reflecting the production flow in the mill.[139] The unit became organised on the basis of product, process, and labora-tories.[140] Researchers were organised into groups. Each group had a leader. Each group was encouraged to look outside the company at trends in steel use and development. In the 1990s, about 70 per cent of the efforts of the research centre were focused on development projects and about 30 per cent on 'trou-ble-shooting'. This is the reverse of the situation in the 1980s.[141]

By 1997, however, as a result of the corporate restructuring, process research activities had been reduced following the extinction of the process research su-perintendency. Some researchers were offered early retirement or even sacked. To avoid a massive loss of highly qualified researchers, the research centre took the initiative of allocating them in the operations units, like the reduction area. As a result, most innovative projects became focused on product development. From 1996, although engaged in product development, the centre had to justify its exis-tence to top management, who tended to see it as a cost burden on corporate fi-nance. As mentioned in interviews in CSN, top management's reduced support for the research centre from 1996 was based on the view that, 'since technology is becoming cheaper and more easily available, it is not necessary to undertake costly investments to develop in-house technological capability'.[142]

As a result, employees in the research centre decreased from 193 in 1991 to 110 in 1997. This reflects the overall reduction in the number of employees in CSN from 22,134 in 1989 to 9,059 in 1997.[143] However, the number of PhDs decreased from three to zero. Between 1991 and 1997, the number of graduate staff decreased from 18 to 12 and the number of technicians from 72 to 59. In 1998 the head of the centre, a highly qualified professional with long experi-ence in both operations and research activities left CSN.[144] Between 1994 and 1995, nearly 70 engineers left CSN. Between 1996 and 1997, about 130 engi-neers were allowed to go as a result of corporate restructuring.[145] The evidence suggests that the reorganisation in CSN did not pay adequate attention to the loss of qualified tacit knowledge from the company. Although the research cen-tre re-organised its activities, and increased efforts on product development, its overall functioning does not seem to have improved substantially in the 1990s over the 1980s, as outlined in Table 8.2.

The plant

USIMINAS   From 1990, experimentation in the operational units was intensi-fied, reflecting the de-concentration and decentralisation from the research cen-tre into the operations sites. Experiments began to be undertaken by teams within the technical divisions of each production unit, under the co-ordination of the metallurgical dept, with the research centre's involvement. Activities took place under well-defined routines of procedures, in other words, in a more

Table 8.2 Evolution of some features of CSN's research centre

Functioning features	1970s–1980s	1990s
• Freedom to do research	indifferent	indifferent → limited
• Perspective on research	technical	technical and commercial
• Research project horizons	short-term	short-term, disconnected from corporate strategy
• Research focus and topics	wide, varied	users' trends and needs
• Coverage of research activities (product, process, energy, raw materials)	diverse	limited (product)
• Interaction with units in the centre	island-type	team-type, limited to product
• Interaction with operations and other support	poor	moderate/good
• Interaction with users	moderate	intense
• Support from top management	weak	Weak → strong → weak and conflictive
• Number of individuals	about 200 (1989)	about 100 (1997)
• Centre's efforts on innovative projects	30 per cent	70 per cent (mostly on product)

Source: Several interviews within CSN.

systematic and interactive way than in the 1980s.[146] One example, was the following experiment in the hot coils unit.

In November 1997, a group of operators from crew B, team DF, of the hot strip mill's technical division prepared the 'Experiment Plan 019/97'. It sought to eliminate a specific type of defect in the pickled and oiled hot coils. The title, objective and specific tasks of the different units involved ('who' should do 'what') were spelt out in the plan. Following implementation, the 'Analysis of Occurrence 061/97' was written. It spelt out the analysis of the activities, the conclusions, and implications for performance improvements. Boosted by their achievement, the team decided to continue with this type of experiment as a way of eliminating other defects on the coils.[147] By doing these experiments, the individuals acquired new tacit knowledge of the intricate properties of the product (hot coils), and the process itself, and reframed their tacit knowledge on the principles underpinning the technology. As the experiment progressed, they gave rise to a series of knowledge-socialisation processes since team members had to interact with one another. It also gave rise to knowledge-codification processes, since individuals had to write their findings, conclusions, and suggestions in an organised way. In the light of Table 3.2, this suggests a strong interaction between the learning processes. While USIMINAS encouraged systematic experimentation, whose results might be uncertain, it also became less tolerant of mistakes relative to these activities: 'Although it is a fashion [to let individuals make mistakes], we have to run a plant 20 per cent above its nominal capacity so we cannot afford to follow this. We keep an eye on the individuals. If things begin to go wrong we have to intervene.'[148]

Nevertheless, USIMINAS seemed aware of the practical benefits deriving from experimentation. Positive action was taken whenever mistakes were made: 'We called attention to the learning acquired from the activity so that future mistakes could be avoided.'[149] In the 1990s, experiments became to be regarded as part of the 'anticipatory view' on process and product activities. This was combined with existing 'never-do-without-knowledge' practice (or 'learning-before-doing'). By the 1990s, this practice had become routinised in the company, as a result of its continuous intensity over the years[150] (see Chapters 6 and 7). This also contributed to minimising mistakes (e.g. in product introduction) that had occurred over past decades.[151] By late-1997, USIMINAS had 78 new patents, 9 of which were overseas, added to its existing 172 patents, a 45 per cent increase leading to a cumulative number of 250.[152] This achievement may reflect: (1) the re-organisation and improved functioning of the internal knowledge-acquisition processes; (2) greater diversity of knowledge-acquisition processes in the 1990s as compared to the 1970s and 1980s; (3) interaction between internal and external knowledge-acquisition and knowledge-socialisation and codification processes.

CSN   As opposed to the previous two phases, CSN's pattern of 'learning-by-doing' in the 1990s was upgraded by the various improvements in production organisation, process, and equipment, made by individuals in different parts of the plant. These were also associated with the implementation of the TQM programme and the QCCs (e.g. improvements in the HSM 2, de-bottlenecking and de-bugging in the hot coils mill, led by a group of operators, and many others). In the light of Table 3.2, the process of carrying out these activities might have contributed to increasing the understanding, by operations individuals, of the technology and its principles, particularly of routine operations.[153] This was probably reflected in the reduction in phone calls for basic problem-solving that operators and foremen used to make to the engineers 'on-call' during the weekends. This was a special service that functioned from 1991 to 1992, as part of the TQM programme.[154]

Therefore, there was an improvement in the internal knowledge-acquisition process in the 1990s in relation to the 1970s–1980s. Until 1997, however, experimentation had not become systematic. It took place intermittently at the initiative of individual employees (e.g. the experiment to improve the gas piping in the BF 2 led by one engineer).[155] Nevertheless, as one of the outcomes of the improved internal knowledge-acquisition processes in the 1990s, 33 new patents were obtained by CSN. This led to a total of 46 patents in the company.

8.2.3 Summary of the Cross-company Differences

In the light of Table 3.2 and the empirical evidence above, this section summarises the differences in the features of the external and internal knowledge-acquisition processes during the liberalisation and privatisation (1990s) of USIMINAS and CSN. These differences in external and internal knowledge-acquisition processes are summarised in Tables 8.3 and 8.4, respectively.

As far as external knowledge-acquisition processes are concerned, Table 8.3 indicates that nine processes were present in USIMINAS and eight in CSN. The increase in variety of processes in CSN indicates an improvement in its knowledge-acquisition processes in the 1990s in relation to the 1970–1980s period.

However, this improvement seemed not enough to achieve the standard attained by USIMINAS. While in USIMINAS the nine processes had diverse content of mechanisms, in CSN most were still limited/moderate. And systematic channelling of external codified knowledge was still absent in CSN. In addition to differences in variety, the eight common processes still differed in their intensity, functioning, and interaction. For example, USIMINAS continued to send engineers overseas on a continuous basis and had the functioning of this process improved to meet the new company's needs. In contrast, although CSN restored this process during the 1990–95 period, from 1996 it had become intermittent. In addition, USIMINAS kept the interaction with teaching and research institutions on a continuous basis. The CSN agreement with UFF started off with good functioning. However, after 1996 its functioning began to deteriorate and it became more intermittent.

As far as internal knowledge-acquisition processes are concerned, Table 8.4 indicates that four processes were present in USIMINAS and three in CSN. The evidence in this chapter suggests that in USIMINAS the process of operating the plant following international efficiency standards had become routinised by the 1990s. In addition, this activity was influenced by the training process associated with the implementation of the fourth quality system – the PAQ. This may have contributed to the deepening and sustaining of its routine capability (Level 2) for processes and production organisation and products. In CSN in the 1990s the concern with efficient operation spread across the plant associated with the TAG and later the TQM training process (as described in Section 8.4). As a result, the way individuals engaged in the routine plant operations might have contributed to completing the accumulation of Levels 1 and 2 routine capability in the plant.

As far as the research centres were concerned, both USIMINAS and CSN reorganised these units in the 1990s. In both companies they became focused more on users' trends. Although USIMINAS paid greater attention to product activities, research in other areas (process improvement, equipment, raw materials, energy) continued. However, the efforts put into them were less than that in the 1970s–1980s period. Nevertheless, this suggests that USIMINAS sought to maintain strategic elements of research in areas other than products.

In CSN the reorganisation of the research centre in the early-1990s contributed to increasing its engagement in development projects rather than trouble-shooting. Later, however, the company engaged in more reductions in the

Table 8.3 Summary of the differences in key features of the external knowledge-acquisition processes during the liberalisation and privatisation phase (1990–97) between USIMINAS and CSN

EXTERNAL KNOWLEDGE-ACQUISITION PROCESSES	VARIETY		INTENSITY		FUNCTIONING		INTERACTION	
	USIMINAS	CSN	USIMINAS	CSN	USIMINAS	CSN	USIMINAS	CSN
Pulling in expertise from outside • Inviting experts for talks • MSc engineers for problem framing • Hiring back retired engineers and/or foremen	Present (diverse)	Present (diverse)	Continuous	Continuous	Good	Moderate	Strong	Moderate
Recruiting operations individuals	Present (diverse)	Present (diverse)	Continuous	Continuous	Excellent	Moderate	Strong	Moderate
Channelling of external codified knowledge • Routinised flow of internal publications	Present (diverse)	Absent	Continuous	–	Excellent	–	Strong	–
Overseas training • Short-term courses • Technical visits, observation tours	Present (diverse)	Present (moderate)	Continuous	Continuous → Intermittent	Excellent	Moderate	Strong	Moderate
Participation in conferences and related events • Attending and/or presenting technical papers	Present (diverse)	Present (diverse)	Continuous	Continuous	Good/Excellent	Good	Strong	Moderate
Using technical assistance and agreements for knowledge acquisition • Technical assistance contracts • Technological agreements with frontier companies	Present (diverse)	Present (moderate)	Continuous	Continuous	Excellent	Moderate	Strong	Moderate
Interaction with suppliers and users • Interactive cells • On-line system for daily problem-solving • Joint innovative projects (eg. steel products design and development; involvement in users' development projects)	Present (diverse)	Present (moderate)	Continuous	Continuous	Excellent	Moderate	Strong	Moderate
Providing scholarship for research and teaching • Tailored MSc and/or PhD	Present (moderate)	Present (limited)	Continuous	Intermittent	Excellent	Moderate/Poor	Strong	Moderate/Weak
Educational infrastructure in the community	Present (diverse)	Present (limited)	Continuous	Intermittent	Good	Good	Strong	Moderate

Source: Own elaboration based on the research.

Table 8.4 Summary of the differences in key features of the internal knowledge-acquisition processes during the liberalisation and privatisation phase (1990–97) between USIMINAS and CSN

INTERNAL KNOWLEDGE-ACQUISITION PROCESSES	VARIETY		INTENSITY		FUNCTIONING		INTERACTION	
	USIMINAS	CSN	USIMINAS	CSN	USIMINAS	CSN	USIMINAS	CSN
Routine plant operation	Present (diverse)	Present (diverse)	Continuous	Continuous	Excellent	Good	Strong	Moderate
Continuous improvements across the plant • Based on cross-functional involvement • Experimentation projects coupled with team-building • Based on TQM tools	Present (diverse)	Present (moderate)	Continuous	Continuous	Excellent	Moderate/Good	Strong	Moderate
Routinised knowledge-process acquisition before engaging in new technical activities	Present (diverse)	Absent	Continuous	–	Excellent	–	Strong	–
Innovative projects within the research centre • Systematic planning of research activities	Present (diverse)	Present (moderate)	Continuous	Continuous → Intermittent	Good/Excellent	Good	Strong	Moderate

Source: Own elaboration based on the research.

process research. This, however, does not seem to have triggered any organised involvement of the operations units in experimentation activities. As a result, by 1997, CSN's research centre was in a weaker position in relation to USIMINAS's to engage in process improvement activities.

As far as learning processes at the plant were concerned, they differed in terms of the variety of their mechanisms. For example, in USIMINAS continuous improvement processes involved systematic experimentation in the operations units and cross-functional involvement. This may reflect the greater integration in the plant during the 1990s in relation to the 1980s. In addition, considering the continuous intensity of improvement activities since the 1960s (Chapters 6 and 7), these had become routinised by the 1990s. In CSN, although there was an increase in continuous improvements activities, they were still limited to the TQM tools within units. And team-based experimentation was still limited and intermittent compared with USIMINAS.

8.3   KNOWLEDGE-CONVERSION PROCESSES

8.3.1   Knowledge-socialisation Processes

In-house training (course-based): basic

USIMINAS   In 1991 the apprentices' training centre was renamed as the professional training centre. Reflecting its continuous intensity over more than thirty years, the basic training for USIMINAS's operators had been routinised by the 1990s. By the mid-1990s, the centre had 21 instructors and 220 trainees aged 16 to 18 years. Quality systems, computing, and environmental protection were added to the contents of existing courses.

CSN   It was not until the early-1990s that CSN began to review the functioning of basic training for operational staff. This review reflected the changes taking place in the plant through TQM implementation. A course for steelmaking operators was created, including the basic TQM concepts. However, after 1996, reflecting the weakening of TQM, ETPC was receiving much less support from CSN. In addition, course contents stopped being updated.[156] This may have had implications for the recruiting process and further qualification of CSN's operators.

In-house training: (course-based)

USIMINAS   The specialised training at the professional training centre consisted of six courses (mechanical, electrical and electronic maintenance, use of lathes, welding, and steelmaking operation). Their content had been reviewed to meet the plant's needs. The practice of joint supervision (centre and site) of trainees had become routine. This functioning led to the acquisition of tacit and codified knowledge by trainees: from studying training modules, where USIMINAS's routinised practices had been documented and through interaction between each other and with instructors, operators and engineers on the site.[157] These activities may have triggered processes of individual knowledge-acquisition and knowledge-socialisation.

Between 1990 and 1997, the ratio of course to individuals per year was, on average, 1.42. This means that, on average, every individual followed more than one course every year. Entry to the training programmes was made more informal and straightforward as opposed to the 1970s and 1980s.[158] Entry to automation courses increased from 16 per cent of those individuals involved with automation activities in 1993 to 33 per cent in 1997. This may reflect USIMINAS's efforts to routinise the integrated process control and PPC systems. In the 1990s, on average, 70 per cent of plant training activities were directed to the operational units, against an average of 55 per cent in the 1980s.[159] This reflects the greater involvement of operational units in innovative projects. By 1997, about 80 options on technical courses were made available. This was 14 more than in 1996 (e.g. including 10 on automation, 28 on leadership, benchmarking, negotiation, USIMINAS's management system, the standard cost system).[160]

On average, 50 per cent of these training programmes were led by in-house instructors (e.g. technicians, engineers). In some cases, the involvement of these instructors reached 100 per cent.[161] This was a powerful mechanism for knowledge-socialisation and codification. The functioning of the 'after-training assessment' became more informal and open among the managers than in the 1980s, leading to quicker decisions on individuals' re-training needs.[162]

In 1995, the self development programme was created as a way of stimulating the self-design of training programmes to meet the needs of different individuals. Between 1996 and 1997, 1,355 people had joined the programme. To support it, USIMINAS made available its large archive of internal videos (e.g. internal seminars given by USIMINAS's individuals in the company's events, USIMINAS's own production techniques like the standard cost system, etc.), a rich reservoir of codified knowledge.[163] Reflecting the intensity of in-house combined with external training, the average number of individuals' school years in USIMINAS, by 1997 was 9.8 against 5 in Brazil as a whole.[164]

As far as communication between plant and corporate office about training activities was concerned, until 1990, 'they used to be like two different companies',[165] From 1991, training practices were unified and the flow of operations individuals into the corporate office became informal. Daily communication became straightforward via direct telephone calls and e-mail. From 1992, an internal network, developed by USIMINAS, permitted an instant flow of information (both voice and image) linking plant employees, corporate office, key users and suppliers. This contributed to a permanent interaction between individuals from different background and parts of the company.[166]

CSN   A key improvement to in-house training in CSN in the 1990s was the wide training between 1990 and 1991 to introduce TQM. This top-down training permitted, for the first time, an integration between corporate management and the plant in training practices. The sequence of the process is described in Table 8.5. This gave rise to a series of knowledge-socialisation interactions among individuals from different backgrounds which had been rare in the 1980s. In the light of Table 3.2, this resulted in greater diversity and improved functioning of in-house training processes in the early-1990s.

Table 8.5   CSN: development of TQM training

Sequence	Participants	Training types and goals
Phase I 12 hours	President, directors, and general managers	Seminars to top management to implement the TQM programme across the corporation.
Phase II 40 hours 120 individuals	Managers, facilitators and 'multipliers'	Course to the managerial level and facilitators to operationalise the TQM.
Phase III 40 hours 550 individuals	Heads of divisions, technicians, and engineers	Course to enable individuals to introduce the TQM in the operations units based on problem-solving.
Phase IV 20 hours 1,400 individuals	Crew heads and supervisors	Course to enable supervisors to identify problems effectively and generate solution following adequate methodology.
Phase V 9,000 individuals	Operators	Diffusion of the methodology through its practical application on the job with real cases.

Source: CSN (*Nove de Abril*, Ano XIII), February, 1990 [internal newspaper]. Interviews with engineers in the quality promotion centre.

The continuous intensity of the company-wide training programme may have contributed to the completion of the accumulation of Level 1 and 2 routine capability for process and production organisation and products in CSN from the early-1990s. However, as a result of changes in top management from the mid-1990s, particularly from 1996, the functioning of the in-house training processes seemed to deteriorate. In 1995, CSN removed the training unit FUGEMMS from the company on a third-party contracting basis. In 1996, the training decisions were decentralised to the business units. However, as managers were under pressure to improve financial results, training decisions tended to be delegated to lower-level staff. Although in some units, training activities were present (e.g. 5 per cent of working hours), its functioning revealed: unclear criteria on choosing individuals for training; incoherence between training programmes and individuals' needs; and absence of any after-training assessment, which prevented the company from knowing whether the training had made any difference in the individuals' performance.[167]

By June 1997, FUGEMMS was brought back into CSN, but weakened. The number of instructors had decreased from 60 in 1990 to 13. FUGEMMS had become a mere co-ordinator of outsourced training activities, but no longer the full executor.[168] The training/individual ratio fell from 1.13 in 1993 to 0.98 in 1997. In other words, by that year, individuals were undertaking less than one course per year. Nevertheless, isolated initiatives were taken in certain operational units – or business units. For example, in the HSM from 1995, a training process was created for operators and supervisors. By using software run on CD-ROM, operators were introduced to every detail of the mill in an animated and 3-D perspective. This was supported by modules and classes. The training was designed by the engineers in the unit helped by a software company. However, this type of initiative was not widespread in the plant.[169]

On-the-job training

USIMINAS Coupled with 'job rotation', OJT became routinised in USIMINAS as a practice to prepare operators before they engaged in routine work. The continuous interaction between individuals of the operations units and the PDC to elaborate training modules, contributed to bringing together different expertise. In addition, the tacit knowledge of various individuals became codified in the modules.[170] This suggests a routinisation of the strong interaction between the knowledge-socialisation and codification processes. The OJT practice became quicker and more effective, as opposed to the 1970s and 1980s, because every production practice had been standardised; the technical level of new operators was good, because most of them had come from the professional training centre; and as a result, they could learn more quickly and even improve existing operating standards.[171]

From 1992, UMSA's new operators were supervised by experienced in-house operators ('Godfathers') or by technicians hired from suppliers. Unlike the iron and steelmaking operators, equipment operators have to manipulate several components to build large equipment (e.g. a sinter boiler). As a result, greater dexterity is needed. OJT became a key mechanism in UMSA whereby new operators could work together, observe, and imitate the experienced operators in order to learn from them. An indication that tacit knowledge was being effectively socialised was when they began to use their own 'language' (e.g. 'dog' for a sticking device; 'bib' for an advanced welding device).[172]

CSN   From 1992, each newly recruited operator worked under the supervision of a 'Godfather' operator on a OJT basis. The 'Godfathers' had to teach the principles of the activity, observe the trainee's progress, correct mistakes and, in particular, transmit their accumulated knowledge. OJT ranged from two to six months depending on the trainee's ability to learn. This daily interaction began to trigger knowledge-sharing within the units contributing to improving the operators' efficiency and also to minimising fatal accidents.[173] In the light of Table 3.2, this suggests a substantial improvement in the diversity and functioning of the OJT process in relation to the 1970s-1980s period.

Training individuals of other steel companies

USIMINAS   This new mechanism reflects USIMINAS's routinised practices of knowledge conversion. Between 1992 and late-1997, about 4,000 USIMINAS's technicians/day were involved in training programmes in different countries. And 6,640 individuals/day from other steel companies in Brazil and in Latin America were trained in USIMINAS in process and production organisation and product-centred activities (e.g. SIDERAR, SIDOR, HUACHIPATO SA).[174] In providing this training, USIMINAS's individuals had their knowledge systematised. They received new insights and viewpoints from interaction with trainees.[175]

CSN   Interviews in the company indicated that this type of training did not take place in CSN.

Shared problem-solving: meetings

USIMINAS   A monthly quality meeting was added to the existing routine daily and weekly meetings. From 1990, they involved the heads of 13 departments, the general plant manager, the marketing manager, the development director, and the president. Stimulated by the president, they discussed problems and proposed actions (e.g. project delays, product development) in an open and even controversial way as long as it led to improvements.[176] After such meetings, each head of department met with the staff within their units down to the level of operators, completing the routine of the functioning of this mechanism: 'The meeting begins with a bottom-up funnelling which then turns into an up-down de-funnelling so that every individual in the company becomes aware of the company's problems, goals, actions. . . so things can go in the same direction.'[177]

CSN   As a result of TQM implementation, the practice of systematic technical meetings began in CSN by 1991. Managers, supervisors, engineers, lead operators met at 7.15 a.m. to review the performance of the previous crews, products, and process problems, etc. This practice, which had been limited over the previous decades, contributed to the socialisation of the individuals' experience, skills, and different perspectives to solve problems in the production lines.[178]

Interactive problem-solving practices

USIMINAS By the early-1990s, the operationalisation of the standard cost system was seen as part of the daily routine in the plant. This mechanism also had become part of the in-house training programmes. In these programmes the long experience in implementation of the system was documented over the years thereby contributing to knowledge-sharing and codification.[179]

CSN During the 1990s, CSN still lacked a systematic process for allowing individuals to interact through problem-solving.

Knowledge-sharing links

USIMINAS Associated with the more integrated technological activities in USIMINAS during the 1990s, the number of researchers informally allocated to the operations units increased from one to five (two engineers and three technicians). They were allocated to the newly created technical divisions from 1992. They were known as 'knowledge-bridging individuals'. The engineering unit also increased the number of engineers in the operational units from one to five.[180] The automation unit intensified this integration by involving the operations individuals in the automation projects and allocating its technicians at the sites (e.g. reduction area).[181]

From 1990, monthly meetings were held to bring together the heads of each operational unit and the team leaders in the research centre to discuss project implementation. These interactions helped to eliminate the risk of developing a product that could not be made in the operations units.[182] They also consisted of greater involvement of the operations individuals in the research centre's projects (e.g. product development, the five-year research plan).[183] The processes were a mix between 'meetings' and 'knowledge-sharing links'. In the light of Table 3.2, this evidence suggests an increase in diversity and continuous intensity of the process for the socialisation of knowledge between individuals across units.

CSN In the early-1990s, CSN created 'product ambassadors'. Their purpose was to improve the relationships between the operational individuals and the researchers for product development. As a result, one researcher, involved in a product development project, stayed in the production units, during the 'transition phase'. The researcher interacted with the operators and technicians to guarantee the efficient manufacturing of new steels.[184] Although the relationship between these types of expertise has improved in the 1990s in contrast to the 1980s, some problems still remained: 'While we are here involved with research in the iron carbide precipitation for product development, the operations individuals are concerned with production volume only. In USIMINAS this is different.'[185]

From 1996, top management in CSN was giving considerable attention to production volume (e.g. incentives, over-emphasis on quantity as CSN's major achievement). The re-emergence of the 'volume-first' practice seems to have limited the efforts on knowledge-sharing for product improvements and development.[186]

Teambuilding

USIMINAS   This process was intensified in the 1990s reflecting the more integrated way in which activities started to be done in contrast to the 1980s. The informal committees and matrix arrangements continued (e.g. product development, patents, engineering, etc.) and new types of teams were created (e.g. experiments in the plant, automation projects, commercial cells, etc.). They contributed to bringing together diverse expertise from different functional areas, thereby triggering knowledge-socialisation. They also contributed to the codification of knowledge by reporting on their activities, elaborating training modules, and project proposals. On the basis of Table 3.2, this reflects a strong interaction between the learning processes. The building of teams became more spontaneous than in the previous phase,[187] indicating an improvement in their functioning in relation to the 1980s: 'Previously, we needed a formal structure to make individuals work in teams. Today that structure no longer exists and they interact intensely. It has become a common job practice.'[188]

CSN   In the early-1990s, staff began to be encouraged to work as teams. This reflected the implementation of TQM and its tools. The CCQ groups for continuous improvements are an example of team-building, particularly from the mid-1990s. Although present, team-building was still incipient in CSN.[189] This may be associated with the limited knowledge-socialisation in the 1940s-1980s period; and the difficulties of integrating individuals across different functional areas (e.g. limited knowledge-sharing links). The progress of team-building across areas may have been further limited by the consolidation of the 'business-units' mindset.

Dissemination of lead operators

USIMINAS   Following the extinction of 'inspection supervisors', by 1992, 'lead operators' began to play an even more critical role in bridging communication between the operators and managers. In addition, by interacting closely,

both formally and informally, with other operators, they could socialise and codify much of their tacit knowledge across the crew during routine and innovative activities (e.g. by leading experiments). Therefore, they began to play an
even more critical role in integrating the knowledge of the crew in a workable
system.[190]

CSN   In the 1990s, as opposed to the 1940s–1980s period, operators began to
play a more active role in their units. In 1990, a campaign initiated by the president encouraged managers and supervisors to talk to and listen to operators.[191]
This initiative, combined with the TQM programme, led to a substantial decline
in the punishments against the operators. These were frequent over the
1960s–1980s period.[192] By 1992, 'Godfather' operators were created. Chosen
on the basis of their accumulated knowledge, their role was to train their peers
and the newly recruited operators. They were also encouraged to be the communication bridge between managers and the rest of the crew in TQM and QCC
implementation.[193] This suggests a substantial improvement in this type of
knowledge-socialisation process in the relation to the previous phase: 'Before
the TQC and the QCCs we never talked to the superintendent. . . We had new
ideas, we knew what had to be changed. . . But we were not supposed to speak.
The TQC gave us a chance to express our knowledge.'[194]

8.3.2   Knowledge-codification Processes

Standardisation practices

USIMINAS   Over time, individuals had become familiar with the practice of
making use of and updating basic technical standards. By the 1990s, these had
become part of the daily routine. As recognised in the plant, several of the improvements in process and production organisation, equipment and products
would not have been possible without them.[195] They were seen as 'intelligent
standards' because they were updated over time as a result of improvements.
This in turn contributed to codifying the tacit knowledge (or parts of it) involved in an innovative activity. The modification of a standard might start with a
suggestion from an operator. But it was put into practice by collective agreement within the unit under the approval of the metallurgical department.[196]
   In the early-1990s, a distinction was made between 'basic technical standards' and 'detailed production procedures'. The technical standards specified
the conditions under which an activity must be carried out ('what', 'why', and
'when'). The 'detailed production procedures' outlined the way the activities
must be carried out ('how' and 'who') in clear and simple language to make it
easily understood by operators. The creation of the production procedures was
associated with a restructuring in the existing quality system through PAQ implementation. In association with the technical standards, the production manu-

als became USIMINAS's 'codified proprietary knowledge'. This also contributed to homogenising knowledge across the crews.[197] From 1992, a group of 50 quality engineers and auditors, trained overseas, began to monitor the daily application and updating of the standards and production manuals.[198] This suggests that this 'codified proprietary knowledge' in the form of standards was being continuously influenced by the processes of external knowledge-acquisition (e.g. overseas training) and internal knowledge-acquisition (e.g. plant experimentation). In other words, there was an interactive relationship between the learning processes.

Another type of standardisation was through the automated systems. From the early-1990s, greater efforts to increase the automated systems for process control led to intense codification of tacit knowledge into different automated systems. For example, the building of the 'specialist system' developed for the BF 3 consisted of codifying and storing the operators' accumulated knowledge to improve control over process parameters. At a certain stage, the system offers a set of decisions to be chosen and made by the operator at the control panel. However, some decisions are made on the basis of the operators' accumulated experience (e.g. interpreting the colour and other aspects of the metallic charge) which are more difficult to document.[199] To cope with this difficulty, other practices (e.g. team-building) are combined to codify individuals' tacit knowledge: 'If we ask individuals to write what they know they will not be able to write everything; but if we bring them together, in teams, to solve a particular problem then they will express their knowledge.'[200]

CSN   In the early-1990s, CSN was engaged in the standardisation of its production practices. This was triggered by the TQM programme and the pursuit of ISO 9000 certification. However, they had a confused start since top management misunderstood the meaning of standards (as 'targets').[201] In 1990, following consultation with USIMINAS's former president and later assisted by a consulting institution, production standardisation began. However, CSN did not know about the adequate scale and level of detail for these standards. 'They [standards] were so detailed that the operators gave up reading.'[202] As a result, about 13,700 standards were cancelled between 1995 and 1997.[203]

When the '5 S programme'[204] began in 1992, its principles were misunderstood by several managers. As a result, they did a radical clean-up in the units, throwing out documents with historical information (e.g. equipment performance records, reports from overseas technical visits).[205] To avoid similar problems, a standardisation office was created to co-ordinate the process and the standardised documents (e.g. technical designs, instructions for inspections, test procedures, job instructions, etc.).[206] The standardisation process consisted of tracing and writing down each of the components involved in units' activities. This involved foremen, supervisors, operators, managers, and engineers.[207] This contributed to codifying different elements of tacit knowledge, a practice that was absent in CSN over the 1940s–80s period.

The production standardisation practice contributed to diminishing, to some extent, the intense knowledge concentration at individual level, as compared to the 1940s–80s period. However, shop-floor evidence reveals that several operators still had limited knowledge of the principles involved in the technology: individuals had not learnt what to do when things diverged from the standards.[208] One example was the application of the standard for washing the converter in the steel shop. At the end of each campaign, the residual slag that had accumulated beneath the converters needed to be removed. In 1996, a group of operators was assigned to this task. Details of this activity which they had to follow had been documented on the 'Basic Oxygen Converter Washing Standard'. The operators began by blowing oxygen up to 300 Nm^3/min. to increase the heat so that the solidified slag could be dissolved; but slag remained because a particular chemical reaction had occurred, therefore another chemical agent had to be used. However, in sticking to the standard, the operators increased the oxygen flow to 750 Nm^3/min. making the heat so intense that the slag leaked into the two converters. This led to a stoppage of several days for repair.[209] This suggests that the operators were able to assimilate the basic routine of operations, but many did not understand 'why' certain things were done.[210] This also suggests a limitation in the functioning of standardisation processes in CSN.

Interviews in CSN revealed that the company recognised the importance of having individual tacit knowledge codified. However, the process of codification seems to have occurred in different ways across the plant. For example, one unit seeking to break free from the old practice of individual knowledge retention had been pressing engineers to codify their knowledge: 'They have to get their knowledge out, otherwise they will be sacked.'[211]

Systematic manipulation of own codified knowledge

USIMINAS From 1996, drawing on the existing knowledge codified in the technical standards and other archives in the operations units, TIC staff engaged in the dissemination of the details of improvements across the company including those in the plant on process and production organisation, products, and equipment from the start-up to the 1990s. To do this, a special database was created to store and disseminate the codified knowledge through the internal network. They created a concise text including the area, authors, date, timing, knowledge involved, and benefits for performance improvement. At the time of the fieldwork for this research, more than 3,600 improvements had been introduced in that network.[212] This illustrates interaction between the knowledge-socialisation and codification processes.

CSN   This process was absent in CSN reflecting the absence of knowledge codification and limited knowledge socialisation in the previous phase.

Reporting from external training

USIMINAS   The stricter criteria for individuals to be sent on local and overseas training was reflected in the way they had to report on it. In the 1980s, they described what had been taught and observed (e.g. plant tours). In the 1990s, however, they were required to generate an innovative proposal (e.g. process, product, etc.) on the basis of their training, in other words, not only 'what' had been learnt but 'how' to apply it in USIMINAS.[213] At the research centre, individuals who had completed their MSc or PhD were encouraged to publish at least two papers per year. They were also asked to circulate the papers among their peers before publishing.[214]

CSN   As result of TAG and TQM implementation in the early-1990s, individuals were encouraged to create improvement projects as a result of their external training. Although this meant an improvement over the 1980s, the interviews in CSN suggested that the process had not become systematic.

Internal seminars

USIMINAS   Reflecting its continuous intensity, this process had been routinised in USIMINAS by the 1990s. Several units by then were running their twentieth symposium. By relating their experiences (e.g. solving problems, leading projects, improving equipment), the staff contributed to a reframing of other individuals' tacit knowledge and mental models. This, in turn, triggered knowledge socialisation processes by instigating further discussions, informal meetings, and debate.[215] The USIMINAS 'Quality Week', a corporate-wide event started in the early-1990s, was attended by staff from the plant and the corporate office, via an interactive teleconference. It involved presentations by groups of engineers, operators, researchers, plant and top managers contributing to spreading and homogenising corporate views: 'USIMINAS's actions have to be directed to improvements. These actions are done by individuals seeking to learn. . . USIMINAS's competitive performance depends on the continuous learning by its individuals. . . This is what our competitors took decades to understand.[216]

CSN   From 1992 to 1995, the building of volunteer QCC groups brought several operators together to discuss and solve problems during 'brainstorming' sessions.[217] The QCC conventions and the annual Technological Week were the main events triggering knowledge socialisation.[218] However, the weakening of the TQM programme probably contributed to decreasing the intensity of knowledge sharing through QCC.

Re-integrating individuals from external training

USIMINAS   As a reflection of stricter criteria for sending individuals on external training, the way they were reintegrated into the company was improved. From 1992, when individuals from the research centre returned from their MSc courses they became team leaders. In parallel, they were a team member in a different project. For example, the steel USI-STAR-350 was developed within one team led by an engineer returning from an MSc in UFMG.[219] In the operational units, these individuals were assigned to be 'new project leaders'.[220]

CSN   Until the 1990s no systematic practice to reintegrate individuals into innovative projects had been developed. This may reflect unclear criteria for sending engineers on overseas training during 1970s–1980s period. As mentioned during one interview, during this period, the training unit 'did not know where to allocate the engineers when they returned from abroad.'[221] As a result, individuals took their own initiative to create new projects.[222]

Learning phrases and symbols

USIMINAS   By the 1990s, the principles of USIMINAS's cycle of technological development had permeated the company's daily technological activities and become routine. The symbol itself was re-designed suggesting the continuous intensity of its use, as shown in Figure 8.1.

Since 1993, as a result of the Daruma Project (TA-4), a massive portrait of Daruma,[223] with one eye painted in and another left blank, hangs in the main hall of the metallurgical department. Daruma dolls, of different sizes and half eyeless, were in various offices throughout the plant. Because they always return to the upright position no matter how they are turned over, these dolls also symbolise the patience, perseverance, and determination of USIMINAS's learning efforts. The second eye would be painted in when the knowledge involved in the recommendations of TA-4 had been internalised by the individuals in the plant. Meanwhile, the blank eye is supposed to instigate, 'bother', and remind everyone that there still is knowledge to be internalised. This triggers knowledge-sharing and codification processes.[224]

CSN   By the early-1990s, top management was able to instil widespread concern about continuous quality improvement. However, the practice of creating learning phrases or symbols to trigger the learning process was absent in CSN. Indeed by the mid-1990s, top management began to spread the view that knowledge had become so available that it could be 'bought' instantaneously.[225]

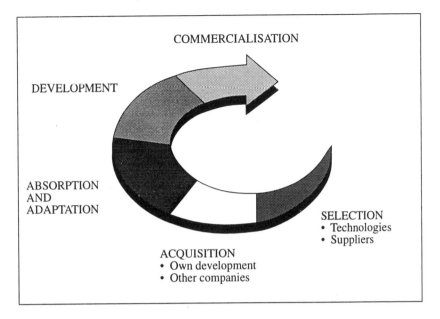

Source: Etrusco, S. C. P. (1997), 'A experiência da USIMINAS no ciclo de transferência de tecnologia e propriedade industrial', *R Esc Minas*, vol. 50, no. 1, pp. 25–29, Jan.–March.

Figure 8.1   USIMINAS's routinised cycle of technological development (1990s)

Thus it seemed to be delivering an opposing message to that of learning processes. It was not until late-1998 that top management began to review this corporate perspective.

8.3.3   Summary of the Cross-company Differences

The cross-company differences in the knowledge-socialisation and codification processes are summarised in Tables 8.6. and 8.7, respectively.

As far as the knowledge-socialisation processes are concerned, eight processes were present in USIMINAS and seven in CSN. Although seven processes were present in both companies, the variety of mechanisms they involved was diverse in all of them in USIMINAS. In contrast, in CSN most of them were of a limited to moderate variety. In addition, they also differed in terms of intensity, functioning and interaction. During the 1990s CSN experienced substantial improvement in the features of its knowledge-socialisation processes. However, the evidence suggests that this improvement was not suf-

Table 8.6 Summary of the differences in key features of the knowledge-socialisation processes during the liberalisation and privatisation phase (1990–97) between USIMINAS and CSN

KNOWLEDGE-SOCIALISATION PROCESSES	VARIETY		INTENSITY		FUNCTIONING		INTERACTION	
	USIMINAS	CSN	USIMINAS	CSN	USIMINAS	CSN	USIMINAS	CSN
In-house training (course-based): basic	Present (diverse)	Present (diverse)	Continuous	Continuous	Excellent	Moderate	Strong	Moderate
In-house training (course-based) • Routinised joint supervision of trainees • Self development programme (e.g. using company's videos; one-day training sessions) • Routinised 'training by in-house individuals' • Company-wide quality training	Present (diverse)	Present (moderate → limited)	Continuous	Continuous → Intermittent	Excellent	Moderate	Strong	Moderate
On the job training (OJT) • Coupled with job rotation • Mixing 'Godfathers' with new operators • Animated CD-ROM training • Blending in-house expertise for OJT modules	Present (diverse)	Present (limited)	Continuous	Continuous	Excellent	Good	Strong	Moderate
Training individuals from other steel companies	Present (diverse)	Absent	Continuous	–	Good	–	Moderate	–
Shared problem-solving: meetings • Cross-functional monthly ('funnelling') • Routine daily/weekly • For innovative projects	Present (diverse)	Present (limited)	Continuous	Continuous	Excellent	Good	Strong	Moderate
Interactive problem-solving practices	Present (diverse)	Absent	Continuous	–	Excellent	–	Strong	–
Knowledge-sharing links • 'Knowledge-bridging' individuals • Mixing links with meetings • 'Product ambassadors'	Present (diverse)	Present (limited)	Continuous	Continuous	Excellent	Moderate	Strong	Moderate
Team building	Present (diverse)	Present (limited)	Continuous	Intermittent	Excellent	Moderate	Strong	Moderate
Disseminating lead operators	Present (diverse)	Present (diverse)	Continuous	Continuous	Excellent	Good	Strong	Moderate/Strong

Source: Own elaboration based on the research.

Table 8.7 *Summary of the differences in key features of the knowledge-codification processes during the liberalisation and privatisation phase (1990–97) between USIMINAS and CSN*

KNOWLEDGE-CODIFICATION PROCESSES	VARIETY		INTENSITY		FUNCTIONING		INTERACTION	
	USIMINAS	CSN	USIMINAS	CSN	USIMINAS	CSN	USIMINAS	CSN
Standardisation practices • Basic standards • Detailed production procedures • Automation systems • Continuous updating of standards • Systematic codification of project engineering designs • 'Quality systems' manuals	Present (diverse)	Present (limited/moderate)	Continuous	Continuous	Excellent	Moderate	Strong	Moderate
Systematic manipulation of own codified knowledge	Present (moderate)	Absent	Continuous	–	Good	–	Strong	–
Reporting from external training • 'Innovative project proposals' • Circulating individuals' technical papers among peers	Present (diverse)	Present (limited)	Continuous	Intermittent	Good	Moderate	Strong	Moderate
Internal seminars • Interactive teleconferencing (plant + head office) • Brainstorming sessions • QCC conventions • Routinised technical symposiums (operations and support units)	Present (diverse)	Present (limited)	Continuous	Continuous	Excellent	Good	Strong	Moderate
Reintegrating individuals from external training • Team leaders • New project leaders	Present (diverse)	Present (limited)	Continuous	Intermittent	Excellent	Good	Strong	Moderate
Learning phrases and symbols	Present (diverse)	Absent	Continuous	–	Excellent	–	Strong	–

Source: Own elaboration based the research.

ficient to consolidate the process. Indeed most processes are still incipient in the company. In contrast, in USIMINAS during the 1990s, most processes had been routinised in an effective way. This may reflect the long period during which they had been present and continuously improved in the company.

As far as knowledge-codification processes are concerned, six processes were present in USIMINAS and four in CSN. Although four processes were present in both companies they differed in terms of the variety of mechanisms they contained. Most processes present in USIMINAS were of a diverse variety. In contrast, the four processes in CSN had a limited variety. For example, in CSN standardisation practices were still incipient. In contrast, they had become diverse and routinised in USIMINAS. In addition, as summarised in Table 8.7 and outlined in the previous subsections, they also differed in terms of intensity, functioning, and interaction.

NOTES

1. Interview with the general manager of the engineering superintendency.
2. Ibid.
3. Interview with an installations and planning engineer in the engineering unit.
4. Interview with the manager of the technical division of the steel shop.
5. Interview with the general manager of the engineering superintendency.
6. Paper by USIMINAS' president: R.C. Soares (1993),'The privatisation experience at USIMINAS', USIMINAS, December.
7. Interview with an installation planning engineer in the engineering superintendency. Also 'Engenharia. Sistema de controle de projetos', in USIMINAS: 22 anos de transferência de tecnologia [internal document, undated].
8. Ibid.
9. 'A experiência da USIMINAS no ciclo de transferência de tecnologia e propriedade industrial', technical assistance co-ordination [internal publication, undated].
10. Interview with USIMINAS's corporate planning manager. Also USIMINAS, Annual reports 1992–94.
11. Interview with USIMINAS's, corporate planning manager.
12. Mr Soares, from 1990 to date. Graduate in metallurgical engineering from the Ouro Preto School of Mines and Metallurgy (1963) holding a PhD in metallurgy from the University of Paris. Former researcher at the Institut de Recherches de la Siderurgie (IRSID). With USIMINAS since February 1971, starting as assistant to the department of industrial engineering then head of the industrial engineering, hot rolling, cold rolling, and metallurgy and inspection, general plant manager and operations director. Former president of the Brazilian Steel Institute (1992–96) and the Brazilian Metallurgy Association (1992). Curator of the Academy of Technological Sciences of Russia. Chosen in 1996 and 1997 as an outstanding corporate leader in Brazil.
13. Interview with the general plant manager.
14. Interview with a planning and installation engineer of the engineering unit.
15. Soares (1993). Interviews and archival records indicate that this replacement of shareholders (banks) involved a series of complex negotiations led by the USIMINAS's, president. Within two years their shares had been sold to other shareholders.

16. Interview with the corporate planning manager. Interview with an organisational analyst. Interview with the plant general manager and with a marketing manager.
17. Soares (1993).
18. Interview with the corporate planning manager. Interview with an organisational analyst. Also USIMINAS, organisational charts 1991, 1996, 1997.
19. Interview with the head of the metallurgical department.
20. The norm 'quality system requirements' was introduced in 1994 by large US car and truck makers (e.g. Chrysler, Ford, GM, Freightliner). The norm consists of requirements of the ISO 9001 added by automotive interpretations and elements of the Malcolm Balbridge Prize.
21. Interview with the head of the metallurgical department. Also paper by three USIMINAS's engineers in 'Aplicação dos requisitos da QS 9000 numa siderúrgica de laminados planos', *Metalurgia e Materiais*, vol. 53, no. 468, pp. 409-411.
22. Interview with the head of the Metallurgical Department. Also presentation given by USIMINAS's president in Tokyo: R. C. Soares (1997), 'A estratégia de desenvolvimento da USIMINAS diante da estabilização da economia brasileira', Tokyo, March 1997.
23. Collective interview with the manager and two engineers of the automation unit.
24. Ibid.
25. Ibid.
26. Interview followed by tour in the control room and the BF area with a technical analyst of the coke ovens and BFs (technical division of the coke ovens and BFs). Notes from direct-site observation.
27. Ibid.
28. Interview with the manager of the technical division of the steel shop. USIMINAS, Annual Reports 1995–96.
29. Interview with the plant general manager. Point also raised in different interviews.
30. Interview with a former process researcher. Head of the Steel Division, Association of engineers of the state of Minas Gerais.
31. Soares (1993). Also internal presentation by three USIMINAS's engineers: 'Ações para o atendimento da indústria automobilística' [copies of overheads, undated]; interview with the development director; interview with the marketing manager.
32. Soares (1993).
33. See 'GM: A ex-estatal está muito mais aberta ao diálogo', Relatório USIMINAS 30 anos, *Gazeta Mercantil*, 29 October 1992.
34. USIMINAS, Special Report, 1997. Annual Reports 1990–97.
35. See paper written by three USIMINAS's engineers, "Aplicação dos requisitos da QS 9000 numa siderúrgica de laminados planos", *Metalurgia & Materiais*, vol. 53, no. 468, pp. 409–11.
36. Interview with the manager of the metallurgical dept; interview with the manager of the research centre.
37. Interview with a product researcher. Also internal presentation by three USIMINAS's engineers: 'Ações para o atendimento da indústria automobilística' [copies of overheads, undated].
38. Interview with a product researcher at USIMINAS's research centre. Also internal presentation by three USIMINAS's engineers: 'Ações para o atendimento da indústria automobilística' [copies of overheads, undated].
39. Internal presentation by three USIMINAS's engineers: 'Ações para o atendimento da indústria automobilística' [copies of overheads, undated].
40. Interview with the marketing manager.

41. Interview with a product resercher at the research centre. Also 'Perspectivas na Siderurgia Brasileira para o Desenvolvimento da Indústria Automobilística no Brasil', ABM, São Paulo, 10 April 1996, mimeo.
42. These are pre-stamped sheets made from plates of different thickness and specifications, welded by laser.
43. Interview with a product researcher at USIMINAS's research centre.
44. Interview with a marketing manager.
45. *Metalurgia ABM*, October, 1997.
46. Interview with an engineer in the steel application unit. Also USIMINAS, Special Report, 1995.
47. Interview followed by plant tour with UMSA's capital goods manager.
48. Ibid.
49. Joint interview with UMSA's capital goods manager and with the quality guarantee manager. In the middle of its crisis, USIMEC lost the right to use its name to a small mechanical company in the state of São Paulo.
50. 'Breve história da USIMEC' [internal publication, undated].
51. 'Relatório USIMINAS 30 anos', *Gazeta Mercantil*, 29 October 1992.
52. Joint interview with UMSA's capital goods manager and the quality guarantee manager. See also 'Relatório USIMINAS 30 anos', *Gazeta Mercantil*, 29 October 1992.
53. Joint interview with UMSA's capital goods manager and the quality guarantee manager.
54. Interview with UMSA's capital goods manager. Also USIMINAS, Special Report, 1995.
55. Interview with USIMINAS's corporate planning manager. Also presentation given by three USIMINAS managers in: 'Ações para atendimento as indústrias automobilística e de autopeças. Enfoque USIMINAS [internal publication, undated].
56. Interview with USIMINAS's corporate planning manager.
57. Interview with a co-ordinator of special projects, CSN's general superintendency of technology.
58. Ibid.
59. Ibid.
60. Interview with an engineer in the technology nucleus. Interview with a co-ordinator of the steel technology nucleus of the general superintendency of technology. Also *Metalurgia ABM*, vol. 47, no. 394, March/April, 1991.
61. Interview with a co-ordinator of special projects of the general superintendency of technology.
62. Interview with a former manager of CSN.
63. Interviews # 1 and 2 with a senior engineer in the steel shop.
64. Interview with a co-ordinator of the steel technology nucleus. Also CSN, Annual Reports 1993–95.
65. CSN, Annual Reports 1993–97.
66. Interview with CSN's former president and other interviews in the company.
67. Metallurgical engineer, former head of the metallurgical division who started his career in CSN.
68. Interview with CSN's former president (who led the creation of the TAG).
69. Ibid.
70. Interview with CSN's former president. Other interviews and casual meetings in the company. Although that programme became known in CSN as 'TQC' it was comprehensive enough to be called TQM as done hereafter.
71. Interview with a manager at CSN's centre of quality promotion.
72. Interview with two managers at the hot strip mill. Also account by the HSM superintendent in 'TQC e piso de fábrica na laminação a quente da CSN' [internal publication, 1992].

73. See 'TQC toma conta da CSN', *Nove de Abril*, ano XII (158), September/October, 1991 [CSN's internal newspaper].
74. Interview with two engineers at the centre for quality promotion. CSN, Annual Reports 1992–96.
75. See 'DO: a qualidade é obrigação de todos', *Nove de Abril*, ano XIII (150), February, 1990.
76. Interview with an engineer in hot strip mill 2; interview with a manager at the hot coils finishing line. This was followed by a tour around the HSM 2 with an engineer. Also 'Laminador de Tiras a Quente # 2: 15 anos primando pela Qualidade', CSN/LTQ2 [internal publication, undated]. Interview with a manager at the hot coils finishing line.
77. Several interviews and casual meetings in CSN.
78. CSN, Annual Report 1996.
79. Interviews with managers in CSN.
80. Interview with a co-ordinator of the TQM implementation in the steel shop.
81. Interview with a product development manager. Interview with a steel shop technician. Interview with two engineers at the quality promotion centre.
82. This is a method to prevent workers' failures. Poka-yoke, or 'mistake proofing' in Japanese, is said to be created by Mr Shigeo Shingo from Toyota.
83. Interview with a manager in CSN.
84. CSN, Annual Report 1996.
85. Interview with a crew supervisor, casual meetings, and notes from direct-site observation. Although in the interview this was not said explicitly, it was possible to 'read between the lines', then confirmed through direct-site observation.
86. Interview with general manager of the research centre and other engineers in the plant.
87. Several interviews in the plant.
88. Interview with a manager and with an operator at the hot coils unit.
89. Interview with a manager at the hot strip mill. Interview with an engineer of the TQC promotion centre. Notes from direct-site observation.
90. Interview with a product development manager..
91. Interview with a manager at the research centre.
92. Interview with a product researcher. Interview with former manager of the research centre. Interviews in the operational units.
93. Interview # 1 with a manager at the research centre.
94. Interviews with managers and engineers in CSN. Also CSN, Annual Reports 1993–97.
95. CSN, Annual Reports 1990–97.
96. Interview with former manager of the research centre.
97. Interview with a manager and a product development researcher at the Research Centre.
98. Interview with a co-ordinator of special projects at CSN's general superintendency of technology. Also 'CSN: 50 anos transformando a face do país', *Metalurgia ABM*, vol. 47, no. 394, 1991. A few interviewees suggested that COBRAPI's deterioration was also influenced by the pressures on SIDERBRAS from local engineering firms.
99. Interview with a manager at the professional development centre.
100. Interview with a manager at the metallurgical department.
101. Interview with the head of the technical assistance unit.
102. Joint interview with UMSA's capital goods manager and with the quality guarantee manager.
103. Interview with CSN's former president.
104. Notes from direct-site observations.
105. Interview with a training analyst in the professional development centre. Also syllabuses of the professional training programmes (1992–97).
106. See CSN, Annual Reports 1992–96. Casual meeting with an engineer of CSN's quality promotion centre.

107. Interview and casual meeting with the former TIC's manager. Letter of resignation by former TIC's manager. Interview with the TIC's manager. Notes from direct-site observation.
108. Interview with the TIC's manager.
109. Ibid.
110. Ibid.
111. Casual meeting with CSN's librarian.
112. USIMINAS, Annual Reports 1990–97.
113. Interview with a manager at the professional development centre.
114. CSN, Annual Reports 1990–97. Several interviews within CSN.
115. Interview with the manager of USIMINAS's technical information centre. Search in the archival records.
116. Interview with a manager at the hot strip mill. Also CSN, Annual Reports 1990–96.
117. Soares (1997).
118. Interview with a manager at the technical division of coke and blast furnace dept. Interviews at the professional development centre.
119. CSN, Annual Reports 1990–97.
120. Interview with a manager at the research centre.
121. Interview with a marketing manager. Interview with a manager at the research centre. Interview with the manager of the technical assistance co-ordination. Also 'Por dentro da USIMINAS: o retrato de uma empresa' [internal publication, 1995].
122. Interview with a product researcher at CSN's research centre.
123. Interview with a manager at the technical assistance co-ordination. Also *Metalurgia ABM*, vol. 47, no. 394, p. 85, March/April 1991.
124. Ibid.
125. USIMINAS, Annual Report 1996.
126. Interview with the adviser to the CSN's superintendent director. Interview with a co-ordinator of the MSc programme at the UFF's School of Metalurgy at Volta Redonda. Also *Metalugia e Materias*, vol. 53, no. 467, p. 37, 1997.
127. Interview with a training analyst at the professional development centre. Also 'Por dentro da USIMINAS: o retrato de uma empresa' [internal publication, 1995].
128. CSN, Annual Reports 1990–97. Interview with a training analyst.
129. USIMINAS, organisational charts (1980s)
130. USIMINAS, organisational charts (1992–97)
131. Interview with the research centre's general manager.
132. Interview with a product researcher.
133. USIMINAS, Annual Report 1996.
134. Interview with the general manager of the research centre. Also copies of internal presentation overheads from the technical assistance co-ordination [1997]. Also *Vida Industrial*, August, 1994
135. Ibid.
136. Interview with a manager at the research centre. Also *Vida Industrial*, August, 1994; paper by an USIMINAS' engineer in Etrusco, S C P. (1997), 'A experiência da USIMINAS no ciclo de transferência de tecnologia e propriedade industrial', *R Esc Minas*, vol. 50, no. 1, pp. 25–9, January/March.
137. Interview with a manager at the research centre. Tour around the laboratories of the research centre with a process researcher.
138. Interview with a product researcher.
139. CSN, organisational charts (1980s).
140. Interview with a manager at the research centre.
141. Ibid.

142. Interview with a manager at the research centre. Interview with a product development manager. Point also raised in other interviews.
143. CSN, Annual Report 1997.
144. Copies from overheads of internal presentation in the research centre.
145. See 'O peso da realidade', Sindicato dos Engenheiros de Volta Redonda, July, 1997.
146. Interviews in the plant.
147. Interview with a technician of the hot strip mill, a crew leader, who enthusiastically described the experiments and its results. Also 'Análise de Ocorrência ILT 167/97' and 'Plano de Experiência ILT 019/97' [These are the experiment plan and report, provided by the technician, from which the confidential information was removed].
148. Interview with the general manager of the plant.
149. Ibid.
150. Interviews with managers, technicians and engineers.
151. Interview with the manager of the metallurgical department.
152. Interview with a manager at the technical assistance co-ordination. Also Etrusco (1997).
153. Interview with a hot strip mill manager. Interview with a manager and an operator of the hot coils mill. Notes from direct site observation.
154. Interviews in CSN.
155. Interview followed by tour in the reduction area with a blast furnace engineer.
156. 'CSN: 50 anos transformando a face do país', *Metalurgia ABM*, vol. 47, no. 394, 1991. Interview with an engineer at the quality promotion centre.
157. Interview with a manager at the professional development centre. Interview # 1 with a training analyst followed by tour in the centre for professional development. Also 'USIMINAS: Gerência de Seleção e Desenvolvimento de Pessoal', November 1997 [internal publication].
158. Interview with a manager at the professional development centre (plant).
159. Syllabuses of the professional training programmes (1997), professional development centre.
160. Syllabuses of the professional training programmes (1992–6).
161. Ibid.
162. Ibid. Interview # 2 with a training analyst.
163. Interview # 2 with a training analyst, followed by a tour around the professional development centre and observation of two videos made by USIMINAS.
164. Interview with a human resource analyst at the human resources management unit, corporate office.
165. Interview with a training analyst. Notes from direct-site observations. Also 'Por dentro da USIMINAS: o retrato de uma empresa' USIMINAS, 1995.
166. Ibid.
167. Casual meeting with an engineer of the corporate co-ordination unit; notes from direct-site observation; interview with a training analyst (CSN).
168. Interview with a senior training analyst.
169. Interview with a HSM manager. Observation of the CD-ROM training presentation given by a technician in that unit; notes from direct-site observation.
170. Interview with a training analyst. Interview with the plant general manager.
171. Interview with the manager of the technical division of the steel shop.
172. Interview followed by plant tour with UMSA's capital goods business manager.
173. Interview with a manager in the hot strip coils mill followed by plant tour with a Godfather operator
174. Interview with a training analyst. Also *USIMINAS Notícias*, November 1997.
175. Interview with a training analyst.
176. Interview with the general plant manager.
177. Interview with a training analyst. Interview with the manager of the metallurgical department.

178. Interview with a manager at the HSM, one of the leaders in the introduction of the TQC in CSN. Notes from direct-site observation from a meeting attended at the hot coils mill. A preparation for a similar meeting was observed in a different day at the cold strip mill.
179. Interview with the manager of the metallurgical dept. Interview with a training analyst at the professional development centre (plant). Also syllabuses of the professional training programmes (1990-97).
180. Interview with the technical assistance manager.
181. Collective interview with the manager and two engineers of the automation unit.
182. Interview with a product development researcher.
183. Interview with a manager at the professional development centre.
184. Interview with a former manager of the research centre.
185. Interview with a product researcher at CSN's research centre.
186. Notes from direct-site observation. Also 'CSN atinge a marca histórica de 100 milhões', *Metalurgia e Materias*, vol. 53, no. 467, July 1997.
187. Several interviews and notes from direct-site observations.
188. Interview with the general plant manager. Also 'USIMINAS. Princípios de Organização' [internal document, undated].
189. Interviews in the company and notes from direct-site observation.
190. Interview with a lead operator at the control cabin of the hot strip mill. Interview with a training analyst at the professional development centre (plant). Also syllabuses of the professional training programmes (1990-96).
191. CSN, *Nove de Abril*, ano XII, n. 160, November–December 1990.
192. Casual meeting with a blast furnace engineer. Also 'CSN atinge a marca dos 100 millhões', *Metalurgia e Materiais*, vol. 53, no. 467, July, 1997.
193. Interview with a manager of the TQM promotion centre.
194. Interview followed by plant tour with a 'Godfather' operator in the hot coils mill.
195. Interview with the manager of the metallurgical department. Several interviews in the plant.
196. Interview and plant tour with two engineers in the hot strip mill.
197. Interview with the manager of the metallurgical department. Interview with a lead operator at the control cabin of the hot strip mill.
198. Interview with an auditing manager.
199. Casual conversation with an operator in BF 3 during a tour guided by a technician of the technical division of the blast furnace and coke ovens dept. Notes from direct-site observation. Also 'USIMINAS, Automação de Alto-Fornos: 22 Anos de Transferência de Tecnologia' [undated].
200. Interview with the general plant manager.
201. Interview with a technician in the steel shop. Interview with USIMINAS's former president.
202. Interview with an engineer at the CSN's quality promotion centre.
203. Interview with na engineer at the CSN's quality promotion centre. Also 'Relatório de Implementação do TQC' [internal document, undated].
204. This is a tool that seeks to provide a neat organisation of the workplace.
205. Several interviews in the plant. Casual meeting with CSN's librarian.
206. 'Manual de Garantia da Qualidade da CSN (Controle de Documentos e Dados)' [internal document, undated].
207. Interview with a technician of the steel shop.
208. Interview with a manager at the HSM 2.
209. Interview with a former foreman of the steel shop.
210. Interview with a manager at the research centre.
211. Interview with a blast furnace engineer.
212. Interview with TIC's manager and notes from direct-site observation.

213. Interview with the manager of the professional development centre.
214. Ibid.
215. Interview with a training analyst (corporate office). Also various programmes of internal seminars 1990–97 [USIMINAS's archival records].
216. Talk given by the general plant manager and notes from direct-site observation at the USIMINAS's Quality Week (November 1997)
217. Interview with a crew supervisor at the steel shop. Interview # 2 with a former foreman of the steel shop and notes from direct-site observation.
218. Several interviews in CSN. Notes from direct-site observation.
219. Interview with a product researcher.
220. Interview with a cold strip mill engineer.
221. Interview with a training analyst at FUGEMSS.
222. Casual meeting with a BF engineer (former researcher).
223. Daruma, or Bodhidharma, was a Buddhist priest from India who lived in the sixth century. He is considered the founder of Zen in the Japanese culture. Darumas are used as votive offerings: the person makes a wish and paints one eye in; when the wish comes true, the other eye is painted in. Daruma is usually portrayed in red, with a scowling face and black beard. Daruma dolls, which are red, armless and legless, have weighted bottoms and always return to the upright position no matter how you throw them. This is said to symbolise patience, perseverance, and determination of the Indian priest. See Geisler, Harlynne (1997), 'Daruma; The Story Behind a Japanese Doll', *Story Bag; A National Storytelling Newsletter* (www.swiftsite.com/storyteller).
224. Interview with the manager of the metallurgical department. Interview with a training analyst. Notes from direct-site observation.
225. Interview with a manager at the research centre.

PART III

Analyses and Conclusions

Introduction

Part II of the book described the paths of technological capability-accumulation and the underlying learning processes in the two steel companies during their lifetime. Both companies have operated the same process technology and produced the same types of products. In addition, they were state-owned from their start-up to the early-1990s, went through similar phases in their lifetime, and since the 1960s, have been operating under the same external conditions. However, they have differed substantially in the way and rate at which they have accumulated levels of capability across different technological functions. The evidence examined in Part II suggests that these differences were strongly influenced by the learning processes used in the companies during their lifetime.

Part III develops analyses and draws conclusions from the evidence on technological capability-accumulation paths and the learning processes examined in Chapters 6 to 8. Part III is organised in three chapters. Chapter 9 explores the influence of the learning processes on the inter-firm differences in paths of technological capability-accumulation. These are interpreted in the light of the framework in Chapter 3. Chapter 10 analyses the implications of paths of technological capability-accumulation for inter-firm differences in operational performance over time. Finally, Chapter 11 closes the book by summarising its conclusions.

9. Cross-company Differences – Technological

9.1 TECHNOLOGICAL CAPABILITY-ACCUMULATION PATHS

Here the differences between USIMINAS and CSN in the paths of technological capability-accumulation across four technological functions are summarised: investments (involving facility user's decision-making and control and project planning and implementation); process and production organisation; product-centred; and equipment. Since Chapters 6 to 8 have already developed a summary of the differences within each phase, this section provides a more concise picture. The differences are summarised here on the basis of rates and consistency of accumulation.

In the light of the framework in Table 3.1, the differences in the rates of accumulation are summarised in Table 9.1. In general, USIMINAS took ten years to accumulate Levels 1 and 2 across all four technological functions. In parallel, USIMINAS proceeded, continuously, to the accumulation of technological capability beyond Level 4. Within 35 years USIMINAS had built up, accumulated, and deepened innovative capability at Level 5 (process and production organisation) and Level 6 (investments, product-centred, and equipment). In contrast, CSN took more than 45 years to complete the accumulation of Levels 1 and 2 routine capabilities, particularly for process and production organisation and product-centred activities. During more than 40 years CSN did not move beyond the accumulation of capability at Level 4, except for product-centred activity. In order to assist in the visualisation of the differences in the consistency of the paths followed by the two companies across different technological functions, this chapter draws on the framework in Figure 3.1. The paths are represented in Figures 9.1 to 9.4. These differences are briefly discussed below.

9.1.1 Investments Capability

USIMINAS followed a path of continuous accumulation of capability from Levels 1 to 6. Although USIMINAS did not move into accumulation of Level 7,

Table 9.1 Differences in the rate of technological capability-accumulation between USIMINAS (1962–97) and CSN (1946–97)[a] (approximate number of years needed to attain each level and type of capability)

Technological Capability Levels	Technological Functions							
	Investments		Process and Production Organisation		Product-centred		Equipment	
	USIMINAS	CSN	USIMINAS	CSN	USIMINAS	CSN	USIMINAS	CSN
Routine								
(1) Basic	10	15	10	45	10	40	10	20
(2) Renewed	10	15	10	50	10	50	10	45
Innovative								
(3) Extra basic	10	20	10	45	10	40	10	15
(4) Pre-intermediate	25	40	25	50	15	45	20	40
(5) Intermediate	30	0	35	0	25	50	30	0
(6) High-intermediate	35	0	0	0	35	0	35	0

Notes: (a) In this case, the initial years refer to the operations start-up year.

Source: Own elaboration based on the research.

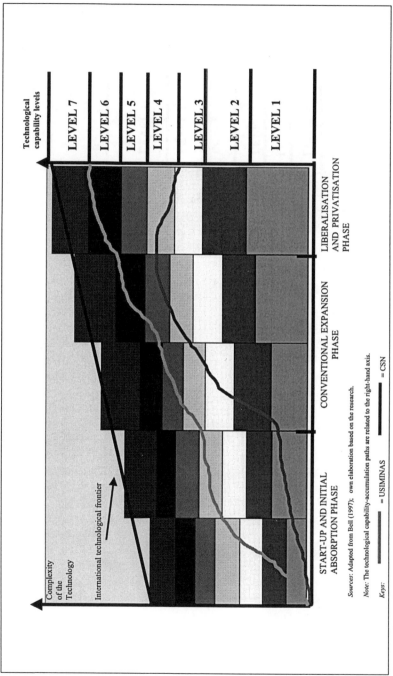

Technological capability levels

LEVEL 7
LEVEL 6
LEVEL 5
LEVEL 4
LEVEL 3
LEVEL 2
LEVEL 1

START-UP AND INITIAL ABSORPTION PHASE

CONVENTIONAL EXPANSION PHASE

LIBERALISATION AND PRIVATISATION PHASE

Complexity of the Technology

International technological frontier

Sources: Adapted from Bell (1997); own elaboration based on the research.

Note: The technological capability-accumulation paths are related to the right-hand axis.

Keys: ▬▬▬ = USIMINAS   ▬▬▬ = CSN

Figure 9.1   Cross-company differences in the paths of accumulation of capability for investments

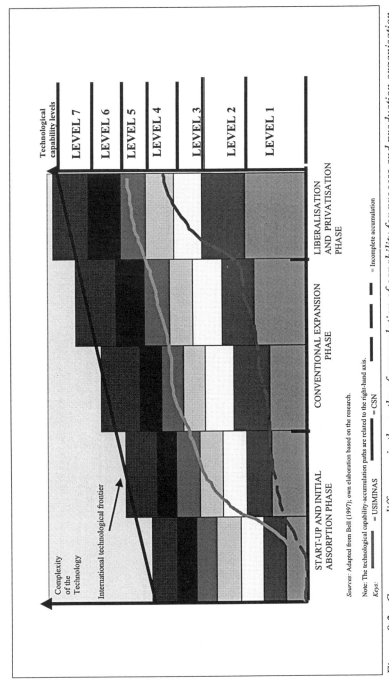

Figure 9.2 Cross-company differences in the paths of accumulation of capability for process and production organisation

Technological capability levels

LEVEL 7
LEVEL 6
LEVEL 5
LEVEL 4
LEVEL 3
LEVEL 2
LEVEL 1

Complexity of the Technology

International technological frontier

START-UP AND INITIAL ABSORPTION PHASE

CONVENTIONAL EXPANSION PHASE

LIBERALISATION AND PRIVATISATION PHASE

Sources: Adapted from Bell ((1997); own elaboration based on the research.

Note: The technological capability-accumulation paths are related to the right-hand axis.

Keys: ▬▬▬ = USIMINAS   ▬▬▬ = CSN   ▬ ▬ ▬ = Incomplete accumulation

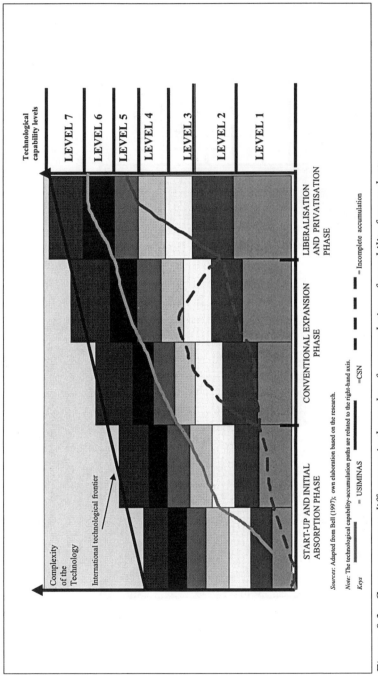

Sources: Adapted from Bell (1997); own elaboration based on the research.

Note: The technological capability-accumulation paths are related to the right-hand axis.

Keys ▬▬▬ = USIMINAS ▬▬▬ = CSN ▬ ▬ ▬ = Incomplete accumulation

Figure 9.3 Cross-company differences in the paths of accumulation of capability for products

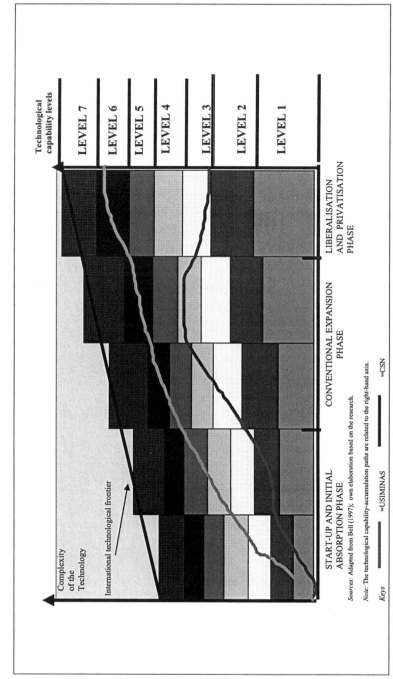

Figure 9.4   Cross-company differences in the paths of accumulation of capability for equipment

Technological
capability levels

LEVEL 7
LEVEL 6
LEVEL 5
LEVEL 4
LEVEL 3
LEVEL 2
LEVEL 1

Complexity
of the
Technology

International technological frontier

START-UP AND INITIAL
ABSORPTION PHASE

CONVENTIONAL EXPANSION
PHASE

LIBERALISATION
AND PRIVATISATION
PHASE

Sources: Adapted from Bell (1997); own elaboration based on the research.

Note: The technological capability-accumulation paths are related to the right-hand axis.

Keys   =USIMINAS   =CSN

230

the company sought to deepen and routinise Level 6 innovative capability. In contrast, CSN accumulated investments capability at Level 4. During the 1990s both companies reduced their in-house investment activities. USIMINAS moved into the deepening of a strategic part of that capability (e.g. basic engineering and overall project management). This permitted the company to be in full control of strategic investment activities and to increase the revenues from technical assistance provided for other companies. The evidence in Part II suggested that by the mid-1980s, CSN seemed to aim for the accumulation of Level 5. However, in the 1990s, CSN went through a more radical reduction in in-house investment activities. As a result, the company moved into a weaker position in relation to USIMINAS as far as full control and execution of strategic investment activities was concerned.

9.1.2 Process and Production Organisation Capability

USIMINAS moved from the accumulation of Levels 1 and 2 to Level 5. However, USIMINAS was not so fast at accumulating capability for processes as it was for products. The evidence in Part II suggests that, in parallel with the accumulation of innovative capabilities, USIMINAS's routine operating capability (Levels 1 and 2) was continuously strengthened over time. As a result, many of the innovative activities (e.g. 'capacity-stretching' or integrated automation) were supported by routine capability for process and production organisation. These activities were also associated with capability for equipment and investments (Level 4). In contrast, in CSN it was not until the early-1990s that the accumulation of Levels 1 and 2 capabilities were completed. This incomplete accumulation of basic operating capabilities must have constrained CSN's efforts during the 1950s–1980s period to move into the accumulation of capability at Levels 3 to 4. It was not until the 1990s that CSN moved into the accumulation of capability at Level 4.

9.1.3 Product-centred Capability

USIMINAS began by accumulating Levels 1 and 2 routine capability. In parallel, it moved into the accumulation and deepening of capability up to Level 6. As it proceeded into the accumulation of innovative capability for products, the company continuously strengthened capability at Levels 1 and 2. The evidence in Part II suggested that USIMINAS would not have achieved such a fast rate of product development capability (Levels 5 and 6) if routine capability at Levels 1 and 2 had not been adequately accumulated. In contrast, in CSN it was not until the 1990s that the accumulation of capability at Levels 1 and 2 for products was completed. Although CSN sought to accumulate innovative capability for products (Level 4 and beyond), this was achieved only slowly and inconsistently. As suggested by the evidence in Part II, one of the reasons for this inconsistency of accumulation in CSN was the absence of adequate accumulation of capability at Levels 1 and 2.

9.1.4 Equipment Capability

Both USIMINAS and CSN engaged in the accumulation of Level 1 routine capability for equipment. However, USIMINAS moved into the accumulation of innovative capability up to Level 6. In contrast, CSN accumulated capability up to Level 4. By the late-1980s both companies were affected by the crisis in Brazil's capital goods industry. As a result, by the early-1990s, both had reduced their in-house equipment activities. However, although USIMINAS (UMSA) reduced the scale of equipment activities in relation to the 1980s, it engaged in deepening the strategic part of its capability (e.g. equipment basic engineering, large equipment manufacturing, technical assistance in revamping engineering). In addition, USIMINAS/UMSA sought to stretch Levels 1 and 2 routine equipment capability into the early stages of car manufacturing. In contrast, CSN adopted a more radical reduction in its innovative equipment activities. As referred to in Chapter 8, by the 1990s, CSN had moved into a weaker position in relation to USIMINAS as far as capability for equipment activities was concerned.

9.1.5 Conclusions

The evidence on the differences in the paths of technological capability-accumulation followed by the two case-study companies suggests that:
1. Although operating within the same industry, using the same process technology and producing the same types of product, being state-owned from start-up to the 1990s, and going through similar phases during their lifetimes, USIMINAS and CSN followed different paths of technological capability-accumulation at differing rates. For instance, USIMINAS followed a consistent path of technological capability-accumulation in which technological capabilities at Levels 1 and 2 routine were built up, accumulated and sustained as the company moved into the building-up and accumulation of Levels 3 to 6 innovative capabilities. In contrast, CSN followed a path characterised by disconnection between the accumulation of routine capabilities (Levels 1 and 2) and innovative capabilities (beyond Level 3). In other words, the company engaged (1970s and 1980s) in (a) the building-up of innovative capabilities without having accumulated the capabilities to operate the plant efficiently; (b) intermittent accumulation of innovative capabilities (e.g. product development capability in the 1970s and 1980s); (c) weakening of innovative capabilities which had already been accumulated (Level 4 ca-

pability for investment and equipment); and (d) as a result, by 1997 CSN had ended up with a narrow range of innovative technological capabilities beyond Level 4 in contrast to the wide range of capability at Level 5 and 6 in USIMINAS.

2. As outlined in Chapter 3, differences in firms' paths of technological capability-accumulation can be expected. This issue has been pointed out in both the LCL (e.g. Lall, 1992, 1994) and the TFCL (Nelson and Winter, 1982; Dosi, 1988; Winter, 1988; Nelson, 1991). This book has added empirical evidence which moves beyond these two bodies of literature by reconstructing the paths followed by two companies. In doing so it has compared and contrasted them to illustrate how technological capability-accumulation differed across the companies in the long term, in terms of rates and consistency of accumulation across different technological functions.

3. The evidence, particularly from USIMINAS, confirms the arguments in the LCL (e.g. Dahlman et al., 1987; Lall, 1992; Hobday, 1995; Kim, 1997a,b) that latecomer companies can start with the barest production capability. On the basis of this, they can build other technological capabilities achieving high levels of technological development. However, since the latecomer company has to cope with a moving technological frontier, this book suggests that looking at the rate of technological capability-accumulation is even more critical than looking at whether or not the accumulation has occurred. This contributes to a more fruitful understanding of the paths of technological capability-accumulation in latecomer companies. This book has indicated that rates of technological capabilityaccumulation differ across technological functions within and between companies. Differences in the way and rate of accumulation of individual technological functions not only illustrates how difficult it is for the latecomer company to move towards the technological frontier. They also have critical implications for the overall characteristics of the paths followed.

For example, USIMINAS's successful entry and achievement of a leading position in the galvanised steel segment was associated with the strengthened Level 6 innovative (high-intermediate) capability for investment, combined with Level 6 innovative capability for products, and Level 5 innovative capability for process and production organisation. And the building of a JIT product delivery system (Level 6 product capability) depended on a highly integrated production system (Level 5 process and production organisation capability). In contrast, CSN's failure to improve the process control parameters in the steel shop may be associated with its weakened Levels 4 to 5 investment capability and the slow rate of accumulation of process and production organisation capability (Levels 1 to 4).

4. In addition, the evidence set out in Chapter 6 strongly suggests that the accumulation of the 'capacity-stretching' capability, or Level 4 capability for

process and production organisation, in USIMINAS had medium and long-term implications for its overall technological capability-accumulation path. This was associated with the fact that this capability was not limited to improvements within the original vintages of plant, but was extended to other new facilities. In other words, the capability accumulated through the operations and improvements in the original facilities contributed to the rapid absorption of the designed capacity and improvements in the new facilities. The cross-vintage accumulation of that capability was associated with (a) the accumulation of Level 4 capability for project engineering (e.g. basic, detailed, and installations engineering), permitting USIMINAS to understand better the intricacies of the process and equipment principles and design, which facilitated improvements in specific parameters; and (b) the accumulation of Level 4 capability for equipment, particularly to undertake large equipment revamping (e.g. blast furnace), permitting USIMINAS to understand in greater detail the 'stretching' possibilities of each new set of facilities and of different types of improvement in process, equipment, and production organisation. In sum, the 'capacity-stretching' capability became a 'platform capability' in USIMINAS with long-term effects. This book, therefore, confirms and extends the findings of previous studies that have explored the issue of 'capacity-stretching' capability (e.g. Dahlman and Fonseca, 1978; Maxwell and Teubal, 1980; Maxwell, 1982). This chapter returns later to the discussion of this issue in Section 9.2.5.

5. The points in (3) and (4) suggest that, at least in the steel industry, the long-term accumulation and sustaining of high-level innovative technological capability (Level 5 and beyond) for individual technological functions are influenced by the way and rate at which other types of capabilities are accumulated and sustained over time. In other words, particularly from Level 5, capabilities become highly interdependent. In addition, it is unlikely that the latecomer company can move further towards the technological frontier without accumulating and sustaining capabilities at the same high level across a wide number of technological functions. Therefore, capabilities across all five technological functions need to be accumulated and sustained in parallel.

6. The evidence in this book also suggests that, at least in the steel industry, the accumulation of routine operating capability (Levels 1 and 2) plays a critical role in the accumulation and sustaining of innovative capabilities. For instance, USIMINAS would not have achieved rapid accumulation of product development capability if it had not developed and strengthened over time Levels 1 and 2 routine capability for products and process and production organisation. Neither would the company have been able to engage successfully in the development of 'capacity-stretching' capability if it had not accumulated Levels 1 to 4 routine capability for investments. Indeed interviews with managers at different levels in USIMINAS even recognised the

presence of two trajectories of capability (operating and innovative) running inside the company. In contrast, in CSN, as indicated in Part II, the slow rate and inconsistent way of accumulation of product development capability was associated with the incomplete accumulation of routine operating capability (Levels 1 and 2) for products and process and production organisation.

However, the existing LCL has not given adequate attention to the accumulation of routine operating capability. The LCL has generally considered this capability as a 'minimum level' whose accumulation seems to take place in any case. Then the attention of the studies moves on to the accumulation of innovative capability. This book has moved further in relation to the LCL (a) by describing how routine operating capability is accumulated and sustained (or fails to be accumulated) over time; and (b) by highlighting the implications of its accumulation (or failure to accumulate) for the accumulation and sustaining of innovative capability.

7. The evidence from USIMINAS suggests that continuous improvement (CI) activities may evolve from the basic level up to the level where they become 'daily routine', as mentioned in different interviews in the company. This is in line with Bessant (1998). In contrast, in CSN these activities were scarce during the 1940s-1980s period. The company seemed to rely on the acquisition of new facilities and the delegation of improvement activities to external technical assistance. It was not until the 1990s that CSN engaged in improvement activities on a more continuous basis following the TAG and the TQM programmes. As pointed out in Caffyn (1997), TQM programmes are not necessarily part of CI, and they may also emerge as a response to crises and/or restructuring. This latter point seems to reflect the case of CSN in the 1990s. The company engaged in continuous process and production organisation improvements through the TQM programme as a response to a series of crises, rather than as a reflection of genuine and long-term concern with CI. This fact may have contributed to the weakening of the TQM programme from 1996, following the loss of top of management support for the programme and CSN organisation into business units.

8. Finally, the evidence also suggests no association between the conventional expansion of physical production capacity and the rate of innovative technological capability-accumulation. For instance, in USIMINAS, a fast rate of accumulation of Levels 1 to 4 innovative capability took place before conventional capacity expansion. In CSN conventional capacity expansion itself did not seem to have exerted any influence on the rate of technological capability-accumulation. Indeed the fastest rates of accumulation of technological capabilities in CSN's lifetime took place during the 1990–97 period, after the conventional expansion phase.

9.2   KEY FEATURES OF THE LEARNING PROCESSES

9.2.1 The Variety of the Learning Processes

The evidence in Part II suggests that, in general, USIMINAS engaged in a more diverse variety of learning processes during its lifetime than did CSN. Table 9.2 summarises the differences in the variety of the learning processes between USIMINAS and CSN.

During the 1956–72 period USIMINAS engaged in external and internal knowledge-acquisition processes and moved into the knowledge-socialisation and codification processes. During the 1950s-1980s period USIMINAS engaged continuously in all four learning processes. In contrast, CSN during the 1940s–1980s period had only engaged in three learning processes. In other words, it had not engaged in systematic knowledge-codification processes. It was not until the early-1990s, nearly 45 years after start-up, that CSN engaged in this process. In the 1990s, USIMINAS engaged in additional mechanisms within the existing processes and increased the variety of knowledge-codification processes. Although similar processes were present in both companies, they differed in terms of the diversity of sub-processes they contained. During the 1960s in USIMINAS, the variety within the learning processes was moderate to diverse. Over subsequent decades, the company developed a more diverse variety for all processes. In contrast, in CSN during the 1950s the variety was limited, and during the 1960s-1980s period the variety was between limited and moderate. It was not until the 1990s that variety within knowledge-acquisition processes increased from moderate to diverse in CSN. Although the variety within knowledge-socialisation and codification processes was still limited to moderate.

Table 9.2   Summarising the differences in the variety of the learning processes between USIMINAS and CSN (numbers in the cells refer to various sub-processes as described in Chapters 6 to 8)

Learning Processes	Start-up and initial absorption phase		Conventional expansion phase		Liberalisation and privatisation phase	
	USIMINAS (1956–72)	**CSN (1938–53)**	**USIMINAS (1973–89)**	**CSN (1954–89)**	**USIMINAS (1990–97)**	**CSN (1990–97)**
External knowledge-acquisition processes	9	6	9	8	9	8
Internal knowledge-acquisition processes	5	3	4	2	4	3
Knowledge-socialisation processes	4	3	8	3	9	7
Knowledge-codification processes	4	0	5	0	6	4
Totals	**22**	**12**	**26**	**13**	**28**	**22**

Source: Own elaboration based on the research.

The evidence suggests that by engaging continuously in a diverse variety of learning processes USIMINAS was able to accumulate technological capabilities across different technological functions. In contrast, CSN engaged in a less diverse variety of learning processes. This must have contributed to its limited accumulation of technological capabilities over time. In USIMINAS the diverse variety of learning processes was associated with fast rates of accumulation of capability for different technological functions. In contrast, in CSN, a limited variety of learning processes was associated with a slow rate of accumulation of technological capability. This evidence is in line with the framework in Chapter 3 whereby a diversity of learning processes is necessary to the building of innovative capabilities.

The evidence in Part II suggests that during the 1956–72 period, USIMINAS engaged in different learning processes that were not only related to the accumulation of capability to operate the plant. Some (e.g. overseas training for product development) were also related to the accumulation of capability to undertake innovative activities independently in the long term. In CSN, however, the concern for and actions related to the long-term development of in-house technological capability were absent during the 1940s–1970s period. Indeed the prolonged absence of critical learning processes (e.g. basic training for operators) must have exerted a negative influence on the rate of accumulation of Levels 1 and 2 routine capability in CSN in contrast to USIMINAS. The accumulation of this capability in turn had different implications for the rate of accumulation of innovative capability in both companies, as discussed earlier in Section 9.1.

9.2.2 The Intensity of the Learning Processes

These cross-company differences had implications for the paths of technological capability-accumulation in USIMINAS (1956-97) and CSN (1938-97). In USIMINAS during the start-up and expansion phases (1956–89), both knowledge-acquisition and knowledge-conversion processes worked on a continuous basis. In contrast, in CSN during these two phases (1938–89) most of the processes worked intermittently. In CSN it was not until the 1990s that these processes became more continuous. Table 9.3 summarises the differences in intensity of the learning processes in USIMINAS and CSN.

As mentioned in Chapter 3, the continuous intensity of learning processes may lead to their routinisation within the company. In USIMINAS the continuous intensity of a diverse variety of learning processes contributed to the routinisation of knowledge-acquisition processes. In parallel, this led to a routinisation of the knowledge-conversion processes. In other words, the continuous intensity of learning processes led to the routinisation of the conversion of individual into organisational learning. In contrast, in CSN during the

Table 9.3 Summarising the differences in the intensity of the learning processes between USIMINAS and CSN

LEARNING PROCESSES	Start-up and initial absorption phase		Conventional expansion phase		Liberalisation and privatisation phase	
	USIMINAS (1956–72)	CSN (1938–53)	USIMINAS (1973–89)	CSN (1954–89)	USIMINAS (1990–97)	CSN (1990–97)
External knowledge-acquisition processes	Continuous	Continuous/ Intermittent	Continuous	Continuous/ Intermittent	Continuous	Continuous/ Intermittent
Internal knowledge-acquisition processes	Continuous	Continuous	Continuous	Intermittent	Continuous	Continuous/ Intermittent
Knowledge-socialisation processes	Continuous	Intermittent	Continuous	Intermittent	Continuous	Continuous
Knowledge-codification processes	Continuous	–	Continuous	–	Continuous	Continuous/ Intermittent

Source: Own elaboration based on the research.

Table 9.4 Summarising the differences in the functioning of the learning processes between USIMINAS and CSN

LEARNING PROCESSES	Start-up and initial absorption phase		Conventional expansion phase		Liberalisation and privatisation phase	
	USIMINAS (1956–72)	CSN (1938–53)	USIMINAS (1973–89)	CSN (1954–89)	USIMINAS (1990–97)	CSN (1990–97)
External knowledge-acquisition processes	Moderate/ Good	Poor/ Moderate	Moderate/ Good	Poor/ Moderate	Good/ Excellent	Good/ Moderate
Internal knowledge-acquisition processes	Good	Poor/ Moderate	Good	Moderate	Good/ Excellent	Good
Knowledge-socialisation processes	Good	Poor	Good	Poor	Good/ Excellent	Moderate/ Good
Knowledge-codification processes	Good	–	Good	–	Good/ Excellent	Moderate/ Good

Source: Own elaboration based on the research.

238

1940s–1980s period, the intermittent nature of the learning processes, in association with their limited to moderate variety, contributed to constraining their routinisation in the company.

The evidence from CSN suggests that during this period, the intensity of knowledge-conversion processes was more intermittent than for knowledge-acquisition processes. In other words, although in CSN the use of some learning processes was intense during that period, they led to the accumulation of learning at the individual rather than organisational level. It was not until the 1990s, about 40 years after operations start-up, that this learning pattern began to change in CSN. However, the evidence from CSN in Part II suggested that the intensity of some learning processes became intermittent from 1996.

The evidence suggests, therefore, that in USIMINAS the continuous intensity of the learning processes contributed to a fast rate of accumulation followed by the sustaining of capability across different technological functions. In contrast, in CSN the intermittent nature of the learning processes, particularly knowledge-conversion processes, contributed to a slow rate of accumulation of capability across different technological functions.

9.2.3 The Functioning of the Learning Processes

The functioning of the learning processes had implications for the paths of technological capability-accumulation in USIMINAS and CSN. The evidence in Part II suggests that although some learning processes were present and some of them worked continuously in both companies, their functioning was different over time. Table 9.4 provides an overall assessment of the functioning of the four learning processes in USIMINAS and CSN over their lifetime.

As demonstrated by the evidence in Part II, in USIMINAS the functioning of the learning processes was continuously improved as the company engaged in more complex technological activities across the three phases (1956–97). In CSN, however, the evidence suggested that little improvement occurred in the functioning of the learning processes over the expansion phase (1954–89) in relation to the previous phase (1938–53). It was not until the 1990s that the functioning of the learning processes was improved in CSN. Even so, they were still different from USIMINAS's.

As indicated in Part II, in USIMINAS another characteristic of the functioning of the learning processes was their relative coherence across corporate levels and over time. In other words, processes were usually implemented at plant level in line with top management expectations (e.g. the training of operators by the Japanese; supervised OJT). In addition, in USIMINAS most of the mechanisms and processes created were further strengthened consistently over time (e.g. the research centre, overseas training, in-house training (course-based: basic, course-based, OJT, training provided to other companies)). This demonstrates consistency in the functioning over time. In contrast, in CSN, particularly during the 1940s–1980s period, some learning processes worked at plant level quite differently from what top management had formally expected (e.g. the training of operators by the Americans; supervised OJT; links between the researchers and engineers from the operations units). In the 1990s particularly from 1996, another inconsistency in CSN was illustrated by the deterioration of the provision of scholarships for research and teaching (e.g. the agreement between CSN and the School of Metallurgy of UFF, as described in Chapter 8). It should be remembered from Chapter 8 that although by 1993 CSN was committed to supporting the postgraduate courses in that school, by 1998 this commitment had diminished.

These discrepancies between the formality and reality of the learning processes in the companies reflect a gap between individuals' rationalised statements and what actually occurs in the firm, as referred to in Argyris and Schön (1978) and also in Dutrénit (2000). This is particularly observed when managers tend to over-estimate the good functioning of learning mechanisms during fieldwork interviews. This book suggests that discrepancies occur not only between managers' statements and reality, but also across the lifetime of the learning processes themselves. In other words, a process can be effectively built and it can work well during the first year, but from the second year, its functioning may deteriorate. Later, the process may disappear. In effect, un-learning may be as important as learning.

In addition, in USIMINAS (1960s) supervised OJT of operators took place in a continuous and rigorous way. In contrast, in CSN (1940s–1950s) supervised OJT functioned poorly, adding to the hostilities between the American instructors and the CSN operators. In addition, while in USIMINAS shared problem-solving was fruitful both between the Brazilians and between them and the Japanese, in CSN such knowledge-sharing (1950s-1980s) was constrained by the knowledge polarisation between engineers and operators and the deliberate retention of tacit knowledge at the engineers' level. Indeed the use of tacit knowledge as a basis for personal competitive advantage in CSN soon became a key constraining factor for knowledge-socialisation and codification processes. Its negative implications for organisational learning were also pointed out in Leonard and Sensiper (1998).

Another key characteristic of the functioning of learning processes in USIMINAS since the late-1950s has been the engagement in knowledge-acquisition processes before engaging in new technological activities. This 'learning-before-doing' process was applied in terms of an overall learning strategy. For example, the intensity of knowledge-acquisition and knowledge-conversion processes over the 1956–72 period seems to have worked as a basis for the company to engage in complex product development and process improvement over the 1970s and 1980s. By the 1990s, that feature of the learning process had been routinised also at the level of day-to-day activities through the 'never-do-without-knowledge' practice, as it was referred to in the company's language. This is in line with evidence from other companies where this type of learning functioning was present and a fast rate of accumulation of technological capability took place (e.g. Hyundai and Samsung) as studied in Kim (1995, 1997a). In CSN, however, this type of functioning was absent, particularly during the 1940s–1980s period.

As far as internal knowledge-acquisition processes are concerned, their different functioning had different implications for technological capability-accumulation in USIMINAS and CSN. In USIMINAS, individuals developed the practice of not only doing routine operational activities, but engaging continuously in innovative and more complex activities. Its positive implications for the fast rate of capability-accumulation was also referred to in other empirical studies (e.g. Ariffin and Bell, 1996; Kim, 1995, 1997a). The implications of these activities for USIMINAS's employees were an increased understanding of the principles involved in the technology (Bell, 1984) and increased confidence and psychological boost in manipulating it (Maxwell, 1982). This must have played a substantial role in the accumulation of capability for routine plant operations and also for continuous improvements in process, production organisation, products, and equipment. In contrast, CSN's individuals went through the expansion phase within the pattern of 'learning-by-doing' routine operational activities. The evidence in Part II has suggested that this functioning of the internal learning process in CSN contributed very little to the accumulation of Levels 1 and 2 routine operating capability during the 1940s–1980s period.

In sum, the functioning of the learning processes in USIMINAS was more effective than in CSN, particularly during the start-up and expansion phases. This difference in the functioning was one of the key contributors to the differences in the paths of technological capability-accumulation followed by the companies during those two phases. In USIMINAS during the 1990s the review and continuous improvement to the functioning of the learning processes are associated with the sustaining of existing capabilities, or their strategic elements. In CSN the improvement in the functioning of some learning processes must have played a critical role in the completion of accumulation of Levels 1 and 2 routine capability for products and process and production organisation.

9.2.4 The Interaction of the Learning Processes

These differences had implications for the paths of technological capability-accumulation in USIMINAS and CSN. Table 9.5 summarises them over time.

Table 9.5 Summarising the differences in the interaction of the learning processes between USIMINAS and CSN

LEARNING PROCESSES	Start-up and initial absorption phase		Conventional expansion phase		Liberalisation and privatisation phase	
	USIMINAS (1956–72)	CSN (1938–53)	USIMINAS (1973–89)	CSN (1954–89)	USIMINAS (1990–97)	CSN (1990–97)
External knowledge-acquisition processes	Moderate	Weak	Moderate/Strong	Weak	Strong	Moderate
Internal knowledge-acquisition processes	Moderate	Weak	Moderate/Strong	Weak	Strong	Moderate
Knowledge-socialisation processes	Moderate	Weak	Moderate/Strong	Weak	Strong	Moderate
Knowledge-codification processes	Moderate	–	Moderate	–	Strong	Moderate

Source: Own elaboration based on the research.

The evidence in Part II and the summary in Table 9.5 suggest that the learning processes interacted in different ways over time in the two companies. This had implications for the conversion of individual into organisational learning and in turn for the paths of technological capability-accumulation followed by USIMINAS and CSN. In USIMINAS since the start-up phase (from 1956), the external knowledge-acquisition processes (e.g. overseas training) began to influence internal learning processes (e.g. continuous efforts on 'capacity-stretching'). In turn, knowledge-socialisation processes (internal seminars) were influenced by internal knowledge-acquisition processes (e.g. presentations on individuals' experience in daily problem-solving) and external knowledge-acquisition processes (e.g. pulling in expertise from outside; channelling external codified knowledge into the company). In addition, the knowledge acquisition-processes triggered knowledge-socialisation processes which, in turn, triggered knowledge-codification processes. As a result, knowledge and skills began to be embedded in the organisational system of the company rather than only in individuals' minds.

In contrast, in CSN, particularly during the start-up and expansion phases (1938–89), there was weak interaction between the learning processes. For example, although CSN had been engaged in external knowledge-acquisition processes (e.g. overseas training; pulling in expertise from outside), these barely influenced the creation of other learning processes. Additionally, they did not trigger knowledge-socialisation and codification. As suggested during several interviews in the company, one of the key characteristic of CSN during the 1940s-1980s period was its limited ability to bring different tacit knowledge into a workable system. In addition, although different learning processes were present in CSN during that period, their weak interaction enabled very little contribution to the conversion of individual learning into organisational learning. This in turn contributed to the slow rate of accumulation of technological capability in the company during that period. It was not until the 1990s, that the interaction between learning processes moved from weak to moderate in CSN.

Although CSN improved interaction between the learning processes from the early-1990s, the evidence from 1996 suggests that certain factors were a threat to that interaction. Among them was the weakening of the TQM programme and the creation of the business units organisation. The mindset of business units, as referred to in Nonaka and Takeuchi (1995), may have contributed to inhibiting the knowledge-socialisation processes.

In contrast, USIMINAS demonstrated an ability to bring different tacit knowledge together into a workable system at different points in time. By the 1990s, the internal knowledge-acquisition processes in USIMINAS had begun to work in a more interactive way compared to the previous phases. This seems to reflect the decentralisation of the research centre's activities into the operations units. For example, by the 1990s the practice of experimentation coupled with team-building had been spread across the plant. In CSN, however, although some experimentation was present, it did not work in an interactive way as happened in USIMINAS. As pointed out in Bessant and Caffyn (1997), the participation of a wide group of individuals in creative problem-solving and innovative activities contributes to accelerating CI within the company. Additionally, in-house experimentation has also been seen as a powerful mechanism for the sustaining of existing capability, while lack of experimentation may work as an inhibitor to the building of new capability (Leonard-Barton, 1995).

In sum, in USIMINAS (1956–97) several learning processes (e.g. internal seminars, channelling of external codified knowledge, in-house training programmes) were created and functioned effectively over time in association with the interaction between these processes. These interactive processes led to the fast conversion of individual into organisational learning and a rapid accumulation and sustaining of capability across different technological functions. In contrast, in CSN, particularly during the 1940s–1980s period, several processes were present but worked in isolation. This contributed to a failure to convert individual into organisational learning and a slow and inconsistent accumulation of technological capability. Therefore, interaction between learning processes is another key feature influencing the paths of technological capability-accumulation

9.2.5 Conclusions

The analysis above suggests that:
1. The differences in the key features of the learning processes between USIMINAS and CSN had implications for the composition of the capabilities that were built and accumulated over time. In other words, during the start-up and expansion phase in USIMINAS (1956-89) technological capability was accumulated in a way that pervaded the whole organisational system of the company. As a result, by the 1990s this pattern of accumulation had become so routinised that USIMINAS's technological capability could hardly be separated from the corporation as a whole. This reflects the effective routinisation of the process of conversion of individual into organisational learning. In contrast, in CSN during the 1940s–1980s period, technological capability was accumulated in a narrower way. In other words, the composition of technological capability being accumulated hardly involved the organisational system of the company. This reflects the failed conversion of individual into organisational learning. It was not until the 1990s that this pattern of accumulation began to change.

   For instance, in USIMINAS technological capabilities were built up and accumulated over time involving the creation, upgrading, re-organisation, and routinisation of different organisational units, organisational and managerial flows and arrangements. These were particularly concentrated on improvements in process and production organisation, equipment, and products, and project engineering. It should be remembered that several innovative activities in equipment, processes and production organisation (e.g. 'capacity-stretching') took place in USIMINAS over the 1960s and early-1970s before the research centre was set up. This suggests that innovative activities can be undertaken successfully in different operations units if production support units and organisational/managerial arrangements in the operations units are built and work effectively, as was the case in USIMINAS. In addition, during the 1970s and 1980s USIMINAS was able to build an effective integration between the research centre and the operational units for product development. From the late-1980s and over the 1990s, the company was able to build up plant-wide integration across the operations units on the basis of automated systems developed in- house.

CSN, on the one hand, did not engage in the building and upgrading of organisational units for improvements in the operational units. On the other hand, although certain organisational units had been built – and formalised on organisational charts – they did not engage in continuous improvement efforts in the production lines. Indeed the absence in CSN of a unit dedicated to automated system control and development contributed to constraining the overall development of process and product capabilities during the 1950s-1980s period. In addition, during this period technological capabilities in CSN were skills residing in engineers and researchers. As a result, capabilities were weakened as these individuals left the company. In contrast to USIMINAS, CSN failed to integrate the operational unit with the research centre for product development. This had negative implications for the rate of product development capability. However, from the early-1990s, CSN began to follow a path of technological capability-accumulation marked by a broader composition of technological capability. This was associated with several improvements in processes and products and the implementation of the TQM programme.

Returning to the discussion of the accumulation of 'capacity-stretching' capability in USIMINAS, developed in Section 9.1.5, the evidence suggests that this capability was accumulated over time and across vintages of plant. Earlier empirical studies (e.g. Hollander, 1965), however, seem to suggest that the capability for continuous improvements within one vintage would not be appropriate for improvements in different vintages. On the basis of the evidence from USIMINAS, this book suggests that, at least in the steel industry, this capability may have a pervasive and long-term effect on improvements across vintages if the company is able: (1) to build on the technical and organisational/managerial improvements undertaken in the previous facilities; (2) to accumulate, in parallel, the capability for different technological functions; and (3) to achieve continuous intensity, good functioning,

246 Technological Learning and Competitive Performance

and interaction of diverse learning processes to convert individual into organisational learning.

Therefore, following earlier studies in the LCL (e.g. Bell, 1982; Tremblay, 1994; Bell and Pavitt, 1995; Dutrénit, 2000) this book confirms the importance of the accumulation of technological capability in a broad way. In other words, it confirms the critical role played by the organisational and managerial components of capability in accelerating the rate of improvements across different technological functions. The importance of these components of capability has also been widely recognised in the TFCL (e.g. Leonard-Barton, 1992a, 1995; Tidd et al., 1997; Bessant, 1998). In addition, the book confirms the importance of integrating different types of capability across units for product development performance (Clark and Fujimoto, 1991; Iansiti and Clark, 1994) and process improvement performance (Pisano, 1997). This is in line with the argument that companies that are able to integrate their capabilities increase competitive advantage (Tsekouras, 1998), an issue that will be explored further in Chapter 10.

2. The evidence in the book suggests a strong association between the features of the learning processes and the way, and rate at which USIMINAS and CSN differed in their paths of technological capability- accumulation. The case of USIMINAS suggests that the company engaged in continuous and deliberate manipulation of key features of the learning processes. Major changes in the learning processes were followed by incremental improvements. This permitted the company to accelerate the rate of accumulation of capability across different technological functions, as summarised in Table 9.1 and represented in Figures 9.1 to 9.4. In contrast, in CSN over four decades (1940–80) little improvement occurred in the learning processes. As a result, CSN proceeded along a slow path of technological capability-accumulation. Specifically, on the basis of the framework in Argyris and Schön (1978) and Senge (1990) the case of USIMINAS suggests that the company combined 'double-loop' or 'generative' with 'single-loop' learning, over time. In contrast, in CSN during the 1940s-1980s period, it seems that only a sort of 'single-loop' learning was dominant. It was not until the 1990s that CSN engaged in 'double-loop' learning by restructuring its learning processes.

Therefore, the move from one level to another of technological capability-accumulation is associated with a deliberate 'discontinuity' in key features of the learning process. In other words, unchanged features of the learning processes over time will contribute very little to a company's ability to move, say, from Level 3 to 4 and then to 5. The earlier the company engages in this deliberate manipulation, as USIMINAS did, the faster the rate of technological capability-accumulation. This also applies to the sustaining of existing technological capability. The evidence in Chapter 8 suggests that if USIMINAS, during the 1990s, had not reviewed the existing learning processes, most of its existing technological capabilities would not have been sustained.

The experience of CSN during the 1990s suggests that the improvement in key features of the learning processes contributed to completing the accumulation of Levels 1 and 2 routine capability. It also contributed to the accumulation of Level 4 innovative (process and product organisation) and Level 5 (products) on a continuous basis. However, as indicated in Chapter 8, from 1996 the company began to face difficulties in being able to guarantee an adequate variety, intensity, functioning, and interaction of some of the learning processes, particularly knowledge-socialisation and codification processes. These difficulties might reflect the length of time during which learning had been accumulated at the individual level. Only in future years will it be possible to know the implications of this fact for the path of capability-accumulation in CSN.

3. The book also suggests that external and internal knowledge-acquisition processes are critical for latecomer companies since they have no previous knowledge base on which to draw. Both knowledge-acquisition and knowledge-conversion processes are critical, complex, and costly for technological capability-accumulation in latecomer companies. However, even when effective, relying only on the former is not sufficient to accelerate the rate of technological capability-accumulation. Therefore, the latecomer company needs to move, in parallel, into knowledge-conversion processes. Their building, however, seems to be even more complex, challenging, and risky for the latecomer firm. The reason is that knowledge-conversion processes do not flow straight from knowledge-acquisition processes. The possibility of failure seems greater for knowledge-conversion than for knowledge-acquisition processes. This seems associated with the difficulties of overcoming internal resistance to ensure coherent functioning and continuous intensity across corporate levels and over time. This is why effective learning processes are associated with deliberate daily in-house efforts by the firm.

4. The experience of the two case-study companies suggests that as different learning processes are built over time inside firms they, and their features, lead to the development of a 'learning system'. This is associated with the intricate and interdependent nature of the learning processes, as examined in this book. The evidence in Part II suggests that learning processes and their features have to be viewed as a 'whole' by the company. This 'organic system' perspective on the learning processes is in line with recent studies in the TFCL (e.g. Senge, 1990; Garvin, 1993; Leonard-Barton, 1990, 1992a, 1992b, 1995; Teece and Pisano, 1994) and the LCL (Dutrénit, 2000). However, the evidence in this book suggests that 'learning systems' may evolve in

an effective way, as in the case of USIMINAS, or in an ineffective way as in the case of CSN (particularly during the 1940–80 period). These different 'learning systems' have different practical implications for the paths of technological capability- accumulation and, in turn, the rate of operational performance improvement. The book reviews the last component of this set of relationships in Chapter 10. The interaction between all three components of this set of relationships is reviewed in Chapter 11.

10. Cross-company Differences in Operational Performance Improvement

10.1 APPROACH AND MAIN PROCEDURES FOR THE ANALYSIS

Operational performance improvement is a critical issue for companies in general. The issue seems even more critical for latecomer companies since they start with levels of performance far below world standards. To catch up with international levels of performance, their rates of performance improvement have to grow faster than the rates of companies operating at the technological frontier. The achievement of world competitive performance depends on how fast they accumulate their technological capability (Bell et al., 1982, 1995). As indicated in Chapter 2, only a few studies in the LCL have systematically investigated the evolution of operational performance improvement associated with the accumulation of technological capability (e.g. Dahlman and Fonseca, 1978; Katz et al., 1978; Bell et al., 1982; Mlawa, 1983; Tremblay, 1994).

This chapter seeks to cover as long a period as possible in the lifetime of the two case-study companies. The analysis draws on different indicators of performance in the two integrated steel plants. The chapter combines descriptive quantitative evidence with qualitative evidence to explain inter-firm differences in operational performance improvement. Most of these differences are expressed in terms of changes in the level of indicators and their rate of improvement. The cross-company comparison of the performance of individual facilities (BF and steel shop) is structured on the basis of time-periods in the facilities' lifetimes rather than phases. The reason for this is that the operational units started at different dates within and across the two companies. However, as demonstrated in Part II, the two companies have followed different paths of technological capability-accumulation during three phases. Therefore, the analysis here also refers to these phases. Specific procedures for comparison are outlined in more detail in each section.

10.2 OPERATIONAL PERFORMANCE IMPROVEMENT

The analysis of cross-company differences in operational performance is based on fourteen different indicators. These are organised in three groups: (1) indicators of the ironmaking process performance (Section 10.2.1 below); (2) indicators of the oxygen steelmaking process performance (Section 10.2.2); and (3) overall plant performance indicators (Section 10.2.3).

10.2.1 Ironmaking Process Performance

This subsection analyses in particular, BF performance, which is normally examined on the basis of three indicators: (1) coke rate (kg/tonne of pig iron); (2) BF productivity (tonnes/m^3/day); and (3) hot metal quality.

In order to provide a clear perspective of the evolution of these indicators, they are initially examined on the basis of two time-periods: (1) the initial 10-year period, covering the start-up year to the tenth year of operation (Y_1 to Y_{10}) approximately; (2) over a longer period (Y_1 to 1989), thus the comparative tables will contain overlapping years. In this way the comparison will be covering roughly the start-up and initial absorption phases. The indicators are then compared during the 1990–97 period. This is related to the liberalisation and privatisation phase. The comparison for coke rate and BF productivity takes into consideration the change in the level of indicators and the average annual rate of reduction and/or increase (per cent per year). Although the three indicators are analysed separately, this section interprets BF performance as a whole not on the basis of individual indicators. In other words, it considers all three indicators to obtain a meaningful inter-firm comparison.

Coke rate (kg/tonne of pig iron)

Coke rate is the amount of coke consumed per tonne of pig iron produced. It should be remembered from Chapter 4 that coke is a critical input into the BF. It represents about 70 per cent of the total cost of raw materials. The coke rate level can be affected by the vintage of the technology embodied in the furnace. Additionally, it can be affected, positively or negatively, by factors such as coal quality, refractory conditions, process activities (e.g. manipulation of the burden preparation and distribution, etc.), and external conditions (e.g. strikes, raw materials supply, energy crisis, etc.). Coke rate tends to increase slowly during the BF campaign as the result of natural alterations in the internal refractories. Although today it would be recommended to assess the fuel rate (coke + injected fuel rates), this subsection analyses coke rates more systematically than fuel rates.

Blast furnace 1   The USIMINAS BF 1 was built in the late-1950s, embodying a later vintage of technology than BF 1 at CSN which had been built in the early-1940s. Consequently, the initial levels of performance differed. However, the key issue here is the rate of change from those differing initial levels. The differences in coke and fuel rates reduction between USIMINAS and CSN for BF 1 are outlined in Table 10.1.

Table 10.1 Differences in coke and fuel rates reduction between USIMINAS and CSN: blast furnace 1

	Reduction in the coke and fuel rates level (kg/tonne)	Average annual rate of reduction (per cent per year)
USIMINAS[a]		
Coke rates BF 1 (1962–72)	731 → 485	–4.02
BF 1 (1962–89)	731 → 446	–1.81
BF 1 (1990–97)	489 → 399	–2.86
Fuel rates BF 1 (1970–76)	561 → 479	–2.59
CSN[a]		
Coke rates BF 1 (1951–60)	792 → 814	+0.30
BF 1 (1951–89)	792 → 565	–0.88
BF 1 (1990–91)	523 → 514	–1.70
Fuel rates BF 1 (1970–76)	674 → 690	+0.39

Note: (a) The years 1946–50 and 1992 are not being considered. As mentioned in the company's documents, during the 1940s CSN had an irregular coal supply. This probably had adverse effects on the coke rate level. In January 1992 the furnace was shut down permanently.
Source: Own elaboration based on the research.

It should be noted that during the 1970–76 period, as indicated in Table 10.1, under the first energy crisis, the fuel rates in USIMINAS fell from 56kg to 479kg or by 2.59 per cent annually on average. In contrast, in CSN the fuel rate *increased* from 674 to 690 or by 0.39 per cent annually.[1] As far as the coke rate over the 1990–97 period is concerned, in USIMINAS it fell from 489 to 399 or by 2.86 per cent on average. In CSN from 1990 to 1991, it dropped from 523 to 514 or by 1.71 per cent. Therefore the differences indicate that USIMINAS was substantially faster than CSN in the rates of coke and fuel rates reduction.

Blast furnace 2   Again the initial levels of performance in USIMINAS's 1965 vintage plant were higher than in CSN's 1954 vintage plant. However, the subsequent rates of improvement differed. The differences in coke rate reduction between USIMINAS and CSN for BF 2 are outlined in Table 10.2.

It should be remembered that during the 1970s both companies were operating under the same energy crises. During the 1980s, the fuel rates in that furnace in USIMINAS (1980–89) dropped from 486 to 446kg or by 0.95 per cent on average. In contrast, in CSN (1982–89) it *increased* from 553 to 562kg or by 0.23 per cent on average. Thus, from the early-1960s to the late-1980s the USIMINAS's BF 2 outperformed CSN's in the rate of reduction of coke and fuel rates. During the 1990s, the coke rate in USIMINAS fell from 491 to 397 or by 2.99 per cent annually. In CSN it dropped from 542 to 412 or by 3.8 per cent

Table 10.2   Differences in coke and fuel rates reduction between USIMINAS and CSN: blast furnace 2

	Reduction in the coke and fuel rates level (kg/tonne)			Average annual rate of reduction (per cent per year)
USIMINAS [a]				
Coke rates BF 2 (1965–74)	612	→	468	–2.93
BF 2 (1965–89)	612	→	446	–1.31
BF 2 (1990–97)	491	→	397	–2.99
Fuel rates BF 2 (1970–74)	571	→	511	–2.73
BF 2 (1970–78)	571	→	479	–2.17
BF 2 (1980–89)	486	→	446	–0.95
BF 2 (1991–97)	486	→	500	+0.47
CSN [b]				
Coke rates BF 2 (1954–64)	809	→	657	–2.06
BF 2 (1954–89)	809	→	562	–1.03
BF 2 (1990–97)	542	→	412	–3.84
Fuel rates BF 2 (1970–74)	612	→	598	–0.57
BF 2 (1970–76)	612	→	629	+0.45
BF 2 (1982–89)	553	→	562	+0.23
BF 2 (1991–97)	494	→	523	+0.95

Notes: (a) The furnace was shut down for revamping during the 1975–77 period.
(b) The furnace was shut down for revamping and reconstruction during the 1977–81 period.
Source: Own elaboration based on the research.

annually. In this case, CSN BF 2 outperformed USIMINAS in the rate of coke rate reduction, although the level of the indicator in CSN was still higher than in USIMINAS.

As far as fuel rates over the 1990s are concerned, the evidence in Table 10.2 indicates that they increased in both companies. However, in USIMINAS (1991–97) they increased from 486 to 500kg or by 0.47 per cent. In contrast, in CSN (1991–97) they increased from 494 to 523kg or by 0.95 per cent. This is twice the rate of increase in USIMINAS. It should be noted that from 1995 both companies were using pulverised coal injection (PCI) technology.[2] This suggests that USIMINAS was more effective than CSN in the yield of the coal fines injection process. In other words, the evidence suggests that during the 1990s USIMINAS was more effective than CSN in managing the fuel rate increase.

Blast furnace 3   In this case the vintages of plant are similar, but the levels and rates of performance improvement differed across the two companies. The differences in coke rate reduction between USIMINAS and CSN for BF 3 are outlined in Table 10.3. In USIMINAS during the four initial years 1975–79 the coke rate fell from 486 to 466kg or by 1.04 per cent annually on average. In contrast, in CSN during the two initial years 1977–79 the coke rate *increased* from 509 to 513 or by 0.39 per cent annually. During the 1980–89 period in

USIMINAS the coke rate fell at the rate of 0.82 per cent annually, while in CSN during the same period it fell at the rate of 0.54 per cent. However, if the longer period of 1975–89 for USIMINAS is considered, coke rate even *increased* from 486 to 491kg or by 0.07 per cent on average.[3] In contrast, in CSN during the 1977–89 period the coke rate fell from 509 to 475kg or by 0.57 per cent on average.

During the 1990–97 period in USIMINAS the coke rate fell from 492 to 407 or by 2.67 per cent annually on average. In contrast, in CSN, over the same period the coke rate fell from 492 to 386 or by 3.4 per cent on average. Therefore, during the 1980s and 1990s CSN BF 3 outperformed USIMINAS BF 3 in the level and rate of coke rate reduction. It should be noted that during the 1970s the USIMINAS BF 3 achieved higher rates of coke rate reduction than CSN. The occurrence of a hard crust in the first campaign of USIMINAS BF 3 might have been one the factors explaining the slow rate of coke rate reduction during the 1980s.

Table 10.3   Differences in coke and fuel rates reduction between USIMINAS and CSN: blast furnace 3[a]

	Reduction in the coke and fuel rates level (kg/tonne)	Average annual rate of reduction (per cent per year)
USIMINAS		
Coke rates BF 3 (1975–79)	486 → 466	−1.04
BF 3 (1975–89)	486 → 491	+0.07
BF 3 (1980–89)	529 → 491	−0.82
BF 3 (1990–97)	492 → 407	−2.67
Fuel rates BF 3 (1975–79)	519 → 508	−0.53
BF 3 (1975–89)	519 → 491	−0.39
BF 3 (1980–89)	533 → 491	−0.90
BF 3 (1995–97)	511 → 510	−0.09
CSN		
Coke rates BF 3 (1977–79)	509 → 513	+0.39
BF 3 (1977–89)	509 → 475	−0.57
BF 3 (1980–89)	499 → 475	−0.54
BF 3 (1990–97)	492 → 386	−3.40
Fuel rates BF 3 (1977–79)	509 → 520	+1.07
BF 3 (1977–89)	509 → 496	−0.21
BF 3 (1980–89)	500 → 496	−0.08
BF 3 (1995–97)	489 → 508	+1.92

Note: (a) For a meaningful comparison, and since the furnaces started in December 1974 in USIMINAS and in May 1976 in CSN, their start-up years are not considered.
Source: Own elaboration based on the research.

It should be noted that during the 1990s, particularly during the 1995–97 period, both companies were using the PCI technology in BF 3. However, in

USIMINAS the fuel rate dropped from 511 to 510 or by 0.09 per cent, while in CSN it *increased* from 489 to 508 or by 1.92 per cent annually on average. Although by 1997 the level of the indicator in the USIMINAS BF 3 was higher than in CSN. However, in USIMINAS the indicator was declining, while in CSN it was increasing. This suggests that, as in the case of BF 2, USIMINAS was more effective than CSN in the process of injecting coal fines in BF 3.

The evidence indicates that USIMINAS had continuously been achieving coke rates below 500kg since 1972.[4] In contrast, it was not until 1992 that CSN began to achieve coke rates consistently below 500kg. From the perspective of the companies' age, in USIMINAS coke rates below that level were achieved from the age of 10. In contrast, in CSN they were achieved only from the age of 46. Figure 10.1 represents the differences in the weighted average coke rates over the companies' lifetimes. In CSN, higher rates of coke rate reduction were only achieved during the 1992–97 period. Indeed over the 1980s, and particularly in the 1990s, CSN BF 3 even outperformed USIMINAS in the rate of coke rate reduction. However, CSN BF 3 did not outperform USIMINAS BF 3 in terms of the rate of fuel rate reduction.

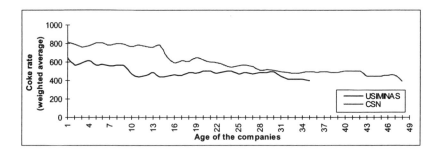

Figure 10.1   Differences in coke rate reduction over the lifetime of USIMINAS and CSN (weighted average for BFs 1, 2, and 3)

The fast rate at which coke rates have fallen in USIMINAS has permitted the company to catch up earlier than CSN with world standards. By the late-1970s, when USIMINAS was 16 years of age, its coke rates were between 430–466kg, while the average coke rates in Japan were 425–430kg and in Germany 490kg (see CEPAL, 1984). In contrast, by the late-1970s, when CSN was 33 years of age its coke rates were still between 513–643kg.[5] By 1988–89, at the age of 27, USIMINAS's coke rate was in line with those of Japan – around 470kg (see Piccinini, 1993). In contrast, in CSN by 1988–89, at the age of 43, it was 487–505kg.

Blast furnace productivity (tonne/m³/day)

This is defined as tonnes of pig iron produced per cubic meters of the internal volume of the furnace in a day. There are other definitions for BF productivity (e.g. useful volume), but the one followed is that used in the case-study companies.

Blast furnace 1 The differences in productivity increase between USIMINAS and CSN for BF 1 are outlined in Table 10.4.

It should be noted that during the 1946–56 period CSN's BF 1 went through three campaigns. In January 1949, its first campaign was interrupted by a crack in the crucible. In October 1949 its second campaign was again interrupted and by a similar problem. Its third campaign (1949–55) was limited by problems in the top and in the crucible.[6] USIMINAS's BF 1 went through one campaign within the initial nine-year period 1962–71[7], without the frequent stoppages that took place in CSN. These stoppages in CSN, which may reflect inadequate process and production organisation and equipment capabilities, clearly had negative implications for BF productivity.

Table 10.4 Differences in BF productivity increase between USIMINAS and CSN: blast furnace 1

	Increase in the productivity level (tonne/m³/day)	Average annual rate of increase (per cent per year)
USIMINAS		
BF 1 (1962–72)	0.57 → 1.45	+9.78
BF 1 (1962–89)	0.57 → 2.47	+5.58
BF 1 (1990–97)	1.90 → 2.24	+2.38
CSN[a]		
BF 1 (1950–60)	0.79 → 0.87	+0.96
BF 1 (1950–89)	0.79 → 1.36	+1.40
BF 1 (1990–91)	1.33 → 1.30	−2.25

Note: (a) In January 1992 the furnace was shut down permanently.
Source: Own elaboration based on the research.

Blast furnace 2 The differences in productivity increase between USIMINAS and CSN for BF 2 are outlined in Table 10.5. In USIMINAS over the 24-year period 1965–89 productivity increased from 0.77 to 2.47 t/m³/day or by 4.97 per cent annually on average. In contrast, in CSN during the 35-year period 1954–89 productivity increased from 0.59 to 0.94 t/m³/day or by only 1.34 per cent annually, less than one third the rate of improvement in USIMINAS. It should be remembered that these periods were equivalent to the start-up and conventional expansion phases in both companies.

During the 1990–97 period in USIMINAS productivity increased from 1.81 to 2.30 t/m³/day or by 3.48 per cent annually. In CSN it was not until 1992, or more than 40 years after the start-up, that the company achieved productivity above 2t/m³/day. From 1990 to 1997 productivity increased from 0.97 to 2.46 t/m³/day or by 14.21 per cent annually on average. In this case, during the 1990–97 period CSN BF 2 substantially outperformed USIMINAS in the rate of productivity increase.

Blast furnace 3 The differences in productivity increase between USIMINAS and CSN for BF 3, a similar vintage plant, are outlined in Table 10.6. Here the performance of BF 3 is also compared under similar time periods.

Table 10.5   Differences in BF productivity increase between USIMINAS and CSN: blast furnace 2

	Increase in the productivity level (tonne/m³/day)		Average annual rate of increase (per cent per year)
USIMINAS			
BF 2 (1965–74)[a]	0.77	→ 1.41	+6.95
BF 2 (1965–89)	0.77	→ 2.47	+4.97
BF 2 (1990–97)	1.81	→ 2.30	+3.48
CSN			
BF 2 (1954–64)	0.59	→ 0.87	+3.96
BF 2 (1954–89)	0.59	→ 0.94	+1.34
BF 2 (1990–97)	0.97	→ 2.46	+14.21

Note: (a) Shut down for revamping between 1975 and 1977.
Source: Own elaboration based on the research.

Table 10.6   Differences in BF productivity increase between USIMINAS and CSN: blast furnace 3[a]

	Increase in the productivity level (tonne/m³/day)		Average annual rate of increase (per cent per year)
USIMINAS			
BF 3 (1975–79)	1.60	→ 1.79	+2.84
BF 3 (1975–89)	1.60	→ 2.56	+3.41
BF 3 (1980–89)	1.69	→ 2.56	+4.72
BF 3 (1990–97)	2.12	→ 2.47	+2.20
CSN			
BF 3 (1977–79)	1.29	→ 1.58	+10.67
BF 3 (1977–89)	1.29	→ 1.61	+1.86
BF 3 (1980–89)	1.74	→ 1.61	−0.85
BF 3 (1990–97)	1.61	→ 2.38	+5.74

Source: Own elaboration based on the research.
Note: (a) Again, for a meaningful comparison the start-up year of the two furnaces are not considered.

During the 1990s in USIMINAS productivity increased from 2.12 to 2.47t/m³/day or by 2.20 per cent on average. In CSN it was not until 1993 that the furnace began to achieve productivity above 2t/m³/day. Productivity in BF 3 increased from 1.61 in 1990 to 2.38t/m³/day in 1997, or by 5.74 per cent on average. Although in the 1990s CSN BF 3 achieved this high rate of increase, CSN could still not catch up with the level of USIMINAS (2.47t/m³/day).

In CSN, as was the case in BF 1, BF 3 suffered from sudden stoppages and unstable production operations, particularly during the 1980s. Indeed 1997 was the first year it operated continuously.[8] This might have been associated with difficulties in stabilising its operations. And this significantly influenced the reduction in productivity during the 1980s.

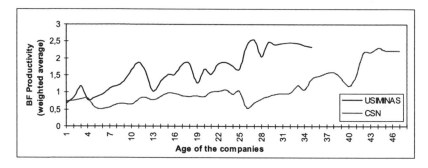

Figure 10.2   Differences in BF productivity increase over the lifetime of USIMINAS and CSN (weighted average for BFs 1, 2, and 3)

By the late-1970s, at the age of 17, USIMINAS had achieved BF productivity of 1.8t/m³/day, while the average productivity in Japan in 1981 was 1.9t/m³/day.[9] By that time, at the age of 33, CSN had achieved BF productivity of only 1.0t/m³/day. From the perspective of the age of the companies, USIMINAS began to operate with BF productivity above 1.5t/m³/day from the age of 10. In contrast, CSN began to operate continuously with productivity above 1.5t/m³/day only from the age of 41, as represented in Figure 10.2.

By 1991 BF productivity in the 29-year old USIMINAS was 2.35t/m³/day on average, while in the Japanese Kiaitsu average productivity was 2.28t/m³/day in 1991 (Gupta et al., 1995). In 1991 average BF productivity in Japan was 2.03t/m³/day (Japan Iron and Steel Federation, 1995). In June 1994, USIMINAS achieved a world record of 2.71t/m³/day.[10] In contrast, in 1991, the 45-year old CSN had achieved average BF productivity around 1.5t/m³/day. The evidence suggests that USIMINAS had been able to achieve and sustain internationally competitive levels of BF productivity over time. In contrast, CSN did not achieve world standards until 1992. Therefore, USIMINAS was faster than CSN in achieving substantial improvements in BF productivity and sustaining them at rising world competitive levels over time.

Hot metal quality

This subsection examines cross-company differences in the hot metal quality on the basis of silicon (Si) content (per cent) in the pig iron. The Si content is associated with the consumption of fuel in the BF. Reduction in Si content is associated with the operational stability of the BF which, in turn, is associated with the good quality of the metallic charge. High hot metal silicon levels have an adverse influence on BF productivity, flux use, and also steelmaking process yield. Thus steel mills seek to reduce Si content. The assessment of hot metal quality may also include the sulphur (S) and phosphorus (P) contents in pig iron, but these are not explored here. The differences between USIMINAS and CSN are summarised in Table 10.7.

Table 10.7   Differences in Silicon (Si) content in the pig iron between USIMINAS and CSN

Periods	Silicon content in the pig iron (per cent): average	
	USIMINAS	**CSN**
1962–69	0.89	n.a.
1970s	0.67	0.80
1980s	0.52	0.80
1990–97	0.41	0.45

Source: USIMINAS and CSN.

In USIMINAS during the initial ten-year period 1962–72, Si content was reduced from 1.28 per cent to 0.75 per cent, being 0.89 per cent on average in that period. Data for CSN during its initial ten-year period 1946–56 are not available. However, evidence from the 1970s suggests that the average for that period would be higher than in USIMINAS. By the late-1970s, when CSN was 32 years old, average Si content was 0.80 per cent. In contrast, when USIMINAS was 16 years of age the silicon content was between 0.58 per cent and 0.54 per cent.

It should be remembered that by the late-1970s both companies were operating under the second world energy crisis. This put pressure on energy consumption world-wide, reflected in the Si content. However, the evidence suggests that the Si content was reduced in USIMINAS but remained unchanged in CSN. During the 1980s in USIMINAS the Si content was between 0.60–0.36 per cent, or 0.52 per cent on average. In CSN, however, the average was 0.80 per cent. By 1991 USIMINAS had reduced Si content to 0.40 per cent, the same as in the Japanese Kiaitsu (Gupta et al., 1995).[11] In CSN in the early 1990s Si content was 0.60 per cent, on average, or 50 per cent higher than in USIMINAS. It was not until 1995 that it dropped to 0.47 per cent.

In sum, during the lifetime of the three BFs USIMINAS was able to combine a fast rate of reduction in coke rate (except in BF 3) with a fast rate of increase in productivity and fast rate of reduction in Si content. These achievements certainly had positive implications for other indicators (e.g. overall energy consumption), as discussed later in Section 10.2.4. In contrast, in CSN, particularly during the 1950s–1980s period, the BFs experienced slow rates of coke rate reduction (except for BF 3), a slow rate of productivity increase, and a slow rate of Si content reduction. This must have had negative implications for overall energy consumption and steelmaking process yield in CSN. It was not until the 1990s, and particularly from 1992, that BF performance improved in CSN. Indeed, in the 1990s CSN even outperformed USIMINAS in terms of the rate of coke rate reduction in BF 2 and BF 3, and productivity increase in BF 2. However, over the long term, and on the basis of the three indicators (including fuel rates), USIMINAS had a substantially better overall BF performance than CSN.[12]

10.2.2 Performance Differences in the Ironmaking Process

The indicators described above can be affected by a number of factors, which may even obscure the role of technological capability. For instance, they can be influenced by, among other things, the vintage of plant, the conditions of the refractories, the quality of the raw materials (coal), the supply of raw materials (local or foreign), energy crises and strikes. The coke consumption may also be reduced through the use of pulverised coal injection (PCI) technology or use of imported coal. Even so, the role of technological capability can be explored.

Following the start-up of the reduction process in USIMINAS (early-1960s), intensive in-house efforts were made to improve raw material preparation. It was realised that an adequate preparation of the raw materials would contribute to improving BF performance. The raw materials dept and the technical dept engaged in these efforts assisted initially by Yawata. At that time, the teams in USIMINAS had little knowledge of the characteristics of the iron ore fines (from Itabira) and its behaviour in sinter manufacturing. As a result, several experiments were undertaken in the pilot sinter machine that had been acquired by USIMINAS. These experiment led to the accumulation of knowledge needed to modify the original sintering process project to adapt it to local iron ore characteristics, and, in particular, to meet BF needs. In the 1960s these efforts consisted of manipulating sinter size, improving the mechanical resistance of the coke and homogenising its ash content, improving sinter quality (e.g. by reducing the variations in silica dioxide (SiO_2)), and reducing the Si content in the pig iron. These improvements had a substantial influence in the reduction in coke rates and the increase in BF productivity.[13]

In CSN during the 1950s in-house incremental efforts to stabilise BF operations, or to manipulate the parameters of the metallic charge, were practically
absent, as indicated by the evidence in Chapter 6. CSN relied on a facility in the
state of Santa Catarina for coal washing and on the use of high-quality imported
coal on a 70 per cent basis, approximately, as a way of reducing the coke rate. It
could be argued that it was not until 1960 that the sinter plant started up in CSN.
Therefore, CSN did not have the means to reduce its coke rate. However, the
sintering process is only one of several ways of reducing the coke rate (e.g. improved burden distribution, changing the high top pressure, manipulating the
slag temperature and viscosity, etc.). However, as demonstrated by the evidence in Part II, in-house continuous process improvement efforts were scarce
in the plant.

During the 1970s USIMINAS intensified its efforts on the operational stabilisation of the BFs. To cope with the emergence of irregularities in the operations
of BF 1 and 2, the company engaged in the removal of a hard crust, as described
in Chapter 6. In addition, a partial straining system was introduced in the silos of
the two furnaces. These were followed by a total straining system one year later.
These changes contributed to improving the quality of the burden. In the
early-1970s, in-house experiments in USIMINAS pointed to the negative influence of non-calcinated carbonates on the coke rate. As a result, operational practices in the BF and the sinter plant were modified to minimise their use.
Additionally, an operational plan was created to constantly monitor the lining
conditions permitting effective interventions within the shortest time. In addition,
this plan sought to improve standardisation in the operations. Improvements in
the loading process were introduced (e.g. modifications to the sampling and raw
materials analysis to establish their characteristics before use, reduction in the
iron ore/coke ratio, modification to the level and sequence of the loading).[14]

From the 1960s to the 1990s USIMINAS continuously improved the conditions of the bellows. In order to reduce stoppages for their replacement, which
had negative influence on the operations of the furnace and on productivity,
USIMINAS engaged in the manufacture of bellows in house. In addition, to increase their lifetime, the refrigeration system and the daily methods of operation were improved. As a result, the lifetime of the bellows increased from 65
days in 1975 to 587 days by 1992.[15] These improvements had positive implications for BF productivity.

In CSN in the 1960s, several improvements to BF 1 were introduced including
enlargement of the crucible, carbon walls, increase in the number of bellows, and
new distributors. However, these improvements were made by Arthur McKee
not by CSN.[16] By the late-1970s, as part of the Stage III of Plan 'D', several improvements were made to BF 1 and 2. For instance, they included increase in the
blast temperature; increase in the top pressure; the substitution in the loading system, among others. However, CSN continued to use firms like Arthur McKee to
undertake these improvements.[17] This suggests that the delegation of the imple-

mentation of improvements to engineering firms may have prevented CSN from assimilating adequately the intricacies of the reduction process. This contributed to limiting CSN's own engagement in further improvements.

In USIMINAS by the early-1980s, mathematical models were developed by the automation unit, the research centre, and the sinter plant. These contributed to reducing sinter dispersion and therefore the coke rate. By the late-1980s, the blast furnace dept, the metallurgical dept and the research centre engaged in joint efforts to improve the reactivity of the coke leading to improvements in silicon content. In addition, from the 1970s USIMINAS introduced the use of sinter with a high content of titanium dioxide (TiO_2) to protect the crucible. This contributed to preventing the BFs from sudden interruptions in their campaigns.[18] In CSN, cracks in the crucible were one of the main factors for interruptions in BF campaigns, particularly over the 1950s–1980s period. These interruptions affected productivity and costs negatively.[19] In USIMINAS from the late-1980s, as indicated in Chapters 7 and 8, more complex and integrated automated process control systems were developed by the automation unit, research centre, and the blast furnace dept. They seemed to have contributed to the sustaining of competitive coke rates and productivity levels during the 1990s.

In CSN during the 1970s and 1980s the BFs operated with irregularities in their functioning. In addition, they experienced unexpected stoppages, and there was a lack of co-ordination with the steelmaking plant. These problems were reflected in the slow rate of productivity improvement in the BFs. They may also reflect the incomplete accumulation of Levels 1 and 2 routine capability. Other factors may have contributed also to low productivity (e.g. frequent strikes in the 1980s). However, interviews with engineers and managers in CSN suggested that if the company had had better operating technological capability, the campaigns of the BFs would have been longer and their productivity higher.

In CSN by the early-1990s, the accumulation of Levels 1 and 2 routine capability was completed. In addition, as the company moved into the accumulation of Level 4 innovative capability for process and production organisation, CSN engaged in more continuous efforts to improve the reduction process parameters in relation to the 1980s. These were undertaken by the process development superintendency, TAG and the TQM programme. The evidence suggests that the period during which CSN did not accumulate adequate in-house technological capability for process and production organisation, particularly at Levels 1 and 2 (1940s–1980s), was associated with slow rate of performance improvement in the iron-making process performance. However, the period in which CSN began to accumulate Levels 1 and 2 routine and then Levels 3 to 4 innovative capabilities for process and production organisation (from the early-1990s) was associated with better performance improvement (see Tables 10.1 to 10.7). USIMINAS, however, had achieved competitive performance earlier and more continuously over its lifetime.

The continuous improvements made by USIMINAS over time contributed to the rapid achievement of competitive coke rates, productivity, and silicon content in the BFs. For instance, by the 1990s in BF 1, a vintage from the 1950s, USIMINAS was achieving coke rates of 400kg and productivity of around 2.20t/m^3/day with silicon levels below 0.45 per cent. In CSN there was an absence of any significant achievements during the lifetime of BF 1. Other factors may have influenced BF performance in USIMINAS. However, this level of performance certainly would not have been achieved if USIMINAS had not accumulated Levels 1 to 5 capability consistently over time. USIMINAS's experience suggests that CSN could have achieved better performance over the 1950s–1980s period if the company had accumulated adequate in-house technological capability. In CSN during the 1990s several factors may have contributed to improving the BF performance (e.g. use of technical assistance, use of imported coal, and the PCI technology). However, the evidence suggests that the completion of the accumulation of basic operating capability (Levels 1 and 2) contributed to the substantial differences in the BF performance over the 1990s in relation to the 1950s–1980s period in CSN.

10.2.3 Steelmaking Process Performance

This subsection analyses the cross-company differences in the oxygen steelmaking process performance in USIMINAS (1963–97) and CSN (1977–97).[20] In USIMINAS there are steel shops 1 and 2. In CSN it is considered as one steel shop. This subsection examines the performance of the steel shop as a whole. The comparison is based on periods over the lifetime of the plants covering 1963–76, 1977–89, and 1990–97 for USIMINAS, and 1977–89 and 1990–97 for CSN. This is to make the comparison more meaningful.

However, one initial comment is important at this stage: the levels of the performance indicators used here are much less determined by the level of technology embodied in the vintage of plant. Indeed the level of the indicators is more influenced by the daily operational practices (technical and organisational) in the steel shop. This subsection reviews three key indicators of the oxygen steelmaking process. Since they are interrelated, the indicators are analysed together and are defined as follows.

1. 'Tap-to-tap' or heat time (minutes). The elapsed time, in minutes, between the heats in the converter.
2. Reblow rate (per cent). The proportion of heats that need to be reblown by oxygen for correction in the steel composition and/or temperature. In a steel shop, following the analysis of the molten steel samples from the converter, it is decided whether to tap the heat or to make corrections. If corrections are needed, they can be made (a) through oxygen reblowing; (b) through the addition of metals (e.g. manganese) preceded by cooling procedures.

3. Hit rate (per cent) or rate of simultaneous achievement of carbon and temperature. The proportion of heats in which the steel composition (carbon) and temperature, desired by the heat order, is simultaneously achieved at the heat end-point. A critical task for any oxygen-based steel shop is to achieve high hit rates to prevent time-consuming and costly oxygen reblows or cooling procedures. These add costly minutes to the heat time reducing the potential productivity of the process.

Differences in the steelmaking process performance: USIMINAS (1963–97) and CSN (1977–97)

The differences across the indicators are summarised in Table 10.8. As indicated there, in USIMINAS during the initial 12-year period 1964–76 the tap-to-tap reduced from 45 to 40 minutes. It should be noted that in USIMINAS during the first month of operation (June 1963) the tap-to-tap was 82 minutes. By the twelfth month (June 1964) it had reduced to 45 minutes, and between 1968 and 1977 the tap-to-tap in USIMINAS was stabilised at around 42 minutes. In the subsequent period of 1977–89 the tap-to-tap reduced from 40 to 32 minutes, and during the 1990s the indicator continued to be stabilised around 32 minutes.

In contrast, in CSN, as indicated in Table 10.8, during the initial 11-year period 1978–89 the tap-to-tap reduced from 62.3 to 38.4 minutes. It should be noted that in CSN during the second year of operation (1978) the tap-to-tap was 62.3 minutes. By the fourth year of operation (1980) the tap-to-tap was 51.3 minutes. It took about ten years (1978–88) for CSN to operate continuously with a tap-to-tap below 40 minutes. In CSN during the 1990–95 period the tap-to-tap varied between 35.1 to 38.1 minutes. It was not until 1997, 20 years after the steel shop start-up, that CSN achieved tap-to-tap of 34 minutes. It should also be noted that CSN's steel shop started with an automated process control technique. This, in principle, would allow the achievement of a 40–45 minutes tap-to-tap within the first year. USIMINAS, however, started with a manual control. This issue is discussed in more detail in the next subsection.

As far as the reblow rate is concerned, in USIMINAS during the initial 13-year period 1963–76, it reduced from 25 to 20 per cent the average during that period being 22.9 per cent. In CSN during the initial 11-year period (1978–89) the reblow rate fell from 36 to 25 per cent and the average rate over the period was 31.8 per cent. In contrast, within a similar period (1977–89), the reblow rate in USIMINAS had fallen from 20 to 10 per cent and the average over the period was 14.6 per cent. In USIMINAS during the 1990–97 period the reblow rate fell from 10 to 6 per cent. In contrast, in CSN from 1990 to May 1997 it fell from 25 per cent to only 16.7 per cent.

Table 10.8   Differences between USIMINAS and CSN in key steelmaking process indicators

	Tap-to-tap (minutes)	Reblow rate (per cent)	Hit rate (per cent)
USIMINAS			
USI (1963-76)	45 → 40	25 → 20	36 → 48
USI (1977-89)	40 → 32	20 → 10	48 → 85
USI (1990-97)	32	10 → 6	85–90
CSN			
CSN (1978-89)	62.3 → 38.4	36 → 25	n.a.
CSN (1990-97)	35 → 34ª	25 → 16.7ª	n.a.

Note: (a) Up to May 1997.
Sources: 'USIMINAS 25 anos', *Metalurgia ABM*, Special Edition, 1987; USIMINAS, 'Aciaria: controle dinâmico de fim de sopro com sublança' [internal publication]. CSN (1996); 'CSN atinge a marca histórica dos 100 milhões', *Metalurgia & Materiais*, 1997, vol. 53, no. 467, July 1997. Interviews in the two companies.

In USIMINAS the improvements over time in the tap-to-tap and reblow rates during the 1963–97 period may reflect the increase in the hit rate from 36 per cent to 85–90 per cent during that period. The evidence from USIMINAS indicates a substantial performance improvement across the three indicators during the lifetime of the steel shop. Although data on hit rates in CSN are not available, the evidence on tap-to-tap and reblow rates suggest that its hit rates would be much lower than in USIMINAS over the lifetime of the steel shop. It also implies that little improvement may have occurred in the hit rate during 1990–97 in relation to the 1978–89 period in CSN. Considering the average ratio of the levels of reblow rates in USIMINAS and CSN for the 1990–97 period, the rates in CSN were 2.7 times higher than in USIMINAS. This implies that CSN must have accumulated higher production costs than in USIMINAS as a result of greater use of oxygen, fluxes, coolants, etc. associated with the reblows.

As mentioned earlier in this chapter, the Si content in the pig iron also has an adverse influence on the steelmaking process yield. While USIMINAS entered into the 1990s with a Si content of 0.36–0.40 per cent, CSN's averaged 0.60 per cent. This suggests that USIMINAS entered into the 1990s with higher yield in the steelmaking process than CSN. Additionally, in USIMINAS the consumption of the refractories (kg/tonne of molten steel) in steel shop 2 in 1989 was 1.9kg/tonnes of steel. By 1993 it had been reduced to 0.69kg/tonnes and was 0.65kg/tonnes in 1994.[21] In contrast, in CSN in 1989 consumption was 9kg/tonnes of steel. By 1993 this had fallen to only 7.3kg/tonnes of steel. This was the lowest level of refractories consumption CSN ever achieved,[22] but it was still more than ten times higher than in USIMINAS. These differences had positive implications for the cost of steel produced in USIMINAS and a negative influence on steelmaking costs in CSN.

Performance differences in the steelmaking process

In USIMINAS, as described in Chapter 6, during the first year of the steel shop, intense efforts were made to achieve efficient operation of the steelmaking process. The steel shop and the metallurgical dept also took the initiative to make systematic comparisons between USIMINAS and a leading Japanese steel mill in the steelmaking performance. By 1963 USIMINAS had set a goal of catching up with the performance achieved in Yawata. Following a series of studies and analyses of samples, within one year USIMINAS had standardised critical heat parameters like sulphur, phosphorus, manganese, and temperature, contributing to stabilising the steel shop operation.[23] This reflects the accumulation of Levels 1 and 2 routine (basic and renewed) capability for process and production organisation. Between 1966 and 1972, under the 'capacity-stretching' projects, continuous and numerous in-house efforts were made in USIMINAS, as described in Chapter 6, led by the metallurgical dept, the steel shop and the engineering unit. These sought to increase the number of heats per day thereby reducing the tap-to-tap rate to increase the capacity utilisation rate.

In CSN during the first year of operation of the steel shop (1977), efforts to operate the unit efficiently and stabilise operations were absent or ineffective, as described in Chapter 7. Instead, CSN seemed to pursue rapid increases in production volume only. This caused a major accident in the steel shop in the first year of operation (see Section 7.1.2) resulting in a 26-day stoppage.[24] As mentioned in different interviews in CSN, during the 1970s and early-1980s, the steel shop had difficulties in making routine steels adequately. In addition, there were frequent conflicts between the continuous casting (CC) mill and the steel shop in relation to the specifications of the steels produced. As described in Chapter 7, during the 1970s and 1980s there was limited co-ordination of production flow between the BFs, the steel shop, and the CC mill. This reflects the limited accumulation of Levels 1 and 2 routine (basic) capability to operate the unit efficiently.

It should be remembered that during the initial four-year period 1963–67, USIMINAS controlled the process parameters manually. During the 1968–75 period that control was based on the 'catch-carbon' strategy.[25] Nevertheless, as described in Chapter 6, several technical and production organisation modifications contributed to improving the process parameters. Drawing on Level 3 and 4 investment capabilities to search and select technologies, whose accumulation was described in Chapters 6 and 7, USIMINAS introduced in 1976 the 'static charge control'.[26] This was in operation until 1982. Drawing on Level 4 innovative capability, USIMINAS developed in-house (by the automation unit, the research centre, and the steel shop) the mathematical models for this control system. This contributed to the rapid reduction in the tap-to-tap and reblow rates in the first year of the introduction of this automated process control system.[27]

In CSN the steel shop started up in 1977 with the 'static charge control' which came together with the acquisition of the whole facility. The mathematical models had been built by the supplier rather than by CSN.[28] In contrast to USIMINAS, CSN had not built an organisational unit dedicated to the development of in-house mathematical models for automated process control (e.g. the automation unit). Instead, CSN relied on the availability of the on-site process control computer programmed with a charge-control model. However, this was not sufficient. As referred to in Chapter 4, to achieve good control of the steelmaking process parameters, accurate information to feed the automated control system (e.g. accuracy of the inputs, consistency of practices and quality of the materials in use, coherence between the computer recommendations and the shop daily practices) is crucial. However, as indicated by the evidence in Chapter 7, by the late-1970s in CSN daily production organisation practices were ineffective or even precarious. This reflects the incomplete accumulation of Levels 1 and 2 capability and the absence of Level 4 innovative capability for process and production organisation compared to USIMINAS.

By 1982, USIMINAS introduced the 'dynamic charge control' in the steel shop.[29] The technology had been acquired from NSC. The introduction of the technology was associated with the previous in-house development of the static mathematical models. In addition, studies were made on the mechanism of the chemical reactions and thermal balance of the steelmaking process by the automation unit and the research centre. This permitted USIMINAS to implement the automated control system effectively and to achieve good results in the first year. During the 1990s, drawing on its own capabilities, USIMINAS introduced the 'supervisory system' combining the principles of the static and dynamic control systems for rigorous control of key process parameters.[30] The improvements over time in the strategies for process control, reflect USIMINAS's Levels 4 and 5 innovative capability for process and production organisation. They also reflect USIMINAS's Levels 5 and 6 innovative capabilities for investments, whose paths of accumulation were described in Chapters 6 to 8. The achievement of low consumption of refractories in USIMINAS was associated with efforts to increase the lifetime of the refractory lining of the converters. These are associated with techniques for preventive and/or corrective maintenance and improvements in the operating conditions, among others. In other words, they reflect the accumulation of capability for equipment and process and production organisation. The evidence from CSN suggests that these techniques had not been developed. In contrast, by the 1990s USIMINAS had been providing technical assistance in steelmaking process control across Latin America.

From 1985 to 1989 the steel shop in CSN was being technically assisted by Japanese companies. The objective was to stabilise operational conditions and to improve the indicators.[31] However, by the mid-1980s, CSN itself had not completed the accumulation of routine operating capabilities. This may be reflected in the slow rate at which the indicators were improved over that period. By 1993 CSN introduced the combined blowing technology which sought to reduce the consumption of ferro-alloys and refractories.

By 1996, CSN introduced the 'dynamic charge control' in the steel shop. The technology was acquired from the German firm Demag which developed the mathematical models. Interviews in the steel shop suggested that the specifications for the control system were not congruent with steel shop characteristics (e.g. product mix). In addition, the production organisation practices of CSN did not seem to meet the conditions for effective implementation of the control system. The result was that by late-1997, the steel shop, drawing on a specialised software company, was still struggling to fix the control system and to overcome its problematic introduction. The evidence suggests that the slow rate of improvement in the steelmaking performance in CSN during the 1970s–1990s period was associated with: (1) lack of organisational units to develop in-house process control systems and mathematical models; (2) poor interaction between the steel shop and the research centre for process problem-solving; and (3) the long time taken to improve the daily production organisation practices in the steel shop, although it was technically assisted during the 1985–89 period. In other words, USIMINAS's experience suggests that a more effective steelmaking process control could have been achieved earlier in CSN if the company had accumulated Levels to 1 to 4 capability for process and production organisation during the 1970s and 1980s.

10.2.4 Overall Plant Performance

Capacity utilisation rate

This is the actual production volume as a percentage of the design capacity of an operational unit. Capacity utilisation rates can be affected by different factors like rates of macroeconomic growth (e.g. recession) and strikes. However, the internal efforts in the plant to increase the utilisation rate of the operations units are another important factor. This subsection outlines the differences in capacity utilisation rates in USIMINAS (1962–97) and CSN (1946–97). The cross-company differences in capacity utilisation rates of those operational units are summarised in Table 10.9.

The evidence in Table 10.9 suggests that, in general, USIMINAS achieved substantially higher capacity utilisation rates than CSN across different facilities over time. In contrast, CSN, particularly during the 1940s–1980s period operated at low capacity utilisation rates. USIMINAS achieved high rates of capacity utilisation during the initial years. For instance, in the steel shop during the initial 1963–72 period capacity utilisation rate increased from 14 per cent to 235 per cent. In contrast, in CSN during the initial 1978–88 period it increased from 53.2 per cent to 87 per cent. In USIMINAS it took three years

Table 10.9   Differences in average capacity utilisation rates between USIMINAS (1962–97) and CSN (1946–97)[a]

Operational units	Average capacity utilisation rate during the 10 initial years of each operational unit (per cent)	Average capacity utilisation rate during the lifetime of each operational unit (per cent)
USIMINAS		
Coke ovens (1963–73)	101.1	
(1963–97)		102.7
Sinter plant (1964–74)	140.2	
(1964–97)		105.0
BF 1 (1963–73)	140.2	
(1963–97)		201.6
BF 2 (1966–74)[b]	158.4	
(1966–97)		202.5
BF 3 (1975–85)	87.4	
(1975–97)		102.5
LD steel shop (1964–74)	128.7	
(1964–97)		112.7
CSN		
Coke ovens (1947–57)	68.1	
(1947–97)		80.2
Sinter plant (1961–71)	78.1	
(1961–97)		82.9
BF 1 (1947–57)	75.9	
(1947–91)[c]		70.3
BF 2 (1955–65)	88.7	
(1955–97)		75.8
BF 3 (1977–87)	84.1	
(1977–97)		90.6
SM steel shop (1947–57)	91.8	
(1947–80)[d]		87.0
Oxygen steel shop (1978–88)	72.1	
(1978–97)		81.0

Notes: (a) The start-up year of each operation unit was not considered for the calculation of the averages. In cases in which the unit was shut down permanently the last year of operation was also not considered.
(b) Shut down for revamping during 1975–77.
(c) Shut down permanently in January 1992.
(d) Shut down permanently in April 1981.

Source: Own elaboration based on the research.

to absorb the design capacity of the steel shop. In contrast, in CSN it took 17 years if molten steel production is considered. The high rates of capacity utilisation of the BFs over time in USIMINAS reflect the increase in their productivity, as examined in Section 10.2.1. In contrast, in CSN it was not until the 1990s that the rates of utilisation of the BFs improved substantially. This may reflect the improvement in the productivity of BFs 2 and 3, as examined in Section 10.2.1.

It might be thought that these differences would reflect differences in the initial plant design relative to market scale – e.g. that USIMINAS was able to achieve high rates of capacity utilisation because it had initial plant design that was well below the size of its (potential) market. However, using Sercovich's (1978) terms, the evidence in Part II does not suggest that CSN's design capacity was above the market absorption capacity over time, or 'over-dimensioned' because: (1) during its start-up phase CSN was the leading company in the flat steel segment; (2) particularly from the early-1950s, there was growing demand for steel products from the automobile, construction, and other industries in Brazil; and (3) during the recession in the 1980s, the company could have explored the export market for its products. However, interviews with managers in CSN indicated that the company had difficulties in increasing exports over the 1980s. This was associated with its poor product quality, reflecting CSN's limited product capability. In turn the evidence in Part II does not suggest that USIMINAS's design capacity was below market capacity, or 'under-dimensioned' over time.

In addition, it could be argued, as in Brumer (1994), that from the late-1970s CSN was given the objective by SIDERBRAS of producing coated sheets. However, as argued, CSN had idle capacity in hot and cold-rolled products. This would be a way of avoiding competition with the products of USIMINAS and COSIPA by limiting the range of products in CSN. This argument, however, falls short as an explanation for the low capacity utilisation rates in CSN. SIDERBRAS was created by the mid-1970s. Even before that, CSN had been operating with low rates of capacity utilisation. The evidence in Part II also demonstrated that continuous efforts to increase the utilisation rates of the facilities, as in USIMINAS, were absent in CSN during the 1950s–1980s period. In addition, the case of the HSM 2 in CSN during the 1980s, as described in Chapter 7, suggests that the low utilisation and availability rates were associated with internal problems in CSN, not only with external ones. In sum, considering that capacity utilisation rates have implications for costs and overall operational income, the evidence implies that USIMINAS was able to achieve lower unit fixed costs and higher operational income than CSN.

Overall energy consumption (Mcal/tonne of steel)

This refers to the overall energy consumption in the plant in relation to tonnes of steel produced. The differences between USIMINAS and CSN are outlined in Table 10.10 which indicates that during the whole period in USIMINAS the indicator was stabilised around 6,200Mcal/tonne.

In contrast, in CSN it varied more frequently, reaching the peak of 7,684Mcal/tonne in 1984. In CSN the consumption in 1979 was considered a 'record' in the company.[32] Although data for the previous period are not available, this suggests that consumption would be higher. The stable trajectory of the indicator in USIMINAS suggests that the company had a more effective energy performance than CSN over time.[33] Taking the ten-year period 1979–89 into consideration, in USIMINAS overall energy consumption reduced from 6,371 to 5,646Mcal/tonne or by 1.20 per cent annually on average. In contrast, in CSN during that same period, the indicator increased from 7,264 to 7,446Mcal/tonne, or by 0.25 per cent annually on average. In addition, the level of the indicator during that period in CSN was, on average, 13 per cent higher than in USIMINAS. It should be noted that during the 1970s and 1980s both companies operated under energy crises and recession. However, as indicated by the evidence, the behaviour of the indicator still differed.

Table 10.10   Differences in overall energy consumption (Mcal/tonne of steel) between USIMINAS (1977–97) and CSN (1979–98)[a]

Year	USIMINAS	CSN	% Difference (CSN in relation to USIMINAS)
1977	6,319	n.a	–
1978	6,222	n.a.	–
1979	6,371	7,264	+14.0
1980	6,461	6,582	+1.9
1981	6,851	6,458	–5.7
1982	6,453	7,092	+9.9
1983	6,225	6,870	+10.4
1984	6,105	7,684	+25.9
1985	6,069	7,085	+16.7
1986	6,349	6,944	+9.4
1987	6,206	6,757	+8.9
1988	5,764	7,138	+23.8
1989	5,646	7,446	+31.9
Average annual rate of reduction (%): 1979–89	–1.20	+0.25	
1990	6,144	7,584	+23.4
1991	5,927	7,360	+24.2
1992	6,071	6,756	+11.2
1993	6,073	6,752	+11.1
1994	6,045	6,749	+11.6
1995	6,138	6,944	+13.1
1996	6,153	6,863	+11.5
1997	6,273	6,634	+5.8
1998	n.a.	6,696	–
Average annual rate of reduction (%): 1990s	+0.29	–1.54	

Note: (a) Data for USIMINAS were provided as Mcal/tonne of crude steel, while for CSN as Mcal/tonne of molten steel. Since production volume expressed in the latter form is slightly higher than the former, this difference would be reflected in the data for CSN.

Sources: USIMINAS: interview with a manager in the energy and utilities unit (energy centre); technical information centre; Annual reports (1976–90); CSN: Interview with the adviser to the director of the Steel Sector; Annual Reports (1975–93).

The year 1990 was marked by the economic recession in Brazil. Therefore production volume fell in the industry. As a result, in USIMINAS the indicator increased from 5,646 in 1989 to 6,144Mcal/tonne in 1990, returning to a level below 6,000Mcal/tonne in 1991. In contrast, in CSN it increased from 7,446 in 1989 to 7,584 in 1990, staying above 7,000Mcal/tonne in the following year. During the 1990–97 period in USIMINAS the indicator increased from 6,144 to 6,273Mcal/tonne or by 0.29 per cent annually on average. However, as during the previous periods, it was again stabilised around 6,200Mcal/tonne. In contrast, in CSN during the 1990–98 period the indicator fell from 7,584 to 6,696 Mcal/tonne or by 1.54 per cent annually on average. Despite the higher rate of reduction of the indicator in CSN in relation to USIMINAS over that period: (1) the variations in indicator level continued although less than during the 1980s; (2) the rate of reduction was not enough for CSN to catch up with the indicator level achieved in USIMINAS. Indeed, during the 1990–97 period the indicator level in CSN was, on average, 13.9 per cent higher than in USIMINAS.

As far as the difference in 1980 in Table 10.10 is concerned, the specific fuel oil consumption in USIMINAS reduced from 35kg/tonne of steel in 1980 to 27.6kg/tonne of steel in 1981 (Piccinini, 1993). In contrast, in CSN the indicator fell from 60 in 1980 to 37kg/tonne of steel in 1981.[34] During the 1981–84 period in USIMINAS the indicator fell from 27.6kg/tonne of steel to 3.6kg/tonne of steel or by 49.2 per cent annually on average (Piccinini, 1993). In contrast, in CSN during the same period it increased from 37kg/tonne to 58kg/tonne, or by 16.1 per cent annually on average.[35] As a result, the differences during the 1980–81 period do not reflect improvement in overall energy performance in CSN. Indeed, they show that the indicator continued improving in USIMINAS.

The evidence in this subsection suggests that the more effective performance of USIMINAS in relation to CSN in overall energy consumption reflects their differences in iron-making and steelmaking processes performance, as examined earlier. In addition, differences in the overall energy performance would have had greater positive implications for operating cost reduction in USIMINAS than in CSN.

Other overall plant performance indicators

This section examines the cross-company differences in six plant performance indicators. Three are examined over the lifetime of USIMINAS and CSN: (1) labour productivity (tonnes/man/year); (2) number of product quality certificates; and (3) number of patents. Others are examined over the 1990s: (4) integrated yield rate; (5) customer complaints rate; and (6) domestic market share (automobile industry).

Labour productivity (tonnes/man/year)   This refers to the number of operational employees in relation to the tonnes of steel produced in a year. The evolution of the indicator is summarised in Table 10.11.

Table 10.11   Comparative evolution of labour productivity in USIMINAS and CSN

	Periods	Evolution of labour productivity (tonnes/man/year)	
	1963–70	15 →	164
USIMINAS	1971–80	185 →	263
	1981–89	182 →	382
	1990–97	300 →	524
	1950–70	32 →	75
CSN	1972–80	73 →	83
	1982–89	97 →	155
	1990–97	160 →	542

Sources: USIMINAS and CSN.

By the late-1980s, at 26 years of age, USIMINAS had achieved labour productivity of 347tonnes/man/year therefore catching up with, and even overcoming, indicators achieved in France (386), the US (429), Japan (351), and Germany (316). In contrast, CSN by the late-1980s was 43 years of age and had achieved productivity of 155tonnes/man/year, while the average in Brazil was 182tonnes/man/year in 1989 (see Peixoto, 1990). It was not until 1996 that CSN achieved productivity of 406tonnes/man/year. By 1996 USIMINAS had achieved labour productivity of 492tonnes/man/year. As referred to in Chapter 8, during the 1990s, both companies reduced their number of employees thereby contributing to increasing the indicator level. For instance, during the 1990–96 period USIMINAS reduced its number of employees from 13,413 to 9,210 or by 31.3 per cent. During that same period CSN reduced its number of employees from 18,222 to 11,086, or by 39.1 per cent.[36]

From 1996 to 1997, the number of employees in USIMINAS reduced from 9,210 to 8,359 or by 9.2 per cent. Interviews in CSN suggested that, particularly from 1996, the company adopted a more radical approach to the reduction of its employees: from 11,086 in 1996 to 9,059 in 1997, or by 18.3 per cent. As referred to in Chapter 8, the approach of CSN to reductions in the number of employees did not seem to consider the loss of qualified and experienced individuals, in other words, the implications for the company of the loss of tacit knowledge.

Number of patents   As indicated in Chapters 6 to 8, during the initial 20-year period 1962–82 USIMINAS had accumulated 83 patents, while CSN had accumulated only one. During the 14-year period 1983–97, USIMINAS had accumulated 250 patents, with 23 overseas across 18 different countries. In contrast, during the 1983–97 period CSN had accumulated 46 patents in Brazil only. This reflects the differences in intensity and variety in the original improvements to production process control, equipment, products, and engineering across the plant. It may also reflect the differences in efforts on knowledge-codification processes, as described in Part II.

Number of product quality certificates   During the 27-year period 1962–89, USIMINAS accumulated 15 product quality certificates. In contrast, no certificate had been accumulated in CSN during the 45-year period 1946–91. During the 1990–97 period USIMINAS obtained 26 new certificates leading to the total of 41 certificates accumulated over its lifetime. In contrast, it was not until 1992 that CSN achieved its first certificate. Only seven certificates were accumulated over the company's lifetime during the 1992–97 period. The number of certificates in USIMINAS during the early years reflects the continuous improvements in in-house quality systems. In other words, it reflects the accumulation and strengthening of Levels 1 and 2 capability for products and processes and production organisation. In CSN the absence of quality certificates in the 1940s–1980s period reflects the incomplete accumulation of those types of capability during that period. However, the award of seven certificates during the 1990s for CSN reflects the improvements across the production lines as a result of the TQM programme.[37] As for USIMINAS, another example of its world-wide recognition in product quality was its nomination by the World Steel Dynamics in 1996 as a steel company competing on the basis of high-valued products for highly-demanding markets. On a scale of 1 to 10 USIMINAS achieved 5, while the South Korean Pohang Iron and Steel Co. (POSCO) and British Steel achieved 7.[38]

Integrated yield rate (%)   This refers to the proportion of molten steel produced that is transformed into final hot and/or cold-rolled steels ready for sale. The difference between the indicator and 100 means loss or scrap. Even small differences in the level of this indicator have substantial implications for overall production costs in a steel company. For example, a 1 per cent loss in 4.5 million tonnes of molten steel production could represent about US$18 million in costs.

In 1984, at the age of 21, USIMINAS achieved an integrated yield of 87.3 per cent, while CSN achieved 78.4 per cent. From 1985 to 1987 the indicator in USIMINAS evolved from 88.5 per cent to 89.9 per cent, while in CSN, during the same period, it evolved from 85.6 per cent to 88.5 per cent. It was not until the early-1990s that CSN was able to stabilise its integrated yield rate around 90 per cent and catch up with USIMINAS. From 1995 both companies were achieving integrated yield rates above 92 per cent. It should be added that USIMINAS entered the 1990s with indices of rejection of hot sheets of less than 1 per cent. In contrast, in CSN by 1990 this indicator was around 4 per cent. By 1992, in CSN the indicator had reduced substantially to 1.9 per cent,[39] but was still higher than in USIMINAS.

Customer complaint rate (%)   In USIMINAS during the early-1980s the customer complaint rate was around 0.10 per cent and was reduced to 0.04 per cent by 1986. From 1991 to 1997 the indicator was reduced from 0.04 per cent to 0.03–0.01 per cent. In contrast, CSN entered the 1990s with a customer complaint rate of 0.38 per cent. In 1991 the indicator went up to 0.58 per cent and then was reduced to 0.19 per cent and 0.12 per cent in 1992 and 1993, respectively. From 1994 to 1997 CSN was able to reduce its customer complaint rates from 0.07 per cent to 0.04–0.03 per cent.[40] The evidence suggests that USIMINAS entered the 1990s with very low customer complaint rates, reflecting the accumulation and strengthening of its capabilities for products at Levels 1 to 5. In contrast, the high rates of customer complaint in CSN up the early-1990s reflect the incomplete accumulation of its product capabilities. Nevertheless, from the mid-1990s the customer complaint rates in CSN were substantially reduced, as a reflection of the company's efforts to build up and strengthen routine and innovative product-centred capabilities.

Domestic market share (automobile industry)   In USIMINAS the market share in the domestic automobile industry increased from 59 to 66 per cent during the 1990–97 period. In contrast, CSN's market share increased from 11 per cent in 1990 to 35 per cent in 1994 then declined to 30 per cent in 1997. This suggests that while USIMINAS strengthened its leading position in this industry, CSN was not able to achieve a more competitive market share.[41] Differences in the number of quality certificates, customer complaint rates and domestic market share are clearly associated with inter-firm differences in routine product quality systems, product strategy, and rate of in-house product improvement development, as examined in Chapters 7 and 8. In other words, their differences reflect the differences in the accumulation of product-centred capability.

Overall plant performance differences
As far as the capacity utilisation rates are concerned, the evidence suggests that the differences are strongly associated with the different ways and rates at which USIMINAS and CSN accumulated technological capability for process and production organisation and equipment. During the initial ten-year period USIMINAS had accumulated Levels 1 and 2 routine and also Levels 3 to 4 innovative capability for process and production organisation and equipment. At Level 4, USIMINAS undertook several 'capacity-stretching' projects across the plant, as in the case of the steel shop (see Box 6.1). For instance, by the late-1980s, USIMINAS had stretched the capacity of the steel shop from 3.5 to 4.2 million tpy of steel, and by the early-1990s the capacity of BF 3 had been stretched from 2 to 2.47 tonnes/m^3/day.

By the 1990s, as mentioned in the interviews in the company, that capability had become part of the company's daily routine (see Chapter 8). In addition, the sustaining of high capacity utilisation rates in the long term may be associated with the accumulation of Level 5 innovative capability to create in-house process control systems to achieve an improved integration between PPC and process control systems (see Chapter 8).

In contrast, the evidence in Chapters 6 to 8 does not suggest that CSN engaged in the systematic accumulation of 'capacity-stretching' capability. It was not until the 1990s that CSN experienced higher capacity utilisation rates on a continuous basis. This may be associated with the completion of accumulation of Levels 1 and 2 routine capability and Levels 3 to 4 innovative capability for process and production organisation (see Chapter 8). Differences in overall energy consumption (Mcal/tonne of steel) clearly reflects the fast accumulation of higher levels of capability for process control in USIMINAS as opposed to CSN. For instance, by the mid-1980s USIMINAS engaged in efforts to build the energy centre within the plant. This organisational unit sought to improve the energy performance of the company. It should be noted that a large part of the structuring of that unit was done by USIMINAS independently.[42] The evidence from CSN suggests that systematic in-house efforts for energy efficiency improvement over the 1970s and 1980s were limited. It was not until the 1990s that the company reorganised and improved its efforts on energy efficiency.

The evidence suggests that the achievements of competitive product-related performance (e.g. number of product quality certificates, integrated yield rate) in USIMINAS are strongly associated with the accumulation of product capability at Levels 1 and 2 (routine) and Levels 3 to 4 (innovative) within ten years (e.g. continuous upgrading of its quality systems), as described in Chapters 6 to 8. In CSN, the improvement in the number of quality certificates and integrated yield rates during the 1990s seems to be strongly associated with the completion of accumulation of Levels 1 and 2 and also innovative Levels 3 to 4 capability for products (e.g. the TQM programme and efforts to improve product quality control in the rolling mills). However, these improvements took place in CSN much later than in USIMINAS.

10.3   CONCLUSIONS

The evidence in this chapter suggests that during its lifetime USIMINAS achieved and sustained competitive operational performance across different indicators. Although the coke rates in BF 3 did not drop substantially, the company achieved substantial improvements in the rate of increase of productivity in this furnace. In addition, USIMINAS achieved substantial improvements in key steelmaking process indicators and overall energy consumption as opposed to CSN. The evidence also suggests that the fast rates at which USIMINAS improved its performance over time are strongly associated with the rates of accumulation and sustaining of capability across different technological functions over time. In contrast, in CSN during the 1940s–1980s period, operational performance improved at a slow rate across different indicators. As demonstrated in Part II, during this period CSN had not completed the accumulation of Levels 1 and 2 capability and had accumulated innovative capability at slow rates.

However, during the 1990s CSN experienced improvements in indicators in the reduction process, capacity utilisation, and integrated yield. Indeed, during the 1990–97 period, CSN even outperformed USIMINAS in the rate of coke rate reduction in BF 2 and BF 3 and in the rate of productivity increase in BF 2. However, these improvements were not followed by any significant improvements in other indicators like those for the steelmaking process and overall energy consumption. The evidence suggests that within a few years (1990–97), CSN sought to capture, as much as possible, the improvements that USIMINAS had been achieving since the early-1960s. However, CSN was not able to catch up with, or outperform, USIMINAS across a wide range of indicators.

In sum, the experience of USIMINAS suggests that if CSN had accumulated technological capability at similar rates to USIMINAS over the 1940s–1980s period, the company could have achieved faster rates of performance improvement earlier and caught up with world competitive levels much more rapidly.

NOTES

1. Data related to fuel rates were obtained from: (i) USIMINAS, 'Dados operacionais dos Altos Fornos', September 1997 [internal publication]; consultation with the technical information centre and interviews in the company; (ii) CSN (1996) and interviews in the company.
2. This technology seeks to reduce the coke consumption.
3. It should be noted that in USIMINAS despite the efforts to improve the metallic charge, there was the occurrence of hard crust in BF 3 (1978–79). In 1981 the furnace went through a revamping. These factors contributed to affecting the coke rate performance during the furnace's first campaign. Despite these factors, productivity performance improved.
4. Except in 1980–83 and 1985–86 for BF 3.
5. It should be noted that by 1960 the average coke rate in Japan was 610kg (see Japan Iron and Steel Industry Federation, 1965). By 1960 CSN was 14 years of age and USIMINAS had not started up. In CSN the coke rates ranged from 760 to 814kg indicating that its BF performance was still below international competitive levels.
6. CSN (1996).
7. USIMINAS (1997).
8. CSN (1996). Interviews in the company.
9. For data on the Japanese mills see Gupta et al. (1995).
10. USIMINAS (1994).

11. Gupta et al. (1995).
12. It should be noted that by 1998 and 1999 USIMINAS was considered to be the most effective steel company in Brazil in combining high yield in the coal fines injection process with high productivity in its BFs. See *Metalurgia & Materiais*, ABM, various numbers, 1998/9.
13. Technical paper by two USIMINAS engineers on the influence of these in-house efforts on the reduction of the coke rate in M.M. Mendes and J.B. Costa (1964), 'Preparação das matérias-primas e seu efeito no alto-forno da USIMINAS. *XIX Congresso Anual da ABM*, São Paulo, July; Also 'USIMINAS: 25 anos', *Metalurgia & Materiais*, Special Edition, 1987.
14. Technical paper by three USIMINAS's engineers, L.C. Abreu et al. (1974).
15. Interview followed by tour in the BF area with a technician from the BF and coke ovens dept. See also technical paper by three engineers from USIMINAS: A. S. Carneiro et al. (1993), 'Fabricação e utilização de ventaneiras para altos-fornos', *Metalurgia & Materiais*, vol. 49, no. 414, pp. 101-5.
16. Interview with a BF engineer. Also CSN, Annual Reports 1962–63.
17. 'CSN: 50 anos transformando a face do país', *Metalurgia & Materiais*, vol. 47, no. 394, 1991. Interview with a BF engineer.
18. 'USIMINAS 25 anos', *Metalurgia & Materiais*, Special Edition, 1987.
19. CSN (1996) and search into the company's archival records.
20. 1963 and 1977 are the start-up years of the oxygen steel shop in USIMINAS and CSN, respectively. CSN used the Siemens-Martin steelmaking process exclusively from the start-up (1946) until 1977. The SM steel shop was shut down permanently in 1981. For a more meaningful comparison this subsection only examines the differences between the oxygen steelmaking process in the two companies.
21. 'Aciaria: aumento de campanha de convertedores' [internal publication, undated].
22. CSN, Annual Report 1993.
23. Paper by four engineers from USIMINAS's steel shop: Fusaro et al. (1965).
24. Interview with a retired steel shop foreman. Also CSN (1996).
25. 'Catch-carbon' is an operational procedure whereby the desired level of carbon is achieved by blowing oxygen to oxidise the carbon during the steelmaking process. As a result, it brings the carbon down to the desired level, say, 0.4 per cent when it is 'caught'.
26. The 'static charge control' strategy is based on statistical, predictive-adaptive control from static models. This control seeks to prescribe the adequate combination of the charge materials (e.g. hot metal, scrap, fluxes, and oxygen) required to meet the endpoint conditions.
27. Interview with the manager of the steel shop technical division. Also 'USIMINAS 25 Anos' *Metalurgia & Materiais*, Special Edition, 1987; 'USIMINAS: 22 anos de transferência de tecnologia', 1994.
28. Interview with a steel shop engineer. Also 'CSN atinge a marca dos 100 milhões', *Metalurgia & Materiais*, 1997
29. The 'dynamic charge control' is based on continuous sampling and analyses of process variables like carbon and temperature. This automated sampling can be done through a sublance (the widely-used method), off-gas analysis, or sonic analysis.
30. Interview with the manager of the steel shop technical division. Also 'USIMINAS 25 Anos' *Metalurgia &Materiais*, Special Edition, 1987; 'USIMINAS: 22 anos de transferencia de tecnologia', 1994.
31. 'CSN alcança a marca de 100 milhões', *Metalurgia e Materias*, 1997.
32. CSN, Annual Report 1979.
33. The more effective overall energy performance of USIMINAS in relation to COSIPA during the 1977-91 period was analysed in Piccinini (1993).
34. CSN, Annual Reports 1980–1984
35. CSN, Annual Reports 1980–1984

36. USIMINAS, Annual Reports, 1990–96; technical information centre; CSN, Annual Reports 1972–97; 'Average operational productivity 1989–97' [overhead of internal presentation, one page, undated].
37. USIMINAS, Annual Reports and Special Reports 1962–97; CSN, Annual Reports 1989-97.
38. R. C. Soares (1997), 'O ambiente de negócios e os fatores de produtividade na siderurgia', *M&M Metalurgia e Materiais*, vol. 53, no. 470, pp. 600–8, October.
39. USIMINAS, Annual Reports 1983-96; 'USIMINAS 25 anos *Metalurgia e Materiais*, 1987, op cit.; interviews in the company. CSN, Annual Reports, 1983-95; *Nove de Abril* (several numbers); interviews in the company.
40. USIMINAS's Annual Report, 1984-85, 1988–97. CSN, 'The Quality Promotion Centre' (1997), mimeo.
41. USIMINAS, Annual Reports and Special Reports 1962–97; CSN, Annual Reports 1989–97.
42. Interview with a manager at the energy centre.

11. Conclusions

This book has focused on key features of the learning processes in two companies in order to explain differences in their paths of technological capability-accumulation and, in turn, in their operational performance improvement. This set of relationships was examined in two large integrated steel companies in Brazil. These issues have been addressed in different ways in both the LCL and the TFCL. However, the problem of how learning processes influence inter-firm differences in paths of technological capability-accumulation and, in turn, the rate of operational performance improvement has not been explored in either body of literature. To examine the paths of technological capability-accumulation this book has drawn on a broad definition and a framework developed in the LCL. To examine the learning processes the book has drawn on available frameworks in the TFCL. The research strategy combined the qualitative with descriptive quantitative approaches and was based on a comparative in-depth case-study. This book has drawn on empirical evidence gathered through detailed fieldwork.

11.1 REVIEWING THE RESEARCH QUESTIONS

This section briefly reviews the three questions addressed in this research:

1. How different were the paths of technological capability-accumulation followed by two large steel companies in Brazil, over time?
2. To what extent can those differences be explained by the key features of the various processes by which knowledge is acquired by individuals and converted into the organisational level – the underlying learning processes?
3. What are the implications of the technological capability-accumulation paths for operational performance improvement in these companies?

The review below begins with question (1) followed by (3) and then returns to question (2).

11.1.1 Differences in Technological Capability-accumulation Paths

This book has reconstructed the paths of technological capability-accumulation followed by USIMINAS and CSN during their lifetime. These paths have been described in an adequate level of detail in the empirical Chapters 6 to 8. Chapter 9 summarised these differences. The paths followed by the two companies were diverse and proceeded at differing rates. In other words, they differed in terms of: (1) rate of accumulation; (2) consistency of accumulation; and (3) composition of technological capability.

Rate of accumulation

Table 9.1 summarised the differences between the two companies. In general, it took 35 years for USIMINAS to accumulate and sustain Levels 5 and 6 capabilities across all five technological functions. In contrast, it took nearly 50 years for CSN to reach Levels 4 and 5 innovative capability for process and production organisation and products, respectively. In addition, it took more than 45 years for CSN to complete the accumulation of Levels 1 and 2 routine operating capability for process and production organisation and products. In contrast, these levels of capability were accumulated within ten years in USIMINAS.

Consistency of accumulation

USIMINAS moved from Level 1 to Levels 5 and 6 capability across all five technological functions in a consistent way. As innovative technological capabilities were accumulated, the company had its routine operating capabilities strengthened over time. This ability to be in control of these two trajectories allowed USIMINAS to undertake innovative activities effectively. By the 1990s USIMINAS engaged in the sustaining and deepening of the technological capabilities (or their strategic elements) that had been accumulated in the previous decades. In contrast, CSN followed an inconsistent path of technological capability-accumulation. During the 1940s–1980s period innovative capabilities were not accumulated beyond Level 4. In parallel, the routine operating capabilities (Levels 1 and 2) were accumulated incompletely during that period. It was not until the early-1990s that CSN completed the accumulation of these levels of capability. In addition, CSN moved into the accumulation of Levels 4 and 5 innovative capability for process and production and products, respectively. However, the existing capability for investments and equipment was weakened. The evidence suggests that for a sustainable engagement in innovative technological activities, the path of accumulation of 'routine operating technological capabilities' has to run in parallel with the accumulation and sustaining of 'innovative technological capabilities'.

Composition of technological capability

In USIMINAS technological capability was accumulated and embodied in both individuals and organisational systems. In other words, it was accumulated in a broad way over time. In CSN, however, particularly during the 1940s–1980s period, technological capability was accumulated as skills and knowledge embodied in individuals. In other words, it was accumulated in a narrower way. It was not until the 1990s that CSN began to accumulate technological capability at both the individual and organisational levels.

11.1.2 Technological Capability-accumulation Paths and Operational Performance Improvement

Chapter 10 of the book examined the differences between USIMINAS and CSN in operational performance improvement over their lifetime. The chapter demonstrated that USIMINAS and CSN differed across the 14 indicators of performance examined. These differences were even more substantial during the 1962–89 period in USIMINAS and the 1946–89 period in CSN. These time-periods are equivalent to their start-up and conventional expansion phases. USIMINAS achieved a fast rate of operational performance improvement from the initial years. As a result, the company was able to catch up more rapidly than CSN with international competitive levels of operational performance.

CSN, in contrast, went through the 1940s–1980s period with a slow rate of operational performance improvement across most of the indicators examined. It was not until the 1990s that CSN experienced substantial performance improvements in some indicators in relation to its previous phases. As indicated in Chapter 10, CSN even outperformed USIMINAS in two indicators in the iron making process, each of them in a different facility. However, despite this improvement in performance during the 1990s, CSN could not catch up with the standard of USIMINAS across most indicators. In addition, Chapter 10 explored the role of technological capability–accumulation in influencing the differences between USIMINAS and CSN across the indicators of operational performance improvement. Chapter 10 suggested two conclusions:

1. The way and rate at which USIMINAS accumulated technological capability contributed to improving several performance indicators at a fast rate. In contrast, Chapter 10 suggested that the slow rate of performance improvement in CSN, particularly during the 1940s–1980s period, was associated with absent and/or slow rates of accumulation of technological capability. However, it was suggested that in CSN the improvement in some indicators during the 1990s was associated with the completion of accumulation of Levels 1 and 2 and Levels 4 and 5 capability, particularly for processes and production organisation and products.

2. Therefore, the book suggests a strong association between rates of operational performance improvement and the rate of accumulation, the consistency over time, and the composition of the paths of technological capability-accumulation.

Some implications for financial performance

Although Chapter 10 did not explore improvements in production costs, it suggested that differences in some indicators of operational performance could have affected production costs differently in USIMINAS and CSN. These effects may have been reflected in the operating income margin (OIM) of these companies.[1] For instance, during the 1980s USIMINAS went through a sequence of positive OIMs. In contrast, during this same period, CSN went through a sequence of seven years of negative OIMs, as indicated in Table 11.1.

Table 11.1   Operating income margin[a] in USIMINAS and CSN during the 1982–2000 period (%)

	USIMINAS	CSN
1982	(0.95)	(18.8)
1983	8.5	(26)
1984	19.9	(8.6)
1985	23.2	(2.5)
1986	2.5	n. a.
1987	13	n. a.
1988	38.3	(44.1)
1989	31.8	(4.48)
1990	35.5	(45)
1991	n.a.	12.3
1992	18.9	49.1
1993	22.2	n. a.
1994	24	40.8
1995	30.3	13
1996	26.5	10.4
1997	13.5	11.5
1998	15.9	25.6
1999	(16.8)	10.5
2000	12.7	52.3

Notes: (a) Operating income/net sales. Numbers in brackets mean negative OIM. n.a. = not available.
Sources: USIMINAS, Annual Reports 1982–2000; CSN, Annual Reports 1982–2000.

As pointed out in the Annual Report of CSN for 1990, the company had an average negative OIM equivalent to US$314 million per year during the eight-year period 1982–90.[2] In contrast, as indicated in Table 11.1, USIMINAS experienced more effective financial performance during that period. By the 1980s USIMINAS had consolidated its leading position in terms of financial performance among the state-owned steel companies in Brazil.[3] It was not until 1991 that CSN engaged in continuous achievement of positive OIMs.[4]

The inter-firm differences in operational performance improvement also seem to have been reflected in the final prices at which the two companies were sold in their privatisation processes. USIMINAS was privatised in October 1991 and CSN in April 1993. On the basis of data from BNDES, the organisation responsible for the privatisation programme in Brazil and the two companies,[5] Table 11.2 outlines the key financial differences between the two companies in their privatisation process.

Table 11.2 Key financial differences between USIMINAS and CSN in their privatisation processes

	USIMINAS[a]	CSN[b]	Ratio USIMINAS/CSN
Sale proceeds (US$ million)[c]	2,059	1,495	1.38
Installed capacity in the year of privatisation (million tonnes)	4.2[d]	4.6[e]	0.91
Crude steel production volume in the year of privatisation (million tonnes)	4.1	4.3	0.95
Sale proceeds/installed capacity (US$ million)	490	325	1.51
Sale proceeds/crude steel production volume (US$ million)	502	347	1.45

Notes: (a) Privatised in April 1991. The second phase of the privatisation process was completed in September 1994.
(b) Privatised in October 1993.
(c) It should be noted that during the main privatisation auction (April 1991), USIMINAS was sold at a price 14.3 per cent higher than the minimum price fixed by BNDES. In contrast, CSN was sold at the minimum price. Value in dollars of 1991 was adjusted to inflation (1991–92).
(d) That refers to the 'stretched' capacity. The nominal capacity was 3.5 million tpy.
(e) That refers also to nominal capacity.
Sources: BNDES (1994); USIMINAS (Annual Reports 1990–91 and the technical information centre); CSN (Annual Reports, 1992–94).

As indicated in Table 11.2, USIMINAS's assets were given a market value 50 per cent higher (US$ 490 million per tonne of installed capacity) than CSN's assets (US$ 325 million per tonne of installed capacity). It is reasonable to consider that a large part of this difference reflected the greater knowledge that was embodied in USIMINAS's physical, human and organisational capital. In a similar vein, one might reasonably argue that if CSN had accumulated knowledge as effectively as USIMINAS over preceding decades, and had embodied it effectively in physical capital, people, organisation and procedures, it might have increased the market value of its assets by as much as US$760 million.

11.1.3   Learning Processes and Paths of Technological Capability-accumulation

This book has explicitly explored the influence of key features of the learning processes on the paths of technological capability-accumulation followed by the two case-study companies over their lifetime. In the light of the framework in Table 3.2, Chapters 6 to 8 described and compared how four learning processes (external and internal knowledge-acquisition and knowledge-socialisation and codification processes) worked within the two companies. This was done on the basis of four features of the learning processes (variety, intensity, functioning, and interaction). Chapter 9 explored the influence of these key features on inter-firm differences in the paths of technological capability-accumulation. As a result, the book suggests three conclusions:

1. There is a strong association between the key features of the knowledge-acquisition and knowledge-conversion processes and the rate, consistency, and composition of the paths of accumulation of technological capability within individual firms.
2. Therefore, the way the learning processes work over time within individual companies is a major contributor to the inter-firm differences in paths of technological capability-accumulation.
3. In addition, the rate of technological capability-accumulation can be accelerated (or slowed) depending on how the company deliberately manipulates the key features of the four learning processes over time. These processes, and their features, may give rise to 'effective' or 'ineffective' 'learning systems' within companies. These 'learning systems' have practical implications for the paths of technological capability-accumulation, and in turn, for the rate of operational performance improvement.

This book has demonstrated that although both companies engaged in learning processes, the key features of these processes differed over time. Indeed the four features of the learning processes in USIMINAS gave rise to an 'effective learning system'. In contrast, in CSN, particularly during the 1940s–1980s period, those features gave rise to an 'ineffective learning system'. A key characteristic of CSN in that period was the failure to convert individual into organisational learning. These 'learning systems' had different implications for the paths of technological capability-accumulation and, in turn, for the rate of operational performance improvement in the two companies. In the light of Figure 3.2 these relationships are summarised in Figure 11.1.

Therefore, the book suggests, on the basis of USIMINAS's experience in comparison with CSN's, that if the key features of the learning processes are deliberately and effectively manipulated over time they produce positive implications for the paths of technological capability-accumulation. This, in turn, has positive implications for the rate of operational performance improvement, and may also have positive implications for financial performance improvement. In other words, continuous and effective in-house efforts on knowledge-acquisition and knowledge-conversion processes do appear to pay off.

Key features of the 'learning systems': processes for converting individual learning into organisational assets	→	Implications for the firms' paths of technological capability-accumulation	→	Implications for the rate of operational performance improvement
USIMINAS (1960s–1990s) 1. Diverse variety 2. Continuous intensity 3. Good functioning 4. Strong interaction	→	Fast rate, consistent, and broad composition of capability accumulated and sustained across all five technological functions (Levels 1 to 5–6)	→	Fast rate of improvement across diverse operational performance indicators. Early catching up with international standards
CSN (1940s–1980s) • Limited variety • Intermittent intensity • Poor functioning • Weak interaction	→	Slow rate, inconsistent, and narrow composition of capability accumulated across a limited number of technological functions (not beyond Level 4)	→	Slow rate of improvement across diverse operational performance indicators
CSN (1990s) • Relatively improved features of the learning processes	→	Completion of accumulation of Levels 1 and 2 capability. Accumulation of Levels 4 to 5 across two technological functions	→	Improvement in a relative number of operational performance indicators. Late catching up with international standards

Source: Own elaboration based on the research.

Figure 11.1   Summary of the relationships between the issues investigated in the book

11.2   OTHER FACTORS AFFECTING THE ISSUES EXAMINED IN THE BOOK

As pointed out in Chapter 3 (Section 3.2.2), this book is aware that other factors may influence the learning processes, the paths of technological capability-accumulation, and the operational performance improvement. As pointed out in Chapter 3, these factors may involve leadership behaviour and external conditions. The book, particularly in Chapters 6 to 8 and Chapter 10, has recognised their presence in the case-study companies. However, their influence is outside the scope of this book and they have been examined very superficially. Nevertheless, a few comments on these two factors are appropriate at this stage.

11.2.1   Corporate Leadership Behaviour

The evidence in Chapters 6 to 8 suggests that the USIMINAS corporate leadership exerted a positive and effective influence on the learning processes over time. For instance, it played a critical role in influencing the creation of key mechanisms like the systematic channelling of external codified knowledge, disciplined and effective OJT, continuous overseas training, and the research centre. This behaviour therefore is congruent with the role of leadership in 'learning organisations', as pointed out in Senge (1990). On other occasions, corporate leadership set challenging goals to push USIMINAS into innovative activities independently (e.g. the planning and implementation of Expansion III). This is congruent with the role of corporate leadership explored in Kim (1995, 1997a,b).

In contrast, in CSN, particularly over the 1940s–1980s period, corporate leadership seemed indifferent to the poor functioning of some learning processes (e.g. importing foreign expertise to lead in-house training; poor functioning of the OJT; poor knowledge links between the researchers and operations engineers). On some occasions, corporate leadership contributed to mitigating the functioning of other mechanisms (e.g. limiting the innovative activities at the research centre; or limiting in-house training programmes). In addition, in CSN from the 1940s, leadership championed the 'Giraffe' reward system. Together with the constant goals to increase production volume at any cost, this system led to the emergence of the 'volume-first practice'. This 'belief' was strong in CSN until the late-1980s. As suggested in Chapter 8, it seemed to re-emerge from 1996. This is in line with the view that features of corporate behaviour may stimulate or constrain the learning processes (e.g. Schein, 1985; Dodgson, 1993; Hedberg, 1981). However, in CSN during the early-1990s, leadership played a critical role in championing the TAG and later the TQM programme thus demonstrating positive and effective behaviour. These actions triggered improvements in knowledge-acquisition and knowledge-conversion processes.

Although more investigation of this issue is needed, the book suggests that behaviour of corporate leadership (including its technical background) may play a critical role in influencing the learning processes. However, positive and effective leadership alone seems not enough to influence firms' paths of technological capability-accumulation if it is not coupled with learning processes. For instance, USIMINAS would certainly not have followed such a path on the basis of corporate leadership behaviour only. In CSN, from time to time, leadership even expected the firm to accumulate technological capability. However, the learning processes as a whole did not work well.

11.2.2 External Conditions

The evidence in Chapters 6 to 8, suggests that external conditions, in particular the IS policy, market protectionism, and the competitive environment, may have exerted some influence on the three issues investigated in the case-study companies over their lifetime. The evidence indicates that even when operating in similar time-periods and under similar conditions the two companies behaved differently. For instance, both USIMINAS and CSN were stimulated by the IS policy to develop in-house technological capability for project engineering and equipment engineering. However, the companies differed in the way and rate at which they built, accumulated, and sustained those capabilities over time, as demonstrated in Chapters 6 to 8 and represented in Figures 9.1 and 9.4.

Indeed during the 1960s–1980s period both USIMINAS and CSN operated under conditions characterised by IS policy and high protectionism of the economy in Brazil. However, USIMINAS continuously sought to accumulate its own technological capabilities to catch up with world competitive standards of operational performance. In contrast, in CSN these efforts were absent. The evidence in Part II suggests that the absence of additional competitive pressure (beyond the presence of USIMINAS in the industry) contributed to influencing that behaviour in CSN. In addition, as demonstrated in Chapter 10, during the 1970s–1980s period both companies operated under the same world energy crises. However, their performance in the reduction and steelmaking processes and, in particular, in overall energy consumption differed substantially. However, during the 1990–97 period, in response to intense competitive pressures and a series of crises, CSN engaged vigorously in in-house efforts to improve its learning processes in order to accumulate technological capability as a way of improving its operational performance.

In addition, during the conventional expansion phase (Chapter 7), particularly during the 1970s, both companies were granted fiscal incentives by government (e.g. Law 6,297 of 1975) to invest in training facilities. USIMINAS used those incentives to upgrade its training centre and improve the functioning of its in-house training programme. In contrast, although CSN built a brand new training facility, the functioning of its in-house training programme was not improved. By the mid-1980s, the training facilities were even shut down as a cost-reduction measure. During the liberalisation and privatisation phase (1990s), as indicated in Chapter 8, both companies were influenced by the 'Brazilian Quality and Productivity Programme' and the 'Industrial Technological Capability Programme' (early-1990s). While USIMINAS responded by routinising its existing process and product quality systems, CSN engaged, for the first time, in the building up of a consistent quality system through the TQM programme. In addition, by the mid-1990s USIMINAS used the incentives of the Industrial Technological Development Programme to upgrade and re-organise its research centre. In contrast, CSN sought to limit the activities of this unit.

On the basis of the evidence in Part II, it is reasonable to argue that in USIMINAS corporate leadership, particularly during the 1960s–1980s period, played a critical role in offsetting the market variable, and pushing the company into the continuous achievement of world competitive performance. In contrast, in CSN during that same period, the absence of similar leadership contributed to opening up the way for the negative influence of a highly protected market on CSN's learning processes, technological capability-accumulation paths, and operational performance improvement.

In sum, although this book has addressed the issue of 'external conditions' very superficially, the evidence does not seem to support the argument in Kim

(1997b) that successful technological capability-accumulation 'requires' an effective national innovation system. In other words, that it depends on external conditions. However, the evidence seems more in line with the argument in Dutrénit (2000) that the learning processes in the firm are a major contributor to its path of technological capability-accumulation. Nevertheless, the evidence does suggest that in addition to the learning processes other factors are also necessary, in particular, an effective and positive corporate leadership and a competitive market environment.

11.3   IMPLICATIONS FOR MANAGEMENT, POLICY AND FUTURE RESEARCH

11.3.1 Implications for Management

The book generates conclusions for managers in large latecomer companies within 'heavy industries', particularly steel. These conclusions may contribute to improving their long-term and/or day-to-day decisions related to three issues: (1) path of technological capability-accumulation (below); (2) learning processes; and (2) rate of operational performance improvement.

Path of technological capability-accumulation
The book has demonstrated that the accumulation of in-house technological capability is a critical task for the latecomer company to overcome its technological barriers and catch up with the technological frontier. In particular, managers should be concerned with the rate, consistency, and composition of technological capability being accumulated. Although catching up with the technological frontier through the accumulation of capability is feasible, the task is complex and costly, demanding continuous, deliberate, and effective efforts from within the firm. It is easy to weaken that technological capability if intra-firm efforts are diminished, and/or neglected or become ineffective.

This book has indicated that the accumulation of technological capability on the basis of a narrow composition (individuals' skills and knowledge) rather than on a broad basis (the firm's own organisational system and managerial arrangements) contributes very little: (1) to accelerating the rate of improvement in processes and production organisation, products, equipment, and project engineering activities; and (2) to accelerating the catching up with the technological frontier. Therefore, decision-making and corporate actions on technological capability development, at least in the steel industry, must be made on a broad basis. In addition, the company should constantly trace its own path to know its behaviour in greater detail. This would permit timely strategic actions to be taken.

In addition, firms should ensure that innovative activities are not concentrated in particular organisational units (e.g. a research centre). Indeed firms should facilitate the involvement of different units and individuals, particularly the production lines, in continuous improvements in process and production organisation, equipment, and products. This perspective is highly supported by the current view on continuous improvement (e.g. Bessant, 1998).

These conclusions are not fundamentally different from the other relevant studies in the LCL (e.g. Dahlman and Fonseca, 1978; Katz et al., 1978; Maxwell, 1981; Bell et al., 1982; Lall, 1987; Hobday, 1995; Kim, 1995, 1997a, 1997b). However, the conclusions here rest on a stronger basis of comparative analysis than these earlier studies, and they draw on a greater level of detail over longer time-periods. As a result, this book has identified inter-firm differences in paths of technological capability in terms of rate of accumulation, consistency over time, and composition of capability. In addition, the book has explored the role of key features of the underlying learning processes in influencing inter-firm differences in the paths of technological capability-accumulation. In doing so, this book has moved beyond the tradition in the LCL of describing the paths of technological capability-accumulation followed by individual firms. In addition, this book has moved beyond the studies in the LCL and also in the TFCL (e.g. Leonard-Barton, 1990, 1992a,b, 1995; Garvin, 1993; Teece and Pisano, 1994; Iansiti, 1998; among others) by exploring the influence of the paths of technological capability-accumulation, and in turn the underlying learning processes, on inter-firm differences in the rate of operational performance improvement over the lifetime of two large companies. The implications of these for the financial performance improvement were also briefly explored.

Learning processes
Key features of the knowledge-acquisition and knowledge-conversion processes are critical in accelerating or slowing down the rate of firms' paths of technological capability-accumulation. More specifically, this study suggests that although knowledge-acquisition processes are critical for the building of capability they are not sufficient. The firm must also engage in parallel processes to convert individual into organisational learning. For example, sending engineers to overseas training in the absence of effective knowledge-socialisation and knowledge-codification processes may contribute very little to accelerating the path of technological capability-accumulation in the firm.

For organisational learning to take place effectively, there must be deliberate and effective daily manipulation of the key features of different learning processes (variety, continuity, functioning, and interaction). Relying on single mechanisms, no matter how powerful they seem to be (e.g. large in-house training facilities and programmes), is unlikely to yield any effective organisational learning. Because of the intricate and systemic nature of the learning processes, a diverse variety of mechanisms needs to be built and continuously improved over time and across corporate levels. This may lead to the routinisation of the conversion of individual into organisational learning. This is necessary for the company to progress rapidly and consistently in the accumulation of capability from Levels 1 to 6.

In sum, this book has generated detailed understanding of the organisational dimensions of the learning processes based on inter-firm comparative analysis. In doing so the book goes further than has yet been derived from studies of the latecomer firm, except probably Kim (1995, 1997a,b) and Dutrénit (2000). In addition, and by exploring the influence of key features of the learning processes on the paths of technological capability-accumulation and, in turn, the operational performance improvement in the long term, the book goes further than studies in the TFCL based on the 'learning organisation' perspective (e.g. Argyris and Schön, 1978; Hedberg, 1981; Cohen and Levinthal, 1990; Garvin, 1993; Huber, 1996a,b; among others). The book also goes further than the studies in the TFCL related to the 'knowledge-building-firm' perspective based on conceptual approaches (e.g. Nelson and Winter, 1982; Dosi and Marengo, 1993; Nonaka, 1994; Teece and Pisano, 1994) and empirical approaches (e.g. Leonard-Barton, 1990, 1992a,b, 1995; Nonaka and Takeuchi, 1995; Iansiti, 1998).

Operational performance improvement
Latecomer companies have to cope with a moving target of international performance indicators. This is not easy. The challenge for the latecomer company is to achieve a fast rate of performance improvement and sustain it over time. This book has indicated that the rate of technological capability-accumulation and, in turn, of the learning processes, play a substantial role in accelerating and/or slowing the rate of improvement in operational performance across different indicators. Since performance indicators, particularly in heavy industries, are very much interrelated, capability for different technological functions must be accumulated. Therefore, the rate at which the latecomer firm can approach the technological frontier depends on the rate and nature of its engagement in in-house technological capability-accumulation and in its knowledge-acquisition and knowledge-conversion processes. In addition, although this study could not develop the point in detail, it has illustrated that large financial benefits are likely to be generated by the firm that manages these learning processes effectively.

The conclusions in this book relative to operational performance improvement confirm, but also move beyond, other studies in the LCL which have tackled the relationship between technological capability and operational performance (e.g. Katz et al., 1978; Dahlman and Fonseca, 1978; Bell et al., 1982; Mlawa, 1983; Tremblay, 1994).

11.3.2 Implications for Policymaking

This book was not designed to address issues concerned primarily with government policy. However, its evidence highlights the importance in latecomer countries of policies designed to stimulate and reinforce the development of technological capability in industrial firms. In the first place, the book suggests, on the basis of CSN's experience, that government policies based on highly protected markets probably contribute to limiting intra-firm learning processes, and in turn, the paths of technological capability- accumulation and operational performance improvement. In other words, the design of government policies that ignores competitive pressures is unlikely to yield fast rates and consistent development of technological capability in industrial firms.

In the second place, as indicated by the evidence in Part II, the two case-study companies behaved differently under similar government policy based on fiscal incentives. The book suggests, therefore, that if fiscal incentives are to be used as a mechanism to stimulate firms' development of technological capability, they should be granted on the basis of firms' rates and levels of capability-accumulation. For example, a firm moving quickly towards the accumulation of Levels 5 and 6 innovative capability across different technological functions should not be granted the same incentive as a firm that accumulated capabilities only slowly up to Level 3 across limited technological functions or has shown a weakening of existing capabilities. Slower firms may need a different incentive combined with exposure to increased international competitive pressures. In order to operationalise this strategy, government could draw more on independent research institutions, with recognised capability, to develop empirical studies on technology and innovation within the industry.

11.3.3 Suggestions for Future Research

This book has shown that most of the studies in the LCL have not linked the issue of paths of technological capability-accumulation with the adequately detailed understanding of issues about the underlying learning processes. Therefore it has explored the relationship between learning processes and paths of technological capability-accumulation and its implications for operational performance improvement. However, this book has addressed only very superficially other issues like the influence of corporate leadership behaviour and external conditions. Therefore it would be fruitful for the LCL if future research could explore the influence of features of corporate behaviour (corporate leadership, beliefs, etc.) on learning processes and paths of technological capability-accumulation in a more systematic way. This would need a stronger link between the LCL and the TFCL. Additionally, future research could explore in greater detail how the internal learning processes in firms are associated with the way firms develop their links with government and/or industrial institutions. Also, the influence of external conditions on learning processes and paths of technological capability-accumulation could be investigated more systematically. In particular, future research could draw on the framework developed in this book and apply it to a comparison of companies in heavy industries, operating in different countries. This would permit a detailed assessment of whether and how different country contexts and policy regimes exert any influence on key features of the learning processes.

This book has examined different indicators of operational performance. However, the book has not addressed production costs or company profitability. Future research could address the improvement in production costs and other financial indicators. This would permit a deeper understanding of the link between technological capability and financial and economic performance. The book has examined the accumulation of capabilities which are technological. Future research could extend the scope of analysis into marketing and/or financial capabilities. This would permit a greater understanding of the paths of capability-accumulation in latecomer firms. Indeed, a comparative analysis would help the understanding of how latecomer firms differ in terms of a broader range of capability-accumulation. If the above suggestions could be implemented, they would contribute to deepening the understanding, as was the primary concern of this book, of how to accelerate the technological capability-accumulation paths in the latecomer firm.

NOTES

1. Operating income margin = operating income/net sales.
2. CSN, Annual Report 1990.
3. The superior financial performance of USIMINAS in relation to COSIPA in the 1977–91 period was compared in Piccinini (1993).
4. CSN, Annual Reports 1991–97; USIMINAS, Annual Reports 1991–97.
5. BNDES (1994); Paula (1998); USIMINAS (consultation at the technical information centre).

Bibliography

ABM (Brazilian Association of Materials and Metallurgy) (1996), 'Perspectivas na Siderurgia para o Desenvolvimento da Indústria Automobilística no Brasil'(Perspectives on the Steel Industry for the Development of the Automobile Industry in Brazil), mimeo, ABM: São Paulo.

Abreu, L.C. (1974), 'Evolução da Produtividade dos Altos Fornos da USIMINAS' (Evolution of productivity of USIMINAS's Blast Furnaces), paper presented at the XXIX Annual Meeting of the Brazilian Materials and Metallurgy Association, July.

Araujo, L.A. (1997), *Manual de Siderurgia (Handbook of Steelmaking)*, vol.I, São Paulo: Arte & Ciência.

Argyris, C. and D. Schön (1978), *Organizational Learning: a Theory of Action Perspective*, Reading, MA: Addison-Wesley.

Ariffin, N. and M. Bell (1996), 'Patterns of Subsidiary-parent Linkages and Technological Capability-building in Electronics TNC Subsidiaries in Malaysia', in K. S. Jomo and G. Felker (eds), *Industrial Technology Development in Malaysia*, London: Routledge, pp. 150-90.

Baer, W. (1969), *The Development of the Brazilian Steel Industry*, Nashville, TN: Vanderbilt University Press.

Baer, W. (1994), *The Brazilian Economy: Growth and Development*, London: Praeger.

Bell, M. (1982), 'Technical Change in Infant Industries: a Review of the Empirical Evidence', mimeo, SPRU, University of Sussex.

Bell, M. (1984), '"Learning" and the Accumulation of Industrial Technological Capacity in Developing Countries', in K. King and M. Fransman (eds), *Technological Capability in the Third World*, London: Macmillan.

Bell, M. (1997), 'Overheads and Notes on Lectures and Seminars', Technology and Development Course, MSc in Technology and Innovation Management Course, SPRU, University of Sussex.

Bell, M. and K. Pavitt (1993), 'Technological Accumulation and Industrial Growth: Contrasts between Developed and Developing Countries', *Industrial and Corporate Change*, **2** (2), 157– 211.

Bell, M. and K. Pavitt (1995), 'The Development of Technological Capabilities', in I. u. Haque (ed.), *Trade, Technology and International Competitiveness*, Washington, DC: World Bank.

Bell, M., D. Scott-Kemmis and W. Satyarakwit (1982), 'Limited Learning in Infant Industry: a Case Study', in F. Stewart and J. James (eds), *The Economics of New Technology in Developing Countries*, London: Frances Pinter.

Bell, M., B. Ross-Larson and L. E. Westphal (1984), 'Assessing the Performance of Infant Industries', *World Bank Staff Working Papers n. 666*, Washington, DC: World Bank.

Bell, M., M. Hobday, S. Abdullah, N. Ariffin and J. Malik (1995) 'Aiming for 2020: a Demand-Driven Perspective on Industrial Technology in Malaysia, Final Report for the World Bank and Ministry of Science, Technology and the Environment, Malaysia, SPRU, University of Sussex.

Bessant, J. (1991), *Managing Advanced Manufacturing Technology. The Challenge of the Fifth Wave*, Oxford: Blackwell.

Bessant, J. (1992), 'Big Bang or Continuous Evolution: Why Incremental Innovation is Gaining Attention in Successful Organisations', *Creativity and Innovation Management*, **1** (2), 59–62.

Bessant, J. (1997), Overheads and Notes on Lectures and Seminars, Inside the Innovating Organisation Course, MSc in Technology and Innovation Management Course, SPRU, University of Sussex.

Bessant, J. (1998), 'Developing Continuous Improvement Capability', *International Journal of Innovation Management*, **2** (4), 409–29.

Bessant, J. and S. Caffyn (1997), 'High-involvement Innovation through Continuous Improvement', *International Journal of Technology Management*, **14** (1), 7–28.

Bessant, J., S. Caffyn, J. Gilbert, R. Harding and S. Webb (1994), 'Rediscovering Continuous Improvement', *Technovation*, **14** (1), 17–29.

Bessant, J. and R. Kaplinsky (1995), 'Industrial Restructuring: Facilitating Organisational Change at the Firm Level', *World Development* (Special Issue), **23** (1), 129–41.

BNDES (The National Bank for Economic and Social Development) (1994), Programa Nacional de Desestatização (National Deregulation Programmme), Activities Report, Rio de Janeiro: BNDES.

Brumer, W. N. (1994), 'The Brazilian Steel Industry', Steel Survival Strategies IX, New York, June.

CEPAL (Economic Commission for Latin America) (1984), La Industria Siderurgica Latinoamericana: Tendencias y Potencial, United Nations, Santiago, Chile.

Caffyn, S. (1997), 'Extending Continuous Improvement to the New Product Development Process', *R&D Management*, **27** (3), 253–67.

Christensen, C. (1997), 'Continuous Casting Investments at USX Corporation', Case 9-697-020, Boston, MA: Harvard Business School Press.

Clark, K. B. and T. Fujimoto (1991), *Product Development Performance*, Boston, MA: Harvard Business School Press.

Cohen, W. M. and D. A. Levinthal (1990), 'Absorptive Capacity: a New Perspective on Learning and Innovation', *Administrative Science Quarterly*, **35** (1), 128–52.

Collinson, S. (1999), 'Knowledge Management Capabilities for Steel Makers: a British–Japanese Corporate Alliance for Organisational Learning', mimeo, Edinburgh: ESRC/University of Edinburgh Management School.

Coombs, R. (1996), 'Core Competencies and the Strategic Management of R&D', *R&D Management*, **26** (4), 345–54.

Coombs, R.W. and R. Hull (1998), 'Knowledge Management Practices and Path-dependency in Innovation', *Research Policy*, **27**, 237–53.

CSN (Companhia Siderúrgica Nacional) (1996), Histórico da Produção: Setor Aço (Production Records: Steel Sector), CSN (SGP/GPP).

Cyert, M. and J. March (1963), *A Behavioral Theory of the Firm*, Englewood Cliffs, NJ: Prentice-Hall.

Dahlman, C. and F.V. Fonseca (1978), 'From Technological Dependence to Technological Development: the Case of the USIMINAS Steel Plant in Brazil', Working Paper, no. 21, IBD/ECLA Research Programme.

Dahlman, C. and L. Westphal (1982), 'Technological Effort in Industrial Development – An Interpretative Survey of Recent Research', in F. Stewart and J. James (eds), *The Economics of New Technology in Developing Countries*, London: Frances Pinter, pp. 105–37.

Dahlman, C., B. Ross-Larson and L. E. Westphal (1987), 'Managing Technological Development: Lessons from the Newly Industrializing Countries', *World Development*, **15** (6), 759–75.

Dodgson, M. (1991), *The Management of Technological Learning: Lessons from a Biotechnology Company*, Berlin: De Gruyter.

Dodgson, M. (1993), 'Organisational Learning: a Review of Some Literatures', *Organisation Studies*, **14** (3), 376–94.

Dosi, G. (1985), 'The Microeconomic Sources and Effects of Innovation. An Assessment of Some Recent Findings', DRC Discussion Paper, n. 33. SPRU, University of Sussex.

Dosi, G. (1988), 'The Nature of the Innovative Process', in G. Dosi, C. Freeman, R. Nelson, G. Silverberg and L. Soete (eds), *Technical Change and Economic Theory*, London: Pinter Publishers.

Dosi, G. and L. Marengo (1993), 'Some Elements of an Evolutionary Theory of Organisational Competences', paper presented at the Tenth World Congress of the International Economic Association, Moscow, August.

Dutrénit, G.B. (2000), *Learning and Knowledge Management in the Firm. From Knowledge Accumulation to Strategic Capabilities*, Cheltenham, UK and Northhampton, MA, USA: Edward Elgar.

Enos, J.L. (1991), *The Creation of Technological Capability in Developing Countries*, London: Pinter Publishers.

Ettlie, J.E. (1988), *Taking Charge of Manufacturing: how Companies are Combining Technological and Organizational Innovations to Compete Successfully*, London: Jossey-Bass.

Fleury, A.C.C. (1977), 'Análise a Nível de Empresa dos Problemas Tecnológicos do Setor de Máquinas-Ferramentas', mimeo, São Paulo: FCAV, mimeo.

Fleury, A.C.C. (1985), 'The Technological Behaviour of State-Owned Enterprises in Brazil', Working Paper, ILO, World Employment Programme Research.

Fusaro, V. S., M. Ramos, A. Assi and C. Silva (1964), 'Resumo de um ano de produção na aciaria LD da USIMINAS' (Summary of one year of production of the LD Steel Shop at USIMINAS), technical paper presented at the XIX Annual Meeting of the Brazilian Metallurgy and Materials Association.

Galimberti, I. (1993), 'Large Chemical Firms in Biotechnology: Case Studies of Learning in Radically New Technology', D.Phil. Thesis, SPRU, University of Sussex.

Garvin, D. A. (1993), 'Building a Learning Organisation', *Harvard Business Review*, **71** (4), 78–91

Girvan, N.P and G. Marcelle (1990), 'Overcoming Technological Dependency: the Case of Electric Arc (Jamaica) Ltd: a Small Firm in a Small Developing Country', *World Development*, **18** (1), 91–107.

Gomes, F.M. (1983), *História da Siderurgia Brasileira (History of the Steel Industry in Brazil)*, Belo Horizonte and São Paulo: Universidade de São Paulo.

Gupta, K. S., S. N. Das and N. Chandra, (1995), 'Indian blast furnace practice: myths, facts and potentials', *Trans. Indian Inst. Met.*, **48** (5), 409–35.

Hedberg, B. (1981), 'How Organisations Learn and Unlearn', in P.C Nystrom and W. Starbuck (eds), *Handbook of Organisational Design*, New York: Oxford University Press, pp. 3–27.

Herbert-Copley, B. (1990), 'Technical Change in Latin American Manufacturing Firms: Review and Synthesis', *World Development*, **18** (11), 1457–69.

Herbert-Copley, B. (1992), 'Technical Change in African Industry: Reflections on IDRC-supported Research', *Canadian Journal of Development Studies*, **13** (2), 231–49.

Hobday, M. (1995), *Innovation in East Asia: the Challenge to Japan*, Aldershot: Edward Elgar.

Hoffman, K. (1989), 'Technological Advance and Organizational Innovation in the Engineering Industries', World Bank IED Working Papers, *Industry Series Paper*, no. 4, Washington, DC:World Bank.

Hollander, S. (1965), *The Sources of Increased Efficiency: a Study of Du Pont Rayon Plants*, Cambridge, MA: MIT Press.

Huber, G. (1996a), 'Organizational Learning: a Guide for Executives in Technology-Critical Organisations', *IJTM Special Publication on Unlearning and Learning*, **11** (7-8), 821–32.

Huber, G. (1996b), 'Organizational Learning: the Contributing Processes and the Literatures', in M. D. Cohen and L. S. Sproull (eds), *Organizational Learning*, London: Sage.

Humphrey, J. (1982), *Fazendo o Milagre*, São Paulo: Vozes.

Humphrey, J. (1993), 'Quality and Productivity in Industry: New Strategies in Developing Countries', *IDS Bulletin*, **24** (2), April.

Humphrey, J. (1995), 'Introduction', *World Development*, **23** (1), 1–7.

Hwang, H-R. (1998), 'Organisational Capabilities and Organisational Rigidities of Korean Chaebol: Case Studies of Semi-Conductor (DRAM) and Personal Computer (PC) Products', D.Phil Thesis, SPRU, University of Sussex.

Iansiti, M. (1998), *Technology Integration*, Boston, MA: Harvard Business School Press.

Iansiti, M. and K. Clark (1994), 'Integration and Dynamic Capability: Evidence from Product Development in Automobiles and Mainframe Computers', *Industrial and Corporate Change*, **33** (3), 557–605.

Iansiti, M. and J. West (1997), 'Technology Integration: Turning Great Research into Great Products', *Harvard Business Review*, **75** (3), 69–79.

IBS (1996), *Anuário Estatístico da Indústria Siderúrgica Brasileira*, Rio de Janeiro: IBS.

IBS (1998), *A Siderurgia Brasileira*, Rio de Janeiro: IBS.

IBS (1999), *IBS Pocket Yearbook*, Rio de Janeiro: IBS

Imai, K. (1987), *Kaizen,* New York: Random House.

IPLAN (The Planning Institute) (1989*), Avaliação dos Planos e Políticas do Setor Siderúrgico Estatal (Assessment of Plans and Policies for the State-owned Steel Industry)*, Brasília*:* Ministério de Indústria e Comércio.

Japan Iron and Steel Federation (1995), 'The Steel Industry of Japan', Tokyo: The Japan Iron and Steel Federation.

Kaplinsky, R. (1994), *Easternisation: the Spread of Japanese Management Techniques to Developing Countries,* London: Frank Cass.

Katz, J. (1976), *Importación de Tecnologia, Aprendizage y Industrialización Dependiente*, México: Fondo de Cultura Economica.

Katz, J. (1985), 'Domestic Technological Innovations and Dynamic Comparative Advantages: Further Reflections on a Comparative Case-Study Program', in N.Rosenberg and C.Frischtak (eds), *International Technology Transfer: Concepts, Measures and Comparisons*, New York: Praeger.

Katz, J. (1987), 'Domestic Technology Generation in LDCs: a Review of Research Findings', in J. Katz (ed.), *Technology Generation in Latin American Manufacturing Industries*, New York: St Martin's Press.

Katz, J., M. Gutkowski, M. Rodrigues and G. Goity (1978), 'Productivity, Technology, and Domestic Efforts in Research and Development', Working Paper n. 14, Buenos Aires, ECLA/IDB/IDRC/UNDP Research Programme on Scientific and Technological Development in Latin America.

Kim, D. (1993), 'The Link between Individual and Organizational Learning', *Sloan Management Review*, Fall, 37–50.

Kim, L. (1995), 'Crisis Construction and Organisational Learning: Capability Building in Catching-up at Hyundai Motor', paper presented at the Hitotsubashi-Organization Science Conference, Tokyo, October.

Kim, L. (1997a), 'The Dynamics of Samsung's Technological Learning in Semiconductors', *California Management Review*, **39** (3), 86–100.

Kim, L. (1997b), *Imitation to Innovation: the Dynamics of Korea's Technological Learning*, Boston, MA: Harvard Business School Press.

Lall, S. (1982), 'Technological Learning in the Third World: Some Implications of Technology Exports', in F. Stewart and J. James (eds), *The Economics of New Technology in Developing Countries*, London: Frances Pinter.

Lall, S. (1987), *Learning to Industrialise: the Acquisition of Technological Capability by India*, London: Macmillan.

Lall, S. (1992), 'Technological Capabilities and Industrialisation', *World Development*, **20** (2), 165–86.

Lall, S. (1994), 'Technological Capabilities', in J.J. Salomon et al. (eds), *The Uncertain Quest: Science, Technology and Development*, Tokyo: UN University Press.

Leonard-Barton, D. (1988), 'Implementation as Mutual Adaptation of Technology and Organization', *Research Policy*, **17** (5), 251–67.

Leonard, D. and S. Sensiper (1998), 'The Role of Tacit Knowledge in Group Innovation', *California Management Review*, **40** (3), 112–32.

Leonard-Barton, D. (1990), 'Implementing New Production Technologies: Exercises in Corporate Learning', in M.A. von Glinow and S.A Mohrman (eds), *Managing Complexity in High Technology Organisations*, New York: Oxford University Press.

Leonard-Barton, D. (1992a), 'Core Capabilities, Core Rigidities: Paradox in Managing New Product Development', *Strategic Management Journal*, **13**, 111–25.

Leonard-Barton, D. (1992b), 'The Factory as a Learning Laboratory', *Sloan Management Review*, **34** (1), 23–38.

Leornard-Barton, D. (1995), *Wellsprings of Knowledge: Building and Sustaining the Sources of Innovation*, Boston, MA: Harvard Business School Press.

Leonard-Barton, D., K. H. Bowen, K. Clark, C. A. Holloway and S. C. Wheelwright (1994), 'How to Integrate Work and Deepen Expertise', *Harvard Business Review*, **72** (5), 121–30.

Malerba, F. and L. Orsenigo (1993), 'Technological Regimes and Organizational Behaviour', *Industrial and Corporate Change*, **2** (1), 45–71.

March, J. and H. Simon (1958), *Organizations*, New York: Wiley.

Maxwell, P. (1981), 'Technological Policy and Firm Learning Efforts in Less Developed Countries: a Case Study of the Experience of the Argentina Steel Firm Acindar SA', D.Phil Thesis, SPRU, University of Sussex.

Maxwell, P.(1982), 'Steelplant Technological Development in Latin America. A Comparative Study of the Selection and Upgrading of Technology in Plants in Argentina, Brazil, Colombia, Mexico and Peru', Working Paper, n. 55. Buenos Aires:ECLA/IDB/IDRC/UNDP Research Programme on Scientific and Technological Development in Latin America.

Maxwell, P. and M. Teubal (1980), 'Capacity-stretching Technical Change: Some Empirical and Theoretical Aspects', Working Paper, No. 36, Buenos Aires: ECLA/IDB/IDRC/UNDP Research Programme on Scientific and Technological Development in Latin America.

McGannon, H. E., N. L. Samways, R. F. Craven and W. Lanford Jr. (1985), *The Making, Shaping and Treating of Steel*, 10th edn, Pittsburgh: Association of Iron and Steel Engineers and United States Steel.

Metalurgia ABM (1988), 'USIMINAS – 25 anos' (USIMINAS – 25 years), Special Edition, October.

Metalurgia ABM (1991), 'CSN: 50 anos transformando a face do país'(CSN: 50 transforming the country's face), **47** (394), March-April.

Metalurgia ABM (1997), 'CSN atinge a marca histórica dos 100 milhões' (CSN reaches the historical record of 100 million tonnes), **53** (467), July.

Meyer-Stamer, J., C. Rauh, H. Riad, S. Schmitt and T. Welte (1991),'Comprehensive Modernisation on the Shop Floor: a Case Study on the Brazilian Machinery Industry', Berlin: GDI.

Miles, M. M. and M. A. Huberman (1984), *Qualitative Data Analysis. A Source of New Methods*, London: Sage.

Mitchell, G. and W. Hamilton (1988), 'Managing R&D as a Strategic Option', *Research Technology Management*, **31** (3), 15–22.

Miyazaki, K. (1993), 'The Dynamics of Competence Building in European and Japanese Firms: the Case of Optoelectronics', D.Phil Thesis, SPRU, University of Sussex.

Mlawa, H. (1983), 'The Acquisition of Technology, Technological Capability and Technical Change: a Study of the Textile Industry in Tanzania', D.Phil. Thesis, SPRU, University of Sussex.

Mody, A., R. Suri and J. Sanders (1992), 'Keeping Pace with Change: Organisational and Technological Imperatives', *World Development*, **20** (12), 1797–1816.

Mukdapitak, Y. (1994), 'The Technology Strategies of Thai Firms', D.Phil Thesis, SPRU, University of Sussex.

Nelson, R. (1987), 'Innovation and Economic Development: Theoretical Retrospect and Prospect', in Katz, J. (ed.), *Technology Generation in Latin American Manufacturing Industries,* New York: St Martin's Press.

Nelson, R. (1991), 'The Role of Firm Differences in a Evolutionary Theory of Technical Advance', *Science and Public Policy*, **18** (6), 347–52,

Nelson, R. and S. Winter (1982), *An Evolutionary Theory of Economic Change*, Cambridge, MA: Harvard University Press.

Nevis, E., A. DiBella and J. Gould (1995), 'Understanding Organizations as Learning Systems', *Sloan Management Review*, Winter, 73–85.

Nonaka, I. (1994), 'A Dynamic Theory of Organisational Knowledge Creation', *Organisational Science,* **5** (1), 15–37.

Nonaka, I. and H. Takeuchi (1995), *The Knowledge Creating Company: How Japanese Companies Create the Dynamics of Innovation*, New York: Oxford University Press.

Pack, H. (1987), *Productivity, Technology and Industrial Development. A Case Study in Textiles*, New York: Oxford University Press.

Patel, P. and K. Pavitt (1994), 'Technological Competencies in the World's Largest Firms: Characteristics, Constraints and Scope for Managerial Choice', STEEP Discussion Paper, No. 13, Brighton, SPRU, May.

Patton, M. Q. (1990), *Qualitative Evaluation and Research Methods*, 2nd edn, Newbury Park, California: Sage.

Paula, G. (1998), 'Privatização e Estrutura de Mercado na Siderurgia Mundial', D.Phil Thesis, Universidade Federal do Rio de Janeiro, Instituto de Economia (UFRJ/IE).

Pavitt, K. (1988), 'Strategic Management in the Innovating Firm', DRC Discussion Paper, No. 61, SPRU, University of Sussex.

Pavitt, K. (1991), 'Key Characteristics of the Large Innovating Firm', *British Journal of Management*, **2**, 41–50.

Pavitt, K. (1998), 'Technologies, Products and Organization in the Innovating Firm: What Adam Smith Tells Us and Joseph Schumpeter Doesn't', *Industrial and Corporate Change*, **7** (3), 433–51.

Peixoto, H.L. (1990), Organização versus Ambiente: o caso da USIMINAS. Das Origens à Privatização, MSc. Dissertation, Belo Horizonte: UFMG.

Penrose, E. T. (1959), *The Theory of the Growth of the Firm*, Oxford: Basil Blackwell.

Pérez, L. and J. Pérez y Peniche (1987), 'A Summary of the Principal Findings of the Case-study on the Technological Behaviour of the Mexican Steel Firm Altos Hornos de Mexico', in J. Katz (ed.), *Technology Generation in Latin American Manufacturing Industry*, New York: St Martin's Press.

Piccinini, M. (1993), 'Technical Change and Energy Efficiency: a Case Study in the Iron and Steel Industry in Brazil', D.Phil Thesis, SPRU, University of Sussex.

Pisano, G. (1997), *The Development Factory: Unlocking the Potential of Process Innovation*, Boston, MA: Harvard Business School Press.

Polanyi, M. (1966), *The Tacit Dimension*, London: Routledge & Kegan Paul.

Prahalad, C. and G. Hamel (1990), 'The Core Competence of the Corporation', *Harvard Business Review*, **90** (3), 79–91.

Rosenberg, N. (1982), *Inside the Black-Box*, Cambridge: Cambridge University Press.

Rosenberg, N. and C. Frischtak (1985), *International Technology Transfer: Concepts, Measures, and Comparisons*, New York: Praeger.

Saravia, E. (1996), 'Proceso de Privatización in Argentina y Brasil: Practicas Utilizadas paral el Ajuste de Personal y Consequencias en Materia de Trabajo y Desempeño Empresarial'. Working Paper, Geneva: International Labour Organization (ILO).

Schein, E. H. (1985), *Organizational Culture and Leadership: a Dynamic View*, London: Jossey-Bass Publishers.

Schroeder, D. and A. Robinson (1991), 'America's Most Successful Export to Japan: Continuous Improvements Programs', *Sloan Management Review*, **32** (3), 67–81.

Scott-Kemmis, D. (1988), 'Learning and the Accumulation of Technological Capacity in Brazilian Pulp and Paper Firms', Working Paper, No. 187, World Employment Programme Research (2-22)

Senge, P. (1990), *The Fifth Discipline: the Art and Practice of the Learning Organisation*, London: Century Business.

Sercovich, F. (1978), 'Ingenieria de Diseño y Cambio Tecnico Endogeno. Un Enfoque Microeconomico Basado en la Experiencia de las Industrias Química y

Petroquímica Argentinas', *Monografia de Trabajo*, No. 19. Buenos Aires: Programa BID/CEPAL sobre Investigación en Temas de Ciencia y Tecnología.

Sherwood, F.P. (1966), 'O aumento do preço do aço da CSN: estudo de um caso', Cadernos de Administração Pública n. 61, Fundação Getulio Vargas/EBAP

Shin, J.S. (1996), *The Economics of the Latecomers. Catching-up, Technology Transfer and Institutions in Germany, Japan and South Korea*, London: Routledge.

Simon, H. (1959), 'Theories of Decision Making in Economics and Behavioral Science', *American Economic Review*, **49** (3), 253–83.

Simon, H. (1961), *Administrative Behaviour. A Study of Decision-Making Process in Administrative Organisation*, 2nd edn, New York: Macmillan.

Simon, H. (1996), 'Bounded Rationality and Organizational Learning', in M.D. Cohen and L.S. Sproull (eds), *Organizational Learning*, London: Sage

Soares e Silva, E. M. (1972), *O Ferro na História e na Economia do Brasil (Iron in the History and in the Economy in Brazil)*, Rio de Janeiro: Biblioteca do Sesquicentenário.

Soares, R.C. (1993), 'The privatisation experience at USIMINAS', Belo Horizonte: USIMINAS.

Soares, R. C. (1997), 'A estratégia de desenvolvimento da USIMINAS diante da estabilização da economia brasileira' (Development strategy of USIMINAS under the stabilisation of the Brazilian economy), Tokyo, March.

Spender, J. C. (1996), 'Competitive Advantage from Tacit Knowledge? Unpacking the Concept and its Strategic Implications', in B. Mosigngeon and A. Edmondson (eds), *Organizational Learning and Competitive Advantage*, London: Sage.

Stalk, G., Evans, P. and Shulman, L.E. (1992), 'Competing on Capabilities: the New Rules of Corporate Strategy', *Harvard Business Review*, **70** (2), 57–81.

Stewart, F. and J. James (1982), 'Introduction', in F. Stewart and J. James (eds), *The Economics of New Technology in Developing Countries*, London: Frances Pinter.

Steel Technology International (1997/8), London: The Sterling Publishing Group PLC,

Suzigan, W. (1986). *Indústria Brasileira: Origem e Desenvolvimento (Brazilian Industry: Origins and Development)*, São Paulo: Brasiliense.

Suzigan, W. and A. Villela (1997), *Industrial Policy in Brazil*, Campinas: UNICAMP/IE.

Teece, D. (1988), 'Technological Change and the Nature of the Firm', in G. Dosi, C. Freeman, R. Nelson, G. Silverberg and L. Soete (eds), *Technical Change and Economic Theory*, London: Pinter Publishers.

Teece, D., G. Pisano and A. Shuen (1990), 'Firm Capabilities, Resources, and the Concept of Strategy: Four Paradigms of Strategic Management', CCC Working Paper, No.94–9, University of California at Berkeley.

Teece, D. and G. Pisano (1994), 'The Dynamic Capabilities of Firms: an Introduction', *Industrial and Corporate Change*, **3** (3), 537–56.

Tidd, J., J. Bessant and K. Pavitt (1997), *Managing Innovation: Integrating Technological Market and Organisational Change*, Chichester: Wiley.

Tiralap, A. (1990), 'The Economics of the Process of Technical Change of the Firm: the Case of the Electronics Industry in Thailand', D. Phil Thesis, SPRU, University of Sussex.

Tremblay, P. (1994), 'Comparative Analysis of Technological Capability and Productivity Growth in the Pulp and Paper Industry in Industrialised and Industrialising Countries', D.Phil Thesis, SPRU, University of Sussex.

Tsekouras, G. (1998), 'Integration, Organisation and Management: Investigating Capability Building', D.Phil Thesis, SPRU, University of Sussex.

USIMINAS/Fundação João Pinheiro (1988), *'USIMINAS conta sua história'* (*USIMINAS tells its history*), Belo Horizonte: USIMINAS/Fundação João Pinheiro.

USIMINAS (1994), 'USIMINAS: 22 anos de transferência de tecnologia' (USIMINAS: 22 years of technology transfer). Belo Horizonte: USIMINAS (engineering superintendency)

USIMINAS (1997), 'Dados Operacionais dos Altos Fornos' (Operational records of Blast Furnaces). USIMINAS (dept of blast furnace and coke ovens).

Viana, H.A.P. (1984), 'International Technology Transfer, Technological Learning and the Assimilation of Imported Technology in a State-owned Enterprise: the Case of SIDOR Steel Plant in Venezuela', D.Phil Thesis, SPRU, University of Sussex.

Voss, C. (1988), 'Implementation: a Key Issue in Manufacturing Technology: the Need for a Field of Study', *Research Policy*, **17** (2), 55–63

Westphal, L. E., L. Kim and C. J. Dahlman (1984), 'Reflections of Korea's Acquisition of Technological Capability', Report DRD77, World Bank Research Department Economics and Research Staff, Washington, DC: World Bank.

Winter, S. (1988), 'On Coase, Competence, and the Corporation', *Journal of Law, Economics, and Organisation*, **4** (1), 163–80.

Womack, J., D. Jones, and D. Roos (1990), *The Machine that Changed the World,* New York: Rawson Associates.

World Development (1984), Special Issue, **12** (5/6), (May–June).

Wortzel, L. H. and H. V. Wortzel (1981), 'Export Marketing Strategies for NIC and LDC-based Firms', *Columbia Journal of World Business*, Spring, pp. 51–60.

Yin, R. K. (1994), *Case Study Research: Design and Methods,* London: Sage.

Index

start-up and initial absorption phase
85–91
knowledge-conversion processes
conventional expansion phase
141–53
liberalisation and privatisation
phase 198–210
start-up and initial absorption phase
94–103
technological capability-accumulation
conventional expansion phase
119–26
liberalisation and privatisation
phase 173–80
start-up and initial absorption phase
78–82
see also cross-company differences
CSN–COR 125, 179
CSN–ZAR 125
customer complaint rate 274
cycle of technological development 153,
209, 210

Daruma Project 168, 187, 209
de-verticalisation, car industry 173
decentralisation, CSN 174
decision-making
CSN 79–80, 120, 174, 177
USIMINAS 69
defensive response, capacity stretching
8
Demag 172, 267
detailed production procedures 205
direct-site observations 59, 184
documentation
of knowledge 102
standardisation 151, 206–7
translating and adaptation 102–3
see also Technical Information Centre
double-loop learning 34, 246
DRAM firms 14
dual phase steels 171
Dwight-Lloyd 126
dynamic charge control 266, 267

early vendor involvement 170
economic recession 115, 123
educational infrastructure in the
community
conventional expansion phase 135–6

liberalisation and privatisation phase
189–90
start-up and initial absorption phase
90
electronic information systems 188
electronics industry 12, 14
emergency course 97–8
enabling capability 31
energy consumption
cross-company differences 269–71,
275
CSN 123
USIMINAS 114
engineering agreement 111–12
engineering superintendency
CSN 120, 174, 181
USIMINAS 112, 130
equipment activities
cross-company differences 230, 232
conventional expansion phase
128
liberalisation and privatisation
phase 183
start-up and initial organisation
phase 84
CSN
conventional expansion phase
125–6
liberalisation and privatisation
phase 180
start-up and initial absorption phase
82
steel industry 49
USIMINAS
conventional expansion phase
117–19
liberalisation and privatisation
phase 171–3
start-up and initial absorption phase
77–8
evolutionary perspective, technological
capability 25–6
expansion plans, USIMINAS 70
Experiment Plan 019/97 193
experimentation
cross-company differences 243–4
USIMINAS 115, 192–4
expertise, importing
conventional expansion phase 128–9,
138